Progress in Mathematics
Volume 84

Series Editors
J. Oesterlé
A. Weinstein

Mappings of Operator Algebras

Proceedings of the Japan–U.S. Joint Seminar,
University of Pennsylvania, 1988

Edited by
Huzihiro Araki
Richard V. Kadison

1991

Birkhäuser
Boston · Basel · Berlin

Huzihiro Araki
Research Institute for
Mathematical Sciences
Kyoto University
Kyoto, Japan

Richard V. Kadison
Department of Mathematics
University of Pennsylvania
Philadelphia, Pennsylvania
U.S.A.

Library of Congress Cataloging-in-Publication Data
Mappings of operator algebras : proceedings of the Japan-U.S. joint seminar. University of Pennsylvania, Philadelphia, Pennsylvania, May 23-27, 1988 / Huzihiro Araki, Richard V. Kadison [editors].
p. cm. — (Progress in mathematics; vol. 84)
Proceedings of the U.S.-Japan Joint Seminar on Operator Algebras.
ISBN 0-8176-3476-2
1. Operator algebras—Congresses. 2. Mappings (Mathematics)–-Congresses. I. Araki, Huzihero, 1932- . II. Kadison, Richard V., 1925- . III. Japan-U.S. Joint Seminar on Operator Algebras (1988 : University of Pennsylvania) IV. Series: Progress in mathematics (Boston, Mass.) ; vol. 84.
QA326.M36 1990
512'.55—dc20 90-35185

Printed on acid-free paper.

ISBN 0-8176-3476-2
ISBN 3-7643-3476-2

Camera-ready copy prepared by the authors and the publisher.
Printed and bound by Edwards Brothers, Inc., Ann Arbor, Michigan.
Printed in the U.S.A.

9 8 7 6 5 4 3 2 1

Shôichirô Sakai

This is dedicated to Shôichirô Sakai, for his brilliant solutions to many of the key problems in the theory of operator algebras that gave us the means and the courage to move ahead and for his inspired leadership in the development of the crucially important theory of unbounded derivations.

With Gratitude and Respect from the Participants of the U.S.–Japan Joint Seminar on Mappings of Operator Algebras

"The results which we shall obtain throw light on an entirely new side of operator theory."

F.J. Murray and J. von Neumann, 1936

Preface

This volume consists of articles contributed by participants at the fourth Japan–U.S. Joint Seminar on Operator Algebras. The seminar took place at the University of Pennsylvania from May 23 through May 27, 1988 under the auspices of the Mathematics Department. It was sponsored and supported by the Japan Society for the Promotion of Science and the National Science Foundation (USA). This sponsorship and support is acknowledged with gratitude.

The seminar was devoted to discussions and lectures on results and problems concerning mappings of operator algebras (C^*-and von Neumann algebras). Among the articles contained in these proceedings, there are papers dealing with actions of groups on C^* algebras, completely bounded mappings, index and subfactor theory, and derivations of operator algebras.

The seminar was held in honor of the sixtieth birthday of Shôichirô Sakai, one of the great leaders of Functional Analysis for many decades. This volume is dedicated to Professor Sakai, on the occasion of that birthday, with the respect and admiration of all the contributors and the participants at the seminar.

H. Araki
Kyoto, Japan

R. Kadison
Philadelphia, Pennsylvania, USA

Contents

On Convex Combinations of Unitary Operators in C*-Algebras

UFFE HAAGERUP

1. INTRODUCTION

Let A be a unital C*–algebra. In [1], Gardner proved that if $x \in A$ and $\|x\| < 1$, then

$$x + U(A) \subseteq U(A) + U(A),$$

where $U(A)$ is the set of unitaries in A. Kadison and Pedersen discovered in [2] that this inclusion together with a simple inductive argument led to the following result:

Let $n \in \mathbb{N}$, $n \geq 3$. If $y \in A$ and $\|y\| < 1 - \frac{2}{n}$, then y is of the form

$$y = \tfrac{1}{n}(u_1 + \dots + u_n)\ , \quad u_i \in U(A)\ ,$$

i.e: y is a mean of n unitaries in A. The main result of this paper is

Theorem

Let A be a unital C^*–algebra and let $n \in \mathbb{N}$, $n \geq 3$. If $y \in A$ and $\|y\| \leq 1 - \frac{2}{n}$, then

$$y = \tfrac{1}{n}(u_1 + \dots + u_n)$$

for some $u_1,\dots,u_n \in U(A)$.

This settles a conjecture of Olsen and Pedersen in [3]. The result is best possible for general C^*–algebras, because if y is a scalar multiple of a non–unitary isometry in a C^*–algebra A and $\|y\| > 1 - \frac{2}{n}$, then y is not the mean of n unitaries in A (cf. [2]). For special C^*–algebras much stronger results hold. Rørdam proved in [5] that if the invertible operators are dense in A, then every operator x in the closed unit ball of A is the mean of n unitaries in A for any $n \geq 3$.

The above Theorem is well known and easy to prove for $n = 4$, and it was proved for $n = 3$ in [3]. Our proof of the general case is obtained by proving the following analogue of Gardner's inclusion:

If $x \in A$ and $\|x\| \leq 1$, then

$$x + 2P \subseteq U(A) + 2P$$

where $P = \{uh \mid u \in U(A), h \in A_+, \|h\| \leq 1\}$.

In the last section of the paper we study the case $n = 3$ more closely. It is proved that for $n = 3$, u_1, u_2 and u_3 can be chosen, such that the spectra of $u_1^*u_2$, $u_2^*u_3$ and $u_3^*u_1$ are contained in the semicircle $\{e^{i\nu} | 0 \leq \nu \leq \pi\}$. Moreover, if $\|y\| < \frac{1}{3}$, then $u_1 \in U(A)$ can be chosen freely and u_2, u_3 are then uniquely determined by the spectral conditions above. This implies that for $x \in A$, $\|x\| < 1$, there is a homeomorphism Φ_x of $U(A)$, such that $\Phi_x^3 = \mathrm{id}$ and

$$x = u + \Phi_x(u) + \Phi_x^2(u)$$

for all $u \in U(A)$.

We wish to thank Gert K. Pedersen for suggesting a major simplification of the proof of the spectral conditions in the key lemma (Lemma 1), which reduced the length of the proof from 8 to 2 manuscript pages.

2. THE MAIN RESULT

Lemma 1

Let A be a unital C*–algebra. Let $x, h \in A$ $\|x\| \leq 1$, $\|h\| \leq 1$, $h \geq 0$. Then

$$x + 2h = u_1 + u_2 + u_3$$

where $u_1,u_2,u_3 \in U(A)$ and $sp(u_1^*u_2)$, $sp(u_2^*u_3)$ are both contained in $\{e^{i\nu} \mid -\frac{\pi}{2} \leq \nu \leq \pi\}$.

Proof. Write $x = a + ib$, where $a,b \in A_{sa}$. Then

$$a^2 + b^2 = \frac{1}{2}(x^*x + xx^*) \leq 1.$$

Particularly $\|a\| \leq 1$ and $\|b\| \leq 1$. By [4, Prop. 1.3.8] the function $t \to t^{\frac{1}{2}}$ is operator monotone on $[0,\infty[$. Therefore

$$|a| \leq (1 - b^2)^{\frac{1}{2}}.$$

Hence

$$-(1 - b^2)^{\frac{1}{2}} \leq a \leq (1 - b^2)^{\frac{1}{2}}.$$

Put

$$c = \frac{1}{2}(2h + a - (1 - b^2)^{\frac{1}{2}}).$$

Then

$$h - (1 - b^2)^{\frac{1}{2}} \leq c \leq h.$$

Particularly $\|c\| \leq 1$, and $c + (1 - b^2)^{\frac{1}{2}} \geq 0$.

Put

$$u_1 = c - i(1 - c^2)^{\frac{1}{2}}$$
$$u_2 = (1 - b^2)^{\frac{1}{2}} + ib$$
$$u_3 = c + i(1 - c^2)^{\frac{1}{2}} .$$

Then $u_1,u_2,u_3 \in U(A)$ and

$$u_1 + u_2 + u_3 = x + 2h.$$

(For $h = 0$, this choice of u_1,u_2,u_3 was used in the proof of [3, Thm. 4.3]).

To prove the conditions on the spectra of $u_1^*u_2$ and $u_2^*u_3$ stated in the lemma, we have to prove that if $\lambda = \cos\theta + i\sin\theta$, $0 < \theta < \frac{\pi}{2}$, then

$$-\lambda \notin sp(u_1^*u_2) \text{ and } -\lambda \notin sp(u_2^*u_3).$$

Let λ be as above and assume that $-\lambda \in sp(u_1^*u_2)$. Then for some faithful representation of A on a Hilbert space H, there is a unit vector $\xi \in H$ for which

$$u_1^*u_2\xi = (-\lambda)\xi.$$

(cf. [4, 4.3.10]). Thus $(u_2 + \lambda u_1)\xi = 0$. Let φ be the vector state on A given by ξ. Then $\varphi(u_2 + \lambda u_1) = 0$, or equivalently

$$\varphi((1-b^2)^{\frac{1}{2}} + ib + (\cos\theta + i\sin\theta)(c - i(1-c^2)^{\frac{1}{2}}) = 0.$$

By considering the real part, we get:

$$\varphi((1-b^2)^{\frac{1}{2}} + \cos\theta\, c + \sin\theta\,(1-c^2)^{\frac{1}{2}}) = 0.$$

By rewriting the equation in the form

$$\cos\theta\, \varphi((1-b^2)^{\frac{1}{2}} + c) + (1-\cos\theta)\varphi((1-b^2)^{\frac{1}{2}})$$
$$+ \sin\theta\, \varphi((1-c^2)^{\frac{1}{2}}) = 0$$

and using $(1-b^2)^{\frac{1}{2}} + c \geq 0$, $(1-b^2)^{\frac{1}{2}} \geq 0$, $(1-c^2)^{\frac{1}{2}} \geq 0$, $0 < \theta < \frac{\pi}{2}$ one gets

$$\varphi((1-b^2)^{\frac{1}{2}} + c) = \varphi((1-b^2)^{\frac{1}{2}}) = \varphi((1-c^2)^{\frac{1}{2}}) = 0.$$

If $(s\xi,\xi) = 0$ for some positive operator $s \in B(H)$, then also $s\xi = 0$. Therefore

$$((1-b^2)^{\frac{1}{2}} + c)\xi = (1-b^2)^{\frac{1}{2}}\xi = (1-c^2)^{\frac{1}{2}}\xi = 0.$$

Particularly $c\xi = (1-c^2)^{\frac{1}{2}}\xi = 0$. This is a contradiction, because

$$\|c\xi\|^2 + \|(1-c^2)^{\frac{1}{2}}\xi\|^2 = 1.$$

The spectral condition on $u_2^*u_3$ can be proved as above by using that

$$sp(u_2^*u_3) = sp(u_3u_2^*).$$

Indeed, if $\lambda = \cos\theta + i\sin\theta$, $0 < \theta < \frac{\pi}{2}$ and $-\lambda \in sp(u_3u_2^*)$, then we can find a vector state φ on A, such that

$$\varphi(u_2^* + \lambda u_3^*) = 0$$

and by taking the real part we get

$$\varphi((1-b^2)^{\frac{1}{2}} + \cos\theta\, c + \sin\theta(1-c^2)^{\frac{1}{2}}) = 0,$$

which is the same equality as in the previous case.

Lemma 2

Let A be a unital C*–algebra and let

$$P = \{uh \mid u \in U(A), h \in A_+, \|h\| \leq 1\}.$$

Then for every $x \in A$, $\|x\| \leq 1$ one has

$$x + 2P \subseteq U(A) + 2P.$$

Proof. Since the spectral condition in Lemma 1 is invariant under multiplication from left (and right) with a fixed unitary, it follows that every $z \in x+2P$ is of the form

$$z = u_1 + u_2 + u_3$$

with $sp(u_1^*u_2)$ and $sp(u_2^*u_3)$ contained in $F = \{e^{i\nu} \mid -\frac{\pi}{2} \leq \nu \leq \pi\}$. The branch of the square root given by $e^{i\nu} \to e^{i\nu/2}$, $-\frac{\pi}{2} \leq \nu \leq \pi$, is continuous on F. Hence there exists $v \in U(A)$ such that

$$u_2^*u_3 = v^2$$

and

$$sp(v) \subseteq \{e^{i\nu} \mid -\tfrac{\pi}{4} \leq \nu \leq \tfrac{\pi}{2}\}.$$

Put $w = u_3v^* = u_2v$. Then

$$u_2 + u_3 = w(v + v^*),$$

and since $sp(v) \subseteq \{s \mid \operatorname{Re} s \geq 0\}$, we have $v + v^* \geq 0$, so that $u_2 + u_3 \in 2P$. This proves lemma 2.

Theorem

Let A be a unital C*–algebra, and let $n \in \mathbb{N}$, $n \geq 3$. If $y \in A$ and $\|y\| \leq 1 - \frac{2}{n}$, then

$$y = \tfrac{1}{n}(u_1 + \dots + u_n)$$

for some $u_1,\dots,u_n \in U(A)$.

Proof. Put $x = \frac{n}{n-2}y$. Then $\|x\| \leq 1$. By repeated use of lemma 2, we get

$$(n-2)x + 2P \subseteq \underbrace{U(A) + \dots + U(A)}_{n-2 \text{ times}} + 2P$$

But $2P \subseteq U(A) + U(A)$, because every selfadjoint operator h of norm less or equal to 1 is the half sum of the two unitaries $h \pm i(1-h^2)^{\frac{1}{2}}$. Hence

$$(n-2)x + 2P \subseteq \underbrace{U(A) + \dots + U(A)}_{n \text{ times}}.$$

Since $0 \in 2P$, we get

$$y = \tfrac{n-2}{n} x \in \tfrac{1}{n}(\underbrace{U(A) + \dots + U(A)}_{n \text{ times}}).$$

3. ON THE SUM OF THREE UNITARIES

Proposition 1

Let A be a unital C*–algebra, and let $x \in A$, $\|x\| \leq 1$. Then

$$x = u_1 + u_2 + u_3 ,$$

where $u_1,u_2,u_3 \in U(A)$, and $sp(u_1^*u_2)$, $sp(u_2^*u_3)$, $sp(u_3^*u_1)$ are contained in $\{e^{i\nu} | 0 \leq \nu \leq \pi\}$.

Proof. Let u_1,u_2,u_3 be as in lemma 1 with $h = 0$, i.e.

$$x = a + ib, \quad a,b \in A_{s.a.}$$

$$c = \tfrac{1}{2}(a - (1-b^2)^{\frac{1}{2}})$$

and

$$(*) \quad \begin{cases} u_1 = c - i(1-c^2)^{\frac{1}{2}} \\ u_2 = (1-b^2)^{\frac{1}{2}} + ib \\ u_3 = c + i(1-c^2)^{\frac{1}{2}} \end{cases}$$

Recall from the proof of Lemma 1 that $a \leq (1-b^2)^{\frac{1}{2}}$. Hence $c \leq 0$, and therefore $sp(u_1)$ is contained in

$$\{-e^{i\nu} | 0 \leq \nu \leq \tfrac{\pi}{2}\}.$$

This implies that

$$\mathrm{sp}(u_3^*u_1) = \mathrm{sp}(u_1^2) \subseteq \{e^{i\nu} \mid 0 \le \nu \le \pi\}.$$

To prove that $\mathrm{sp}(u_1^*u_2)$ and $\mathrm{sp}(u_2^*u_3)$ are also contained in $\{e^{i\nu} \mid 0 \le \nu \le \pi\}$ we can proceed as in the proof of lemma 1. From lemma 1 we already know that the two spectra are contained in $\{e^{i\nu} \mid -\frac{\pi}{2} \le \nu \le \pi\}$, so it remains to be proved that if $\lambda = \cos\theta + i\sin\theta$ for $\frac{\pi}{2} \le \theta < \pi$, then

$$-\lambda \notin \mathrm{sp}(u_1^*u_2) \text{ and } -\lambda \notin \mathrm{sp}(u_2^*u_3).$$

As in the proof of prop. 1, $-\lambda \in \mathrm{sp}(u_1^*u_2)$ or $-\lambda \in \mathrm{sp}(u_2^*u_3)$ would imply that

$$(*) \qquad \varphi((1-b^2)^{\frac{1}{2}} + \cos\theta\, c + \sin\theta\, (1-c^2)^{\frac{1}{2}}) = 0.$$

Since $(1-b^2)^{\frac{1}{2}} \ge 0$, $(1-c^2)^{\frac{1}{2}} \ge 0$, $c \le 0$, $\cos\theta \le 0$ and $\sin\theta > 0$ it follows that

$$\varphi((1-b^2)^{\frac{1}{2}}) = \cos\theta\, \varphi(c) = \sin\theta\, \varphi(1-c^2)^{\frac{1}{2}} = 0.$$

Particularly

$$\varphi((1-b^2)^{\frac{1}{2}}) = \varphi((1-c^2)^{\frac{1}{2}}) = 0.$$

Since $-(1-b^2)^{\frac{1}{2}} \le a \le (1-b^2)^{\frac{1}{2}}$, we have

$$0 \le -c \le (1-b^2)^{\frac{1}{2}}.$$

Hence also $\varphi(-c) = 0$. We can assume that φ is a vector state $(\cdot\, \xi,\xi)$ for faithful representation of A on a Hilbert space. Then

$$(1-c^2)^{\frac{1}{2}}\xi = (-c)\xi = 0,$$

because $(1-c^2)^{\frac{1}{2}} \ge 0$ and $(-c) \ge 0$. This contradicts that

$$\|c\xi\|^2 + \|(1-c^2)^{\frac{1}{2}}\xi\|^2 = 1.$$

Hence $\mathrm{sp}(u_1^*u_2)$ and $\mathrm{sp}(u_2^*u_3)$ are both contained in $\{e^{i\nu} \mid 0 \le \nu \le \pi\}$.

Proposition 2

Let A be a unital C*–algebra, and let $x \in A$, $\|x\| < 1$. Then for every $u_1 \in U(A)$ there exists one and only one pair of unitaries $u_2, u_3 \in U(A)$, such that

$$x = u_1 + u_2 + u_3$$

and $sp(u_1^*u_2)$, $sp(u_2^*u_3)$, $sp(u_3^*u_1)$ are contained in $\{e^{i\nu} \mid 0 \le \nu \le \pi\}$.

Proof. Let $u_1 \in U(A)$. Since $\|x\| < 1$, $x - u_1$ is invertible in A. Hence $x - u_1$ has a polar decomposition

$$x - u_1 = w(2s),$$

where $w \in U(A)$, $s \in A_+$, $\|s\| \le 1$. Put

$$u_2 = w(s - i(1-s^2)^{\frac{1}{2}})$$

$$u_3 = w(s + i(1-s^2)^{\frac{1}{2}}).$$

Then clearly $u_2, u_3 \in U(A)$ and $x = u_1 + u_2 + u_3$ (this is Gardner's construction in [1]).

Let $a, b \in A_{s.a.}$ be the two selfadjoint operators determined by

$$-w^*x = a + ib.$$

Note that $\|x\| < 1$ implies that $\|a\| < 1$ and $\|b\| < 1$. We prove next that $s = \frac{1}{2}((1-b^2)^{\frac{1}{2}} - a)$.

Put

$$v_i = -w^*u_i, \quad i = 1,2,3.$$

Then

$$v_1 = 2s - w^*x = (a + 2s) + ib$$

$$v_2 = -s + i(1-s^2)^{\frac{1}{2}}$$

$$v_3 = -s - i(1-s^2)^{\frac{1}{2}}$$

and

$$v_1 + v_2 + v_3 = - w^*x.$$

Since v_1 is unitary,

$$(a + 2s)^2 + b^2 = \frac{1}{2}(v_1^*v_1 + v_1v_1^*) = 1 .$$

Thus

$$|a + 2s| = (1-b^2)^{\frac{1}{2}}.$$

Since $\|b\| < 1$, $(1-b^2)^{\frac{1}{2}}$ is invertible. Hence $a - 2s$ has a polar decomposition

$$a + 2s = u(1-b^2)^{\frac{1}{2}},$$

where $u \in U(A)$, $u = u^*$ and u commutes with $(1-b^2)^{\frac{1}{2}}$. We may assume that A is faithfully represented on a Hilbert space H. Then

$$H = H_+ \oplus H_- ,$$

where H_+ and H_- are the eigenspaces for u corresponding to $+1$ and -1, respectively. For $\xi \in H_-$

$$((a + 2s)\xi,\xi) = -((1-b^2)^{\frac{1}{2}}\xi,\xi)$$

or

$$(2s\xi,\xi) + (((1-b^2)^{\frac{1}{2}} + a)\xi,\xi)) = 0.$$

As in the proof of lemma 1 one gets

$$-(1-b^2)^{\frac{1}{2}} \leq a \leq (1-b^2)^{\frac{1}{2}}.$$

Hence both s and $(1-b^2)^{\frac{1}{2}} + a$ are positive, so we get

$$s\xi = ((1-b^2)^{\frac{1}{2}} + a)\xi = 0 .$$

But $s = \frac{1}{2}|x-u_1|$ is invertible and therefore $\xi = 0$. This proves that $H_- = \{0\}$, i.e. $u = 1$. Hence

$$a + 2s = (1-b^2)^{\frac{1}{2}}$$

or

$$s = \tfrac{1}{2}((1-b^2)^{\frac{1}{2}} - a).$$

Particularly

$$0 \leq s \leq (1-b^2)^{\frac{1}{2}}.$$

Moreover,

$$v_1 = (1-b^2)^{\frac{1}{2}} + ib.$$

This shows that the three unitaries

$$v_3 = -s - i(1-s^2)^{\frac{1}{2}}$$

$$v_1 = (1-b^2)^{\frac{1}{2}} - ib$$

$$v_2 = -s + i(1-s^2)^{\frac{1}{2}}$$

written in this order are exactly the three unitaries in the decomposition of $-w^*x = a + ib$ given in the start of the proof of Proposition 1 (the operator c in the proof of Proposition 1 is $-s$).

Hence by Proposition 1, the spectra of $v_3^*v_1$, $v_1^*v_2$ and $v_2^*v_3$ are all contained in

$$\{e^{i\nu} | 0 \leq \nu \leq \pi\}.$$

Since $u_1^*u_2 = v_1^*v_2$, $u_2^*u_3 = v_2^*v_3$ and $u_3^*u_1 = v_3^*v_1$, we have proved the existence of u_2 and u_3 in Proposition 2.

To prove uniqueness, let $u_1 \in U(A)$, and assume that $u_2, u_3 \in U(A)$ satisfies the conditions in prop. 2. The condition on the spectrum of $u_2^*u_3$ implies that there is a $v \in U(A)$, such that

$$v^2 = u_2^*u_3 \text{ and } sp(v) \subseteq \{e^{i\nu} | 0 \leq \nu \leq \tfrac{\pi}{2}\}.$$

Put $w = u_2 v = u_3 v^*$. Then

$$w^*(x - u_1) = w^*(u_2 + u_3)$$
$$= v^* + v.$$

Put $s = \frac{1}{2}(v + v^*)$ and $t = \frac{1}{2i}(v - v^*)$. The condition on $sp(v)$ implies that $s \geq 0$ and $t \geq 0$. Therefore

$$(x - u_1) = w(2s)$$

is the unique polar decomposaition of the invertible operator $x - u_1$. Since $v = s + it$ is unitary and $t \geq 0$, we have $t = (1-s^2)^{\frac{1}{2}}$. Hence w and $v = s + it$ are uniquely determined from u_1. This completes the proof, because $u_2 = wv^*$ and $u_3 = wv$.

Corollary

Let A be a unital C*–algebra, and let $x \in A$, $\|x\| < 1$. Then there is a unique homeomorphism Φ_x of $U(A)$, such that

1) $\Phi_x^3 = id$
2) $x = u + \Phi_x(u) + \Phi_x^2(u)$ for all $u \in U(A)$
3) $sp(u^*\Phi_x(u)) \subseteq \{e^{i\nu} | 0 \leq \nu \leq \pi\}$ for all $u \in U(A)$.

Proof. For $u_1 \in U(A)$ we put $\Phi_x(u_1) = u_2$, where

$$x = u_1 + u_2 + u_3$$

is the unique decomposition in Proposition 2. Since the condition on the spectra of $u_1^* u_2$, $u_2^* u_3$ and $u_3^* u_1$ in Proposition 2 is invariant under cyclic permutations of (u_1, u_2, u_3) we have

$$\Phi_x(u_2) = u_3 \, , \ \Phi_x(u_3) = u_1 \, .$$

Hence $\Phi_x^3 = \mathrm{id}$, and

$$x = u + \Phi_x(u) + \Phi_x^2(u) \quad , \quad u \in U(A).$$

Also 3) above follows from Proposition 2. It is easily seen that the polar decomposition is norm–continuous on the invertible elements of A. Hence from the explicite formula for $u_2 = \Phi_x(u_1)$ given in the proof of Proposition 2 one gets that Φ_x is continuous, and since $\Phi_x^3 = \mathrm{id}$ it is a homeomorphism. The uniqueness of Φ_x is clear from Proposition 2.

REFERENCES

[1] L.T. Gardner, An elementary proof of the Russo–Dye Theorem. Proc. Amer. Math. Soc. 90 (1984), 171.

[2] R.V. Kadison and G.K. Pedersen, Means and convex combinations of unitary operators. Math. Scand. 57 (1985), 249–266.

[3] C.L. Olsen and G.K. Pedersen, Convex combinations of unitary operators in von Neumann algebras. Journ. Funct. Anal. 66 (1986), 365–380.

[4] G.K. Pedersen, C*–algebras and their automorphism groups. Academic Press 1979.

[5] M. Rørdam, Advances in the theory of unitary rank and regular approximation. Annals of Math. 128 (1988), 153–172.

Institut for Matematik og Datalogi, Odense Universitet,
Campusvej 55, DK–5230 Odense M, Denmark.

Approximately Inner Derivations, Decompositions and Vector Fields of Simple C*-Algebras

PALLE JORGENSEN

TABLE OF CONTENTS

Research supported in part by NSF.

§1. INTRODUCTION

Given a mathematical structure, one of the basic associated objects is its automorphism group. In differential geometry, manifolds are equipped with a differentiable structure and the objects of study are the diffeomorphism group and its subgroups. In quantum mechanics, the observables form a non–Abelian algebra. In some of the more accessible cases, this algebra is a simple C^*–algebra, and the objects of study are the associated groups of automorphisms. These are the quantum mechanical symmetry groups.

In this generality, the problem is hopelessly difficult:

(i) simple C^*–algebras are not classified and generally poorly understood.

(ii) the group of all automorphisms has so far not lent itself to computation.

In this paper, we restrict attention to a family of C^*–algebras which was first studied by Slawney [Sla] in connection with an attempt to unify the theory of the CAR and the CCR algebras (those on the anti (respectively, the commutation) relations of quantum mechanics), and to the Cuntz algebras.

There is a simple C^*–algebra $\mathfrak{A}$ associated to an antisymmetric nondegenerate form in two variables, and $\mathfrak{A}$ is well understood from the point of view of invariants. But it does not come equipped with a differentiable structure. Yet the differential geometric tensors and the Yang–Mills functional have been introduced in [Con 3], [C–R] and [Rie 2]. We set out to define a differentiable structure in terms of the canonical torus action on $\mathfrak{A}$ (which is an ergodic action and determines $\mathfrak{A}$).

Smooth actions are then defined relative to the chosen differentiable structure. We ask if the same differentiable structure results when a different smooth Lie group action is used to generate the differentiable structure. We show that the answer is affirmative if the nondegenerate form is assumed to satisfy a certain generic condition (which will be defined). It is a diophantine condition familiar in number theory.

The automorphism groups are nonlinear, and we show that the study of smooth Lie group actions (in the generic case) may be partially linearized by introducing the infinitesimal Lie algebra representation. We show, in particular, that every smooth Lie algebra exponentiates to a smooth Lie group action, thus generalizing a familiar result in differential geometry which is proved by solving a system of first order differential equations. The analogous differential equations do not make sense in the noncommuative setting.

In the study of the smooth Lie algebras, we show (again in the generic case) that the Levi decomposition diagonalizes in a particularly nice and simple way. This allows us to catalogue the actions in terms of the semisimple ones and the solvable ones; and we showed in [BEGJ] that the first ones are compact and inner, while the second class exhibits an interesting dichotomy. The dichotomy is defined in terms of exact sequences of Lie algebras developed further in Section 5 below. One side of the dichotomy corresponds to a trivial Ext–group, while the second side yields nondegenerate central extensions, and the Heisenberg Lie algebras in particular.

Finally, we classified in [BEGJ] the Heisenberg actions up to smooth equivalence and expressed the classification parameters in terms of the moduli space for the Yang–Mills problem studied by Connes and Rieffel [C–R].

In Section 2 below, the noncommuative tori are reviewed, and the main problems are formulated. An early result of Sakai [Sak 5] is a main source of inspiration. We recall it here in a slightly more general form than the original one, cf. [BR, p. 262]: Let $\mathfrak{A}$ be an approximately finite–dimensional C^*–algebra (AF or *matricial* are alternative names). Let δ be a closed derivation with dense domain $D(\delta)$ in $\mathfrak{A}$, i.e.,

$$\delta(ab) = \delta(a)b + a\delta(b), \qquad a,b \in D(\delta).$$

Then there is an ascending nest $\{\mathfrak{A}_n : n = 1,2,\cdots\}$ of finite dimensional C^*–subalgebras such that

(i) $$\mathfrak{A}_n \subset D(\delta)$$

and

(ii) $$\mathfrak{A}_0 := \bigcup_{n=1}^{\infty} \mathfrak{A}_n \text{ is norm dense in } \mathfrak{A}.$$

Condition (i) allows us to choose matrix units in the domain, and, for each n, find an element $h_n \in \mathfrak{A}$ such that

(iii) $$\delta(a) = h_n a - a h_n, \quad a \in \mathfrak{A}_n.$$

It follows that

(iv) $$\lim_{n \to \infty} (\mathrm{ad}\, h_n)(a) = \delta(a)$$

holds for all $a \in \mathfrak{A}_0$ where $\mathrm{ad}\, h_n$ is the inner derivation, defined by

$$(\mathrm{ad}\, h_n)(a) = h_n a - a h_n, \quad a \in \mathfrak{A}, \tag{1.1}$$

and the limit in (iv) is relative to the C^*–norm.

To summarize, Sakai showed that every closed derivation is approximately inner on some dense subalgebra in the domain. As a corollary, we get that closed derivations with dense domain in $C(K)$, where K is a Cantor set, must vanish identically. By analogy, we say that AF algebras have 0–dimensional differentiable structure.

Sakai asked in [Sak 7] for an analysis of simple C^*–algebras (other than UHF) with nontrivial differentiable structure. A partial answer was obtained in [Ta] for the irrational rotation algebras, and a systematic study was supplied in [BEJ] for all of the noncommutative tori.

In Section 3, we show that the problem may be attacked effectively with the use of a new inner product on the space of all derivations with a given domain. The inner product is defined for the derivations of the noncommutative tori, studied in [BEJ], and the results from [BEJ] are extended.

In Section 7, we introduce the inner product on the derivations of $\mathcal{O}_n$, studied in [BEvGJ], and the results from that paper are extended in Sections 8–12.

Let $\{s_i\}_1^n$ be the basis isometries generating $\mathcal{O}_n$, and let $\mathscr{P}_n$ be the (polynomial) *–algebra generated by the s_i's. The derivation δ_g defined by

$$\delta_g(s_i) = (\sqrt{-1})s_i, \quad i = 1,\cdots,n, \tag{1.2}$$

is called the gauge–generator. Let $\mathrm{Der}(\mathscr{P}_n,\mathcal{O}_n)$ denote the derivations from $\mathscr{P}_n$ into $\mathcal{O}_n$. We show in Section 12 that every $\delta \in \mathrm{Der}(\mathscr{P}_n,\mathcal{O}_n)$ decomposes uniquely:

$$\delta = \lambda\delta_g + \tilde{\delta} \tag{1.3}$$

where $\lambda \in \mathbb{C}$, and where $\tilde{\delta}$ is singular, or approximately inner *in mean.* (Both notions to be defined.) Let $\|\cdot\|_2$ be the Hilbert–norm defined by the inner product on $\mathrm{Der}(\mathscr{P}_n,\mathcal{O}_n)$. (This is the inner product from Section 7 below.) Then the approximation property of $\tilde{\delta}$ referred to in (1.3) is as follows: There is a sequence $(h_p) \subset \mathcal{O}_n$ such that

$$\lim_{p\to\infty} \|\tilde{\delta} - \mathrm{ad}\, h_p\|_2 = 0. \tag{1.4}$$

§2. THE NONCOMMUTATIVE TORI

In this section, we introduce the noncommutative tori and state the decomposition theorem which is the basis for our analysis of the smooth structures.

Let Γ be a discrete abelian group, and let ρ be a function on $\Gamma \times \Gamma$ with values in the circle group $\mathbb{T}^1$ such that

(i) ρ is antisymmetric, i.e.,

$$\rho(\xi,\gamma) = \rho(\gamma,\xi)^{-1}, \quad \xi,\gamma \in \Gamma.$$

(ii) $\rho(\cdot,\gamma)$ is a (multiplicative) character in Γ for all $\gamma \in \Gamma$, i.e.,

$$\rho(\xi_1 + \xi_2,\gamma) = \rho(\xi_1,\gamma)\rho(\xi_2,\gamma), \quad \xi_1, \xi_2 \in \Gamma.$$

(iii) ρ is nondegenerate, i.e., $\rho(\cdot,\gamma)$ is a nontrivial character for all $\gamma \in \Gamma\backslash\{0\}$.

Let $K = \hat{\Gamma}$ be the compact group of all characters on Γ. We shall be mainly interested in the case $\Gamma = \mathbb{Z}^d$ (the d–dimensional integer–lattice), and in this case $\hat{\Gamma} \simeq \mathbb{T}^d$ (the d–torus).

There is a familiar construction of a C^*–algebra dynamical system $(\mathfrak{A},K,\alpha)$ where $\mathfrak{A}$ is a simple C^*–algebra, and

$$\alpha : K \longrightarrow \mathrm{Aut}(\mathfrak{A}) \tag{2.1}$$

is a representation of K by $*$–automorphisms of $\mathfrak{A}$. The construction is given in [OPT] and [BEJ], and we review details: The C^*–algebra $\mathfrak{A}$ is a "universal" C^*–algebra on a set of generators, denoted $\{u(\gamma) : \gamma \in \Gamma\}$ such that $u(\gamma)^* = u(\gamma)^{-1}$, and

$$u(\xi)u(\gamma) = \rho(\xi,\gamma)u(\gamma)u(\xi), \quad \xi,\gamma \in \Gamma.$$

The duality between Γ and K will be denoted $\langle \xi,k \rangle$, $\xi \in \Gamma$, $k \in K$, and the formula

$$\alpha_k(u(\xi)) = \langle \xi,k \rangle u(\xi) \tag{2.2}$$

can be shown to define an action of K on $\mathfrak{A}$. It is strongly continuous, and *ergodic.* Ergodicity means that the fixed–point subalgebra $\{a \in \mathfrak{A} : \alpha_k(a) = a, \ k \in K\}$ reduces to the scalars times the identity. Finally, $\mathfrak{A}$ has a unique trace which will be denoted τ and may be specified by the integral

$$\tau = \int_K \alpha_k \, dk \tag{2.3}$$

with respect to normalized Haar measure on K. The nontrivial fact is the result of [OPT] that every *ergodic* C^*–dynamical system $(\mathfrak{A},K,\alpha)$ with a C^*–algebra $\mathfrak{A}$ and a compact abelian group K is associated to some antisymmetric bicharacter ρ on the dual (discrete) character group $\Gamma \simeq \hat{K}$, and, moreover, $\mathfrak{A}$ is simple iff ρ is nondegenerate. Moreover, the compact bicharacter group $(\Gamma \wedge \Gamma)\hat{}$ parametrizes the isomorphism classes of dynamical systems as above.

In the sequel, $\mathfrak{A}$ will be a simple C^*–algebra associated to some given ρ on $\Gamma = \mathbb{Z}^d$. We shall need two dense subalgebras $\mathfrak{A}_0$ and $\mathfrak{A}^\infty$ of $\mathfrak{A}$.

We define $\mathfrak{A}_0$ as the algebra spanned by the $u(\gamma)$'s, $\gamma \in \Gamma$, and we define $\mathfrak{A}^\infty$ as the C^∞–elements for the ergodic action α of $\mathbb{T}^d$, i.e.,

$$\mathfrak{A}^{\infty} := \{a \in \mathfrak{A} : (k \longrightarrow \alpha_k(a)) \in C^{\infty}(\mathbf{T}^d, \mathfrak{A})\}. \tag{2.4}$$

Sakai asked in [Sak 7] for classes of simple C^*–algebras with specified dense subalgebras such that the derivations, with domain equal to the given subalgebra, have a unique decomposition into a sum of a Lie–algebra part and an approximately inner part. The Lie algebra part was meant to represent the differential geometry. Sakai's question was based on an earlier result for closed derivations with dense domain in a UHF (or more generally, AF) C^*–algebra. Modulo an (important) technical point regarding the domain, Sakai [Sak 5] showed, in the UHF–case, that all closed derivations are approximately inner. In the abelian case, the corresponding result [Sak 6] states that every closed derivation with dense domain in the C^*–algebra of continuous functions on the Cantor set must be zero. (See also [Bat].) The UHF algebras are thought of as noncommutative Cantor sets, zero dimensional with trivial differentiable structure.

We showed in [BEJ] that the decomposition theorem holds for derivations in the simple C^*–algebras associated to bicharacters (called noncommutative tori).

We will begin by introducing a new positive definite sesquilinear inner product on the space of all derivations. We then show that the known decomposition results may be obtained as a Hilbert space decomposition. New results are then obtained for $\mathcal{O}_n$.

This allows us, in addition, to get new results on derivations, and differentiable structure. The inner product (to be described below) was motivated by the Yang–Mills functional introduced by Connes and Rieffel, see [C–R] and [Rie 2].

An analogous inner product may be defined on the derivations of other simple C^*–algebras. In Section 7, we do this for Cuntz's algebra $\mathcal{O}_n$.

The main reference for Secs. 3–4 is [BEJ], for Secs. 5–6 it is [BEGJ(a)–(b)], and for Sec. 7 it is [BEvGJ].

§3. THE INNER PRODUCT

We first define an inner product on the maps from $\mathfrak{A}_0$ into $L^2(\mathfrak{A},\tau)$, the trace Hilbert space, as follows. Let $\{e_i : i = 1,2,\cdots,d\}$ be the usual basis for $\Gamma = \mathbb{Z}^d$ given by $e_i = (0,\cdots,0,1,0,\cdots,0)$ with one on the i–th place. Recall that $\mathfrak{A}_0$ is spanned by the set of unitary elements $\{u(\xi) : \xi \in \Gamma\}$ satisfying

$$u(\xi)u(\gamma) = \rho(\xi,\gamma)u(\gamma)u(\xi), \quad \xi,\gamma \in \Gamma, \tag{3.1}$$

where ρ is the given bicharacter. It follows that $\mathfrak{A}_0$ is generated, as an algebra, by the smaller set $\{u(e_i) : i = 1,\cdots,d\}$ of unitaries. We shall use the shorter notation, $u_i = u(e_i)$, $i = 1,\cdots,d$, for convenience. If δ_1,δ_2 are mappings from $\mathfrak{A}_0$ to $L^2(\mathfrak{A},\tau)$ where τ is the trace, we define a sesquilinear form,

$$(\delta_1,\delta_2) := \sum_{i=1}^{d} \tau(\delta_1(u_i)^*\delta_2(u_i)). \tag{3.2}$$

It is linear in the second variable, and satisfies $(\delta_1,\delta_2) = \overline{(\delta_2,\delta_1)}$. We also have $(\delta,\delta) \geq 0$ by the positivity of the trace τ. But, in general, it does not follow that $\delta = 0$ if $(\delta,\delta) = 0$. But if δ is furthermore assumed to be a derivation, and $(\delta,\delta) = 0$, then we may infer that $\delta = 0$. If $(\delta,\delta) = 0$, then $\delta(u_i) = 0$, $i = 1,\cdots,d$. Since the elements $\{u_i\}$ generate $\mathfrak{A}_0$ as an algebra, we can show from this, using Leibniz' rule, that then $\delta(a) = 0$ for all $a \in \mathfrak{A}_0$.

We now recall the basis derivations with domain $\mathfrak{A}_0$.

For $x \in \mathbb{R}^d$, the derivation δ_x is defined as follows,

$$\delta_x(u(\xi)) = \sqrt{-1}\, x\cdot\xi\, u(\xi), \quad \xi \in \Gamma = \mathbb{Z}^d, \tag{3.3}$$

where $x\cdot\xi$ refers to the (usual) dot–product in $\mathbb{R}^d$.

For $\gamma \in \Gamma$, the *inner* derivation $\delta_\gamma := \mathrm{ad}(u(\gamma))$ is defined by

$$\delta_\gamma(a) = u(\gamma)a - au(\gamma), \quad a \in \mathfrak{A}_0. \tag{3.4}$$

We shall need the following

Lemma 3.1. *The following orthogonality relations are satisfied for the inner product* $(\cdot,\cdot)$ *and the basis derivations* $\{\delta_x : x \in \mathbb{R}^d\}$ *and* $\{\delta_\gamma : \gamma \in \Gamma = \mathbb{Z}^d\}$:

(i) $$(\delta_x, \delta_y) = x\cdot y\,, \quad x,y \in \mathbb{R}^d;$$

(ii) $$(\delta_x, \delta_\gamma) = 0\,, \quad x \in \mathbb{R}^d\,, \quad \gamma \in \Gamma;$$

(iii) $$(\delta_{\gamma_1}, \delta_{\gamma_2}) = 0 \quad \text{if } \gamma_1 \neq \gamma_2 \ \textit{in } \Gamma;$$

(iv) $$(\delta_\gamma, \delta_\gamma) = \sum_{i=1}^{d} |1 - \rho(e_i, \gamma)|^2\,, \quad \gamma \in \Gamma.$$

Proof. Since the proof amounts to only straightforward computations, we shall omit details.

We add two comments: If $\gamma = 0$ in $\Gamma = \mathbb{Z}^d$, then $\delta_0 = \mathrm{ad}(u(0))$. But $u(0)$ is the identity element in the algebra $\mathfrak{A}_0$, and in the C^*–algebra completion $\mathfrak{A}$. So $\delta_0 = 0$.

Further, the bicharacter ρ is assumed to be nondegenerate. So if $\gamma \neq 0$ in Γ,

$\rho(e_i,\gamma) \neq 1$ for some i, and it follows that $(\delta_\gamma,\delta_\gamma) > 0$ from the expression (iv) in the lemma.

We also need

Lemma 3.2. *The completion of the space* $\mathrm{Der}(\mathfrak{A}_0)$ *of all derivations from* $\mathfrak{A}_0$ *to* $\mathfrak{A}_0$ *in the Hilbert norm defined by the inner product* $(\cdot,\cdot)$ *on* $\mathrm{Der}(\mathfrak{A}_0)$ *is the space of all module derivations from* $\mathfrak{A}_0$ *to* $L^2(\mathfrak{A},\tau)$.

Proof. The details are quite straightforward and will be omitted.
We just recall that a derivation δ from $\mathfrak{A}_0$ into $L^2(\mathfrak{A},\tau)$ is a linear mapping

$$\delta : \mathfrak{A}_0 \longrightarrow L^2(\mathfrak{A},\tau) \tag{3.5}$$

satisfying the natural Leibniz rule when $L^2(\mathfrak{A},\tau)$ is viewed as a bimodule over the algebra $\mathfrak{A}_0$. For a discussion of module derivation domains, and closability in the respective graph norms, we refer to [Con 3], and [Jo 5], [PSa 1], and [BR, Sec. 3.2.4].

Definitions 3.3. (a) The linear space $\{\delta_x : x \in \mathbb{C}^d\}$ is a commutative family of derivations $\delta_x : \mathfrak{A}_0 \longrightarrow \mathfrak{A}_0$, and it will be denoted

$$A = \{\delta_x : x \in \mathbb{C}^d\}.$$

(b) A derivation $\delta : \mathfrak{A}_0 \longrightarrow \mathfrak{A}$ is said to be *approximately inner* if there is a sequence of elements $\{h_n : n = 1,2,\cdots\}$ in $\mathfrak{A}$ such that

$$\lim_{n\to\infty} ||\delta(a) - \mathrm{ad}(h_n)(a)|| = 0 \tag{3.6}$$

for all $a \in \mathfrak{A}_0$, where the norm $\|\cdot\|$ is now *the* C^**-norm* of the C^*–algebra $\mathfrak{A}$, and where

$$\mathrm{ad}(h_n)(a) := h_n a - a h_n, \quad a \in \mathfrak{A}_0.$$

(c) A derivation,

$$\delta : \mathfrak{A}_0 \longrightarrow L^2(\mathfrak{A},\tau)$$

is said to be *approximately inner in the* L^2*–norm* if there is a sequence $\{h_n : n = 1,2,\cdots\}$ in $\mathfrak{A}$ such that

$$\lim_{n\to\infty} \|\delta(a) - \mathrm{ad}\, h_n(a)\|_2 = 0 \tag{3.7}$$

for all $a \in \mathfrak{A}_0$ where $\|\cdot\|_2$ now refers to the trace–norm given by

$$\|b\|_2 = \tau(b^*b)^{\frac{1}{2}}, \quad b \in \mathfrak{A}, \tag{3.8}$$

where again τ is the canonical trace on $\mathfrak{A}$.

<u>**Theorem 3.4**</u> (Bratteli–Elliott–Jorgensen [BEJ]). *Let* $\Gamma = \mathbb{Z}^d$, *and let* ρ *be a nondegenerate antisymmetric bicharacter on* Γ, *and let* $\mathfrak{A}$ *be the corresponding simple* C^**–algebra obtained by completing the set of generators* $\{u(\xi) : \xi \in \Gamma\}$ *on the relations* (3.1). *Let* $\mathfrak{A}_0$ *be the dense subalgebra generated by* $\{u(\xi)\}$.

(a) *Then every derivation,* $\delta : \mathfrak{A}_0 \longrightarrow \mathfrak{A}$, *decomposes uniquely in the form*

$$\delta = \delta_x + \mathfrak{F} \tag{3.9}$$

where $\delta_x \in A$ *and* $\mathfrak{F}$ *is approximately inner.*

(b) *Every module derivation* $\delta : \mathfrak{A}_0 \longrightarrow L^2(\mathfrak{A},\tau)$ *decomposes uniquely in the form* (3.9) *as specified in* (a) *with the modification that the second component* $\mathfrak{F}$ *is approximately inner in the* L^2*—norm only.*

__Proof.__ We sketch a proof below which is different from the one which was originally given in [BEJ], see also [Jo 6] for module derivations. The original proof was based on a detailed harmonic analysis of the (canonical) ergodic action,

$$\alpha : \mathbb{T}^d : \longrightarrow \mathrm{Aut}(\mathfrak{A}), \tag{3.10}$$

of the d—torus which is generated by the abelian Lie algebra $A_{\mathbb{R}} = \{\delta_x : x \in \mathbb{R}^d\}$.

Let $\Gamma = \mathbb{Z}^d$ be realized as the dual of the compact group $\mathbb{T}^d$, and let the pairing be given by the notation $\langle\xi,k\rangle$, $\xi \in \Gamma$, $k \in \mathbb{T}^d$. We recall that the action in (3.10) is determined by

$$\alpha_k(u(\xi)) = \langle\xi,k\rangle u(\xi), \ \ k \in \mathbb{T}^d, \ \ \xi \in \mathbb{Z}^d. \tag{3.11}$$

The proof, which is given below, has the virtue of applying simultaneously to (a) and (b) in the theorem, and further identifies the decomposition as a pure Hilbert space *orthogonal* decomposition.

Let $\mathscr{H}$ be the Hilbert space of all module derivations

$$\delta : \mathfrak{A}_0 \longrightarrow L^2(\mathfrak{A},\tau). \tag{3.12}$$

Then $A = \{\delta_x : x \in \mathbb{C}^d\}$ is a d–dimensional complex linear subspace of $\mathcal{H}$. As a result, it is closed (relative to the Hilbert norm defined by the inner product on $\mathcal{H}$, given by (3.2) above). Let P_A denote the corresponding *orthogonal* projection of $\mathcal{H}$ onto A, and define the two components of δ as follows:

$$\delta_x := P_A(\delta) \tag{3.13}$$

and

$$\tilde{\delta} := \delta - P_A(\delta). \tag{3.14}$$

It remains to show (the main problem!) that $\tilde{\delta}$ is approximately inner as specified in (a), and (b), respectively. The uniqueness part is then immediate from the (basic) projection theorem of Hilbert space, applied to the orthogonal projection P_A.

We shall need two additional lemmas:

__Lemma 3.5__. *The ergodic action,* $\alpha : \mathbb{T}^d \longrightarrow \mathrm{Aut}(\mathfrak{A})$, *given by formula* (3.11), *implements a strongly continuous unitary representation* U *of* $\mathbb{T}^d$ *on* $\mathcal{H}$ *as follows:*

$$U_k(\delta) := \alpha_k \circ \delta \circ \alpha_k{-1}\ , \quad \delta \in \mathcal{H},\ k \in \mathbb{T}^d, \tag{3.15}$$

and the corresponding spectral subspaces are

$$\{\delta \in \mathcal{H} : U_k(\delta) = \delta, k \in \mathbb{T}^d\} = \{\delta_x : x \in \mathbb{C}^d\} \tag{3.16}$$

and

$$\{\delta \in \mathcal{H} : U_k(\delta) = \langle \gamma,k\rangle \delta\ ,\ k \in \mathbb{T}^d\} = \mathbb{C}\delta_\gamma \tag{3.17}$$

for all $\gamma \in \Gamma = \mathbb{Z}^d$.

Lemma 3.6. *The set*

$$\{\delta_x : x \in \mathbb{C}^d\} \cup \{\delta_\gamma : \gamma \in \Gamma\}$$

(which is orthogonal relative to the inner product on $\mathcal{H}$*) spans a dense subspace of* $\mathcal{H}$ *relative to the norm defined by the inner product.*

Proofs. It is clear that $U_k(\delta_x) = \delta_x$ and $U_k(\delta_\gamma) = \langle\gamma,k\rangle\delta_\gamma$ hold for all $k \in \mathbb{T}^d$, $x \in \mathbb{C}^d$, and $\gamma \in \Gamma$. We claim that $\{U_k : k \in \mathbb{T}^d\}$ is a unitary representation on $\mathcal{H}$, and that the corresponding spectral subspaces are spanned by the respective vectors δ_x and δ_γ.

This follows from the known fact that $\{\alpha_k : k \in \mathbb{T}^d\}$ implements a unitary representation on $L^2(\mathfrak{A},\tau)$ with known spectral subspaces. The latter fact, in turn, is a corollary of the GNS–representation applied to the canonical trace τ on $\mathfrak{A}$. This trace is unique (since the bicharacter ρ is assumed nondegenerate) which implies that τ preserves every automorphism of $\mathfrak{A}$.

The second lemma (Lemma 3.6) now follows from the first by a direct application of the spectral theorem (in its SNAG (Stone–Naimark–Ambrose–Godemont) form [Am], [Ma]) applied to the unitary representation $\{U_k : k \in \mathbb{T}^d\}$ on $\mathcal{H}$.

Note. The following convenient notation will be used in the sequel:

$$U_k(\delta)= \mathrm{Ad}(\alpha_k)(\delta) = \mathrm{Ad}(k)(\delta) := \alpha_k \circ \delta \circ \alpha_k{-1}$$

defined for $k \in \mathbb{T}^d$ and $\delta \in \mathcal{H}$.

<u>**Proof of Theorem 3.4**</u> (continued). For $x \in \mathbb{C}^d$ and $\gamma \in \mathbb{Z}^d$ we have $x = \sum_i x_i e_i$, $\gamma = \sum_i \gamma_i e_i$ with $x_i \in \mathbb{C}$, $\gamma_i \in \mathbb{Z}$. Now define $|x| = (\sum |x_i|^2)^{1/2}$ and $|\gamma| = (\sum_i \gamma_i^2)^{1/2}$. Further, let $\Gamma_n = \{\gamma \in \Gamma = \mathbb{Z}^d : |\gamma| \le n\}$ and $\mathcal{H}_n = \text{span}\{\delta_\gamma : \gamma \in \Gamma_n\}$. Since Γ_n, $n = 1,2,\cdots$, is finite, each subspace $\mathcal{H}_n$ is finite–dimensional and closed. Let P_n be the corresponding *orthogonal* projection of $\mathcal{H}$ onto $\mathcal{H}_n$. It follows from the spectral theorem (i.e., the SNAG–theorem applied to $(U,\mathcal{H})$) that

$$I = P_A \oplus (\lim_{n\to\infty} P_n) \tag{3.18}$$

where I denotes the identity operator of $\mathcal{H}$.

In each of the two cases, (a) and (b) discussed in Theorem 3.4, we have $\delta \in \mathcal{H}$ where δ is the given derivation.

It now follows from (3.18) that

$$\lim_{n\to\infty} \|\delta - P_A(\delta) - P_n(\delta)\|_{\mathcal{H}} = 0 \tag{3.19}$$

where $\|\cdot\|_{\mathcal{H}}$ is the Hilbert norm on $\mathcal{H}$ defined by the inner product. We have already shown (Lemma 3.6) that $P_A\delta$ is of the form δ_x for $x \in \mathbb{C}^d$, and (Lemmas 3.5–6) that $P_n(\delta)$ is spanned by the finite set of inner derivations given by $\{\delta_\gamma : \gamma \in \Gamma_n\}$ where $\delta_\gamma = \text{ad}(u(\gamma))$.

This completes part (b) of the theorem since

$$\tilde{\delta}(a) := \delta(a) - \delta_x(a) = \lim_{n\to\infty} (P_n\delta)(a) \tag{3.20}$$

holds for $a \in \mathfrak{A}_0$ where the limit is in the L^2–norm on $L^2(\mathfrak{A},\tau)$ (which, as noted, is defined by the trace τ, cf. formula (3.2) and (2.3)).

In part (a), it remains to show that, for a given derivation $\delta : \mathfrak{A}_0 \longrightarrow \mathfrak{A}$ of the C^*–algebra, the limit (3.20) is also valid *relative to the* C^**–norm.*

This is *not* quite true as stated. The limit, relative to the C^*–norm, holds only in the *Cesàro sense.* But this is enough to show that the derivation (difference), $\gamma = \delta - \delta_x$ is approximately inner. The Cesàro limit is

$$\gamma(a) = \lim_{N \to \infty} \frac{1}{N} \sum_{n=1}^{N} (P_n \delta)(a) \tag{3.21}$$

for $a \in \mathfrak{A}_0$. This limit does exist, relative to the C^*–norm, for all $a \in \mathfrak{A}_0$.

The proof of the last assertion may be completed in (at least) two ways: We can get it as a consequence of the pointwise ergodic theorem of G. Birkhoff [Wal, Sect. 1.6]; or, we may obtain it, as in the original paper [BEJ], by applying the Fejér kernel to the ergodic $\mathbf{T}^d$–action. We refer to [BEJ] for details on this last point.

From the Cesàro–mean construction in the above proof we also read off the following,

Corollary 3.7. *For derivations* $\delta \in \mathrm{Der}(\mathfrak{A}_0,\mathfrak{A})$ *the following are equivalent:*

(i) *There is a sequence* (h_n) *in* $\mathfrak{A}$ *such that*

$$\lim_{n \to \infty} \|\delta(a) - (\mathrm{ad}\, h_n)(a)\| = 0, \quad a \in \mathfrak{A}_0,$$

holds in the C^**–norm.*

(ii) *There is a sequence* (k_n) *in* $\mathfrak{A}$ *such that*

$$\lim_{n\to\infty} \|\delta - \mathrm{ad}(k_n)\|_2 = 0$$

where $\|\cdot\|_2$ *denotes the Hilbert–norm on the derivations.*

Corollary 3.8. *Let* $\mathfrak{A}$ *be the simple* C^**–algebra based on a given antisymmetric nondegenerate bicharacter, and let* $\{u(\xi) : \xi \in \mathbb{Z}^d\}$ *be the generating unitary elements, cf.* (3.1) *above. Let* $\mathfrak{A}_0$ *be the algebra generated by the* $u(\xi)$'s *and let* τ *be the trace on* $\mathfrak{A}$.

Then a given derivation,

$$\delta : \mathfrak{A}_0 \longrightarrow \mathfrak{A}$$

is ***norm*** *approximately inner if and only if*

$$\tau(u(\xi)^* \delta(u(\xi))) = 0\ , \quad \xi \in \mathbb{Z}^d. \tag{3.22}$$

Proof. If $\{h_n\}$ is a sequence of elements in $\mathfrak{A}$ such that

$$\lim_{n\to\infty} \|\delta(a) - [h_n, a]\| = 0$$

for all $a \in \mathfrak{A}_0$ (where $\|\cdot\|$ is the C^*–norm), then

$$\tau(u(\xi)^* \delta(u(\xi))) = \lim_{n\to\infty} \tau(u(\xi)^*[h_n, u(\xi)]) = 0$$

for all ξ, by virtue of the trace property. This is the easy implication of the corollary. Both implications follow, in fact, from Theorem 3.4 where the approximately inner derivations are characterized as the orthogonal complement of $A = \{\delta_x : x \in \mathbb{C}^d\}$. It therefore suffices to show that $(\delta_x,\delta) = 0$ for all $x \in \mathbb{C}^d$ if and only if condition (3.22) is satisfied.

We have

$$0 = (\delta_x,\delta) = \sum_i \tau(\delta_x(u_i)^*\delta(u_i)) = -\sqrt{-1}\sum_i(\overline{x}\cdot e_i)\tau(u_i^*\,\delta(u_i))$$

for all $x \in \mathbb{C}^d$. This is equivalent to the identity

$$\tau(u_i^*\,\delta(u_i)) = 0\,, \quad i = 1,\cdots,d.$$

But the mapping

$$\xi \longrightarrow \tau(u(\xi)^*\delta(u(\xi))) \tag{3.23}$$

is a homomorphism on $\mathbb{Z}^d$ so it vanishes identically iff it vanishes on the basis vectors e_i, $i = 1,\cdots,d$. The corollary now follows from Theorem 3.4 as noted above.

§4. SMOOTH DERIVATIONS

Let ρ be a given antisymmetric nondegenerate bicharacter on the discrete group $\Gamma = \mathbb{Z}^d$, let $\{u(\xi) : \xi \in \Gamma\}$ be the generating unitaries satisfying (3.1). Let $\mathfrak{A}$ be the corresponding simple C^*–algebra, and let

$$\alpha : \mathbf{T}^d \longrightarrow \mathrm{Aut}(\mathfrak{A})$$

be the canonical ergodic action of the d–torus, cf. (3.11).

The algebra $\mathfrak{A}^\infty$ of smooth elements in $\mathfrak{A}$ is defined as

$$\mathfrak{A}^\infty = \{a \in \mathfrak{A} : (k \longrightarrow \alpha_k(a)) \in C^\infty(\mathbf{T}^d,\mathfrak{A})\}. \tag{4.1}$$

It is well known that $a \in \mathfrak{A}^\infty$ iff, for all $n \in \{1,2,\cdots\}$, there is a constant C_n such that

$$|\tau(u(\xi)^*a)| \le \frac{C_n}{(1+|\xi|)^n} \quad \text{for all } \xi \in \mathbb{Z}^d \tag{4.2}$$

where τ denotes the trace on $\mathfrak{A}$.

We proved in [BEJ] and [BGJ] that every derivation $\delta : \mathfrak{A}_0 \longrightarrow \mathfrak{A}^\infty$ extends uniquely, by closure, to a derivation from $\mathfrak{A}^\infty$ to $\mathfrak{A}^\infty$, and further that $\mathfrak{A}_0$ is a core for the extension to $\mathfrak{A}^\infty$. A derivation of the ring $\mathfrak{A}^\infty$ is said to be a *smooth derivation.* A derivation of the ring $\mathfrak{A}_0$ is said to be a *finite derivation.* Since $\mathfrak{A}_0 \subset \mathfrak{A}^\infty$, it follows that finite derivations are smooth. Both classes of derivations form a Lie algebra (infinite-dimensional) under the commutator bracket,

$$[\delta_1,\delta_2] := \delta_1\delta_2 - \delta_2\delta_1. \tag{4.3}$$

<u>**Corollary 4.1**</u> (Jorgensen [Jo 5]). *The approximately inner derivations in* $\mathrm{Der}(\mathfrak{A}^\infty)$ *form an ideal under the commutator bracket.*

<u>**Proof**</u>. Let δ_1,δ_2 be two smooth derivations and assume that δ_2 is approximately inner. The assertion is that $\delta := [\delta_1,\delta_2]$ is also approximately inner. To verify this, it is enough, by virtue of Corollary 3.8 above, to check that δ satisfies condition (3.22). Since the Lie algebra

$$A = \{\delta_x : x \in \mathbb{C}^d\} \tag{4.4}$$

is abelian, the result is equivalent to the assertion that the commutator of *any* pair of smooth derivations is approximately inner.

In the computation, we shall use the following fact (which is immediate from Theorem 3.4 in the above formulation): For every derivation δ defined on $\mathfrak{A}_0$, or on $\mathfrak{A}^\infty$, we have

$$\tau(\delta(a)) = 0 \tag{4.5}$$

for all elements a in the domain.

Furthermore, there is a 1–1 correspondence between derivations δ, defined on $\mathfrak{A}_0$, and cocycles, defined on Γ with values in $\mathfrak{A}$. If the derivation δ is given, then the cocycle $c = c_\delta$ is the function, $c : \Gamma \longrightarrow \mathfrak{A}$, defined by

$$c(\xi) := u(\xi)^*\delta(u(\xi)) , \quad \xi \in \Gamma. \tag{4.6}$$

It satisfies

$$c(\xi+\gamma) = u(\gamma)^* \, c(\xi)u(\gamma) + c(\gamma), \quad \xi,\gamma \in \Gamma. \tag{4.7}$$

This is called the cocycle identity. Conversely, every cocycle defines a derivation by virtue of formula (4.6).

Let δ_1,δ_2 be a pair of derivations with cocycles c_1,c_2, respectively. Then the cocycle c_δ of the commutator $\delta=[\delta_1,\delta_2]$ is given by

$$c_\delta(\xi)=\delta_1(c_2(\xi))-\delta_2(c_1(\xi))+[c_1(\xi),\,c_2(\xi)],\text{ for all }\ \xi\in\Gamma \tag{4.8}$$

With this, the verification of (3.22) is easy. We have:

$$\tau(u(\xi)^*\delta(u(\xi)))=\tau(c_\delta(\xi))=\tau(\delta_1(c_2(\xi)))-\tau(\delta_2(c_1(\xi)))+\tau([c_1(\xi),c_2(\xi)])=0$$

for all $\xi\in\Gamma$ where we use the formula for the cocycle c_δ, and the property of trace τ. Corollary 3.8 then applies, and we conclude that δ is approximately inner.

<u>Corollary 4.2</u>. *Let $\delta\longrightarrow\delta^*$ be the involution defined by*

$$\delta^*(a):=-\delta(a^*)^*\,,\quad a\in\mathfrak{A}_0.$$

Then

$$(\delta_1,\delta_2)=(\delta_2^*,\delta_1^*)$$

for all $\delta_1,\delta_2\in\mathrm{Der}(\mathfrak{A}_0,\mathfrak{A})$. It follows that the inner product is real valued on derivations δ satisfying $\delta^*=-\delta$; and that the mapping J defined by $J(\delta):=\delta^*$, $\delta\in\mathscr{H}$, is an involutory conjugation on the derivations.

<u>Proof</u>. Compute!

We now turn to the smoothness considerations. The bicharacter ρ is said to be *generic* if, for all $\gamma \in \Gamma$, there is a constant $C = C_\gamma$, and an integer N, such that

$$|(1-\rho(\gamma,\xi))^{-1}| \leq C|\xi|^N. \tag{4.9}$$

The antisymmetric bicharacters form a compact group (denoted $(\Gamma \wedge \Gamma)^\wedge$) and it is known, see [Sch] and [Ell]) that the generic bicharacters are dense relative to the Haar measure of this group.

We now have the next corollary:

<u>**Corollary 4.3**</u> ([BEJ] and [Con 7]). (a) *All finite derivations decompose uniquely as*

$$\delta = \delta_X + \mathfrak{d} \tag{4.10}$$

where $\delta_X \in A$ *and* $\mathfrak{d}$ *is inner by an element in* $\mathfrak{A}_0$.

(b) *Suppose* ρ *is generic. Then every smooth derivation* δ *decomposes as in* (4.10) *where now the inner component* $\mathfrak{d}$ *is inner by an element in* $\mathfrak{A}^\infty$.

<u>**Proof**</u>. The derivation δ is given and we assume that one of the assumptions (a) or (b) is satisfied. Let P_A be the orthogonal projection, relative to the Hilbert space $\mathscr{H}$ of all derivations, onto $A_{\mathbb{C}} \approx \mathbb{C}^d$; and for each $\gamma \in \Gamma$, let P_γ be the projection onto the one–dimensional space spanned by δ_γ in $\mathscr{H}$. Recall

$$\delta_\gamma := \mathrm{ad}(u(\gamma)).$$

We then have the orthogonal expansion

$$\delta = P_A(\delta) + \sum_\gamma P_\gamma(\delta) \tag{4.11}$$

where

$$P_\gamma(\delta) = (\delta_\gamma,\delta)(\delta_\gamma,\delta_\gamma)^{-1}\delta_\gamma = \left(\frac{\sum_i \tau(u(\gamma)\,\delta(u_i)u_i^*)(1-\rho(\gamma,e_i))}{\sum_i |1-\rho(e_i,\gamma)|^2} \right) \delta_\gamma \tag{4.12}$$

which may be summarized as

$$P_\gamma\delta = D(\gamma)\delta_\gamma$$

where the number $D(\gamma)$ is given as the above fraction. It follows that

$$\mathfrak{\delta} = \mathrm{ad}(h)$$

where

$$h = \sum_\gamma D(\gamma)u(\gamma), \tag{4.13}$$

and we claim that this sum is finite if δ is assumed finite. The sum is convergent in the C^*–norm, and $h \in \mathfrak{A}^\infty$, if δ is smooth and ρ satisfies the generic condition.

First apply the Cauchy–Schwarz inequality to the expression for $D(\gamma)$. We get

$$|D(\gamma)| \leq \left(\frac{\sum_i |\tau(u(\gamma)\delta(u_i)u_i^*)|^2}{\sum_i |1-\rho(e_i,\gamma)|^2} \right)^{1/2} \tag{4.14}$$

If δ is finite, the elements $\delta(u_i)u_i^*$, $i = 1,\cdots,d$, are contained in a finite–dimensional linear subspace of $\mathfrak{A}_0$ so the coefficients $\gamma \longrightarrow \tau(u(\gamma)\delta(u_i)u_i^*)$ must vanish for γ outside some finite subset of Γ. This shows that the expression in (4.13) is a finite sum and the proof of (a) is completed.

With the assumptions from (b) there are $N \in \{1,2,\cdots\}$ and $C_N < \infty$ such that

$$\left(\sum_i |1-\rho(e_i,\gamma)|^2\right)^{-1/2} \leq C_N|\gamma|^N \quad \text{for all } \gamma \in \Gamma.$$

For every $k \in \{1,2,\cdots\}$, there is a constant C_k such that

$$\left(\sum_i |\tau(u(\gamma)\delta(u_i)u_i^*)|^2\right)^{-1/2} \leq C_k(1+|\gamma|)^{-k}.$$

So, for $\ell \in \{1,2,\cdots\}$, we have

$$|\gamma|^\ell |D(\gamma)| \leq C_N C_k (1+|\gamma|)^{-k+N+\ell}.$$

But the sum

$$\sum_\gamma (1+|\gamma|)^{-k+N+\ell}$$

is finite if $k > N + \ell + \frac{d}{2}$. It follows that the expression (4.13) for the implementing element h is convergent in $\mathfrak{A}^\infty$, and the proof of (b) is completed.

In the next section we develop some purely Lie algebraic results which will be used subseqently in the analysis of smooth Lie actions on the noncommutative tori. The results from Sec. 5 were stated and proved in [BEGJ] for the specialized setting of derivation Lie algebras and diophantine assumptions. We show here that the setting is, in fact, pure Lie algebra theory, and the results are extended somewhat.

§5. EXTENSIONS OF LIE ALGEBRAS

Let A and M be Lie algebras over the same ground field $\mathbb{R}$ or $\mathbb{C}$. Let Der(M) be the Lie algebra of all Lie derivations of M. It follows from the Jacobi–identity that Der(M) is again a Lie algebra under commutator bracket. Let

$$F : A \longrightarrow \mathrm{Der}(M) \tag{5.1}$$

be a representation of A into Der(M). We have, in particular, the identity

$$F_{[a_1,a_2]}(m) = [F_{a_1},F_{a_2}](m) = F_{a_1}(F_{a_2}m)-F_{a_2}(F_{a_1}m) \text{ for all } a_1,a_2 \in A \text{ and } m \in M.$$

Then there is a universal extension $E_A(M)$ of A by M. The elements in $E_A(M)$ may be described as pairs (a,m), $a \in A$, $m \in M$; and the Lie bracket in $E_A(M)$ is defined by

$$[(a,m),(a',m')]= ([a,a'],F_a m' - F_{a'} m + [m,m']) \tag{5.2}$$

for $a,a' \in A$ an $m,m' \in M$.

We then get a natural short exact sequence

$$0 \longrightarrow M \xrightarrow{\ h\ } E_A(M) \xrightarrow{\ f\ } A \longrightarrow 0$$

in the category of Lie algebras where the connecting Lie homomorphisms

$$h: M \longrightarrow E_A(M) \quad \text{and} \quad f : E_A(M) \longrightarrow A$$

are defined by

$$h(m) = (0,m)\,, \qquad m \in M \tag{5.3}$$

and

$$f(a,m) = a\,, \qquad a \in A. \tag{5.4}$$

If the two Lie algebras A and M carry involutions, then we shall assume that the representation F of A in Der(M) respects the involutions. Specifically, we assume

$$F_a(m^*) = -F_{a^*}(m)^*\,, \qquad a \in A, \quad m \in M. \tag{5.5}$$

We then define an involution on $E_A(M)$ by

$$(a,m)^* = (a^*,m^*)\,, \qquad a \in A, \quad m \in M,$$

and note that then

$$[x_1,x_2]^* = -[x_1^*,x_2^*] \tag{5.6}$$

holds for all $x_1,x_2 \in E_A(M)$.

Whenever involutions exist on A and M, we shall always assume that the compatibility condition (5.5) is satisfied for the representation of A in Der(M).

<u>**Theorem 5.1**</u>. *Let* L *be a finite-dimensional Lie subalgebra of* $E_A(M)$. *Let*

$$f : E_A(M) \longrightarrow A$$

be the canonical homomorphism defined by

$$f(a,m) := a\,, \qquad a \in A, \quad m \in M.$$

Let

$$L_0 := f(L) \quad \text{and} \quad L_1 := \ker(f|_L). \tag{5.7}$$

We assume that the adjoint representation of L *on the ideal* L_1 *is completely diagonalizable, and further that* L_0 *is abelian.*

(a) *Then one of the following two conditions must hold:*

(i) *The short exact sequence*

$$0 \longrightarrow L_1 \longrightarrow L \longrightarrow L_0 \longrightarrow 0$$

splits in the category of Lie algebras.

(ii) L *contains a Lie subalgebra which is isomorphic to the* 3*−dimensional Heisenberg algebra.*

(b) *The two properties* (i) *and* (ii) *are mutually exclusive.*

(c) *If* z *is a nonzero element in* [L,L], $z = [x_1,x_2]$, $x_i \in L$, *and if* z *commutes with* x_i, $i = 1,2$, *then it follows that* z *is in the center of* L.

(d) *If* L *contains a copy of the Heisenberg Lie algebra, then* $\dim L_0 \geq 2$.

__Corollary 5.2__. *Let L be a finite−dimensional solvable (non−nilpotent) Lie subalgebra of* $E_A(M)$ *such that the adjoint representation of* L *on* L_1 *is completely diagonalizable and* L_0 *is abelian. Then the short exact sequence*

$$0 \longrightarrow L_1 \longrightarrow L \longrightarrow L_0 \longrightarrow 0$$

splits if and only if the derived ideal $[L,L]$ *is abelian.*

Corollary 5.3. *Let* $L \subset E_A(M)$ *be a finite-dimensional Lie subalgebra, and let*

$$0 \longrightarrow L_1 \longrightarrow L \longrightarrow L_0 \longrightarrow 0$$

be the associated short exact sequence. Assume that L_0 *is abelian and that the restriction to* L_1 *of the adjoint representation of* L *is completely diagonalizable.*

(a) *Then there is an (abelian) subalgebra* T *of* L *such that*

$$0 \longrightarrow (T \cap L_1) \longrightarrow T \longrightarrow f(T) \longrightarrow 0$$

is split, and the following maximality condition holds: Whenever T' *is another such subalgebra such that* $f(T) \subset f(T')$, *then* $f(T) = f(T')$, *or equivalently,* $T' \subset T+L_1$.

(b) *Furthermore, there is a linear subspace* U *of* L *such that* $[U,U] \subset \mathscr{Z}(L)$ and $f(U) = L_0$, *where* $\mathscr{Z}(L)$ *denotes the center of* L.

(c) *If* T *is chosen as in* (a), *then* $f(T)$ *is a* ***proper*** *Lie subalgebra of* L_0 *if, and only if,* $[U,U] \neq 0$ *for every subspace* U *satisfying the conditions in* (b).

Suppose now that the two Lie algebras A and M carry compatible involutions which we shall denote $*$ in both cases. It follows that $E_A(M)$ inherits the natural involution which is defined by

$$(a,m)^* = (a^*,m^*) , \quad a \in A, \quad m \in M.$$

We shall say that a Lie subalgebra L of $E_A(M)$ is *real* if

$$x^* = -x \qquad \text{for all } x \in L. \tag{5.8}$$

__Corollary 5.4__. *Suppose the Lie subalgebra of* $E_A(M)$ *satisfies the conditions in Theorem* 5.1, *and suppose further that* L *is real relative to a compatible involution. Assume further that the adjoint representation of* L *on the complexification* $L_1^{\mathbb{C}}$ *is completely diagonalizable with purely imaginary roots.*

Then the conclusions in Theorem 5.1 *hold with the modification of* (i) *that the exact sequence*

$$0 \longrightarrow L_1 \longrightarrow L \longrightarrow L_0 \longrightarrow 0$$

is split in the category of ***real*** *Lie algebras.*

__Proof of Theorem 5.1__. (Due to [BEGJ] in the case of operator Lie algebras with diagonal adjoint representation.) We first consider the case when the ground field is assumed to be $\mathbb{C}$, the complex numbers.

Let T be a linear subspace of L which is commutative, i.e., $[x_1,x_2] = 0$ for all $x_1,x_2 \in T$. We shall choose T such that the dimension of f(T) (as a linear subspace of L_0) is maximal. If T is so chosen (i.e., satisfying the maximal condition), we may have $f(T) = L_0$, or else f(T) a proper subspace of L_0. In the first case, the exact sequence (5.10) splits, and, in the second case, L contains a copy of a Heisenberg–Lie subalgebra.

Consider the adjoint representation of T on L. Since T is abelian, and the ground field is $\mathbb{C}$, L is the direct sum of its generalized eigenspaces relative to the representation. The weight–subspaces will be denoted $\Phi(\psi) \subset L$ where ψ runs over the nonzero weighs. For the zero weight, we have

$$\Phi(\psi) = \{y \in L : \forall\, x \in T\ \exists\, k \text{ s.t. } (\mathrm{ad}x - \psi(x))^k(y) = 0\},$$

$$\Psi(0) = \{y \in L : \forall\, x \in T\ \exists\, k \ \text{ s.t. } (\mathrm{ad}x)^k(y) = 0\}$$

where we recall that ψ is a linear functional on T.

We have

$$L = \Phi(0) + \sum_{\psi}^{\oplus} \Phi(\psi) \tag{5.9}$$

as a direct sum, i.e., with linearly independent terms.

We first note that the spaces $\Phi(\psi)$ for $\psi \neq 0$ are contained in the ideal L_1. For if $\psi(x) \neq 0$, $x \in T$ where ψ is a weight, then (for $y \in \Phi(\psi)$),

$$\begin{aligned} 0 &= (\mathrm{ad}x - \psi(x))^k(y) \\ &= \sum_{i=0}^{k} \begin{bmatrix} k \\ i \end{bmatrix} (-\psi(x))^i (\mathrm{ad}x)^{k-i}(y) \\ &= (-\psi(x))^k y + k(-\psi(x))^{k-1}[x,y] + \cdots + (\mathrm{ad}x)^k(y). \end{aligned}$$

Since all but the first term $(-\psi(x))^k y$ are in $[L,L]$ we conclude that $y \in [L,L]$. But we assumed that L_0 from the exact sequence

$$0 \longrightarrow L_1 \longrightarrow L \longrightarrow L_0 \longrightarrow 0 \tag{5.10}$$

is abelian, and it follows from this that $[L,L] \subset L_1$.

It follows that

$$f(L) = f(\Phi(0)) \simeq \Phi(0)/(\Phi(0) \cap L_1)$$

where we have used the canonical isomorphism

$$L_0 \simeq L/L_1 \tag{5.11}$$

implicit in the given exact sequence (5.10). If the containment

$$f(T) \subset L_0 = f(L) = f(\Phi(0))$$

is proper, there is a vector $f(x')$, $x' \in \Phi(0)$, which is not in $f(T)$. This means that x' is not in $T+L_1$.

For every $x \in T$, there is a positive integer k such that $(\mathrm{ad}x)^k(x') = 0$ since $x' \in \Phi(0)$.

We can now consider the weight–space decomposition of L_1 relative to the adjoint representation of L on L_1. Let the weight spaces be denoted $L_1(0)$, and $L_1(\varphi)$ corresponding to nonzero weights φ. Recall, the weights φ are linear functionals on L, and

$$L_1(0) = \{y \in L_1 : \forall\, u \in L\ \exists\, k \text{ s.t. } (\mathrm{ad}u)^k(y) = 0\},$$

$$L_1(\varphi) = \{y \in L_1 : \forall\, u \in L\ \exists\, k \text{ s.t. } (\mathrm{ad}u-\varphi(u))^k(y) = 0\},$$

and

$$L_1 = L_1(0) + \sum_{\varphi}^{\oplus} L_1(\varphi). \tag{5.12}$$

We shall now consider the element $[x,x']$ relative to the latter of the two root space decompositions, i.e., (5.12). Let

$$[x,x'] = y_0 + \sum_{\varphi} y_{\varphi} \tag{5.13}$$

with $y_0 \in L_1(0)$ and $y_\varphi \in L_1(\varphi)$. Then

$$0 = (\mathrm{ad}x)^k(x') = \sum_{\varphi} \varphi(x)^{k-1} y_\varphi \tag{5.14}$$

if $k \geq 2$. It follows that $\varphi(x)^{k-1} y_\varphi = 0$ for all φ, and therefore

$$\varphi(x)y_\varphi = 0. \tag{5.14'}$$

Substitution of this back into (5.14) yields $(\mathrm{ad}x)^2(x') = 0$.

For an arbitrary pair of elements $x_1, x_2 \in T$, we have

$$[x_i, x'] = y_0^i + \sum_{\varphi} y_\varphi^i, \quad i = 1,2,$$

where the weight space decomposition (5.12) is used on the two elements $[x_i, x'] \in L_1$, $i = 1,2$. But we also have $[x_1, x_2] = 0$, so

$$\begin{aligned} 0 &= [[x_1,x_2],x'] \\ &= [[x_1,x'],x_2] + [x_1,[x_2,x']] \\ &= \sum_{\varphi}(\varphi(x_1)y_\varphi^2 - \varphi(x_2)y_\varphi^1). \end{aligned}$$

We conclude that

$$\varphi(x_1)y_\varphi^2 - \varphi(x_2)y_\varphi^1 = 0 \text{ for all } \varphi \tag{5.15}$$

by virtue of the linear independence of the generalized eigenspaces $L_1(\varphi)$ in the decomposition (5.12).

Let $U = \{\varphi : \varphi(x') \neq 0\}$, $U' = \{\varphi : \varphi(x') = 0\}$, and choose elements $x_1, \cdots, x_r$ in T such that $\{f(x_i)\}_{i=1}^{r}$ is a linear basis for f(T). If

$$[x_i, x'] = y_0^i + \sum_{\varphi} y_{\varphi}^i, \tag{5.16}$$

define new elements by the formula

$$x_i' = x_i + \sum_{\varphi \in U} \frac{1}{\varphi(x')} y_{\varphi}^i \tag{5.17}$$

and note that then

$$[x_i', x_j'] = \sum_{\varphi \in U} \frac{1}{\varphi(x')} (\varphi(x_i) y_{\varphi}^j - \varphi(x_j) y_{\varphi}^i).$$

It follows, by virtue of (5.15), that $\{x_i'\}_{i=1}^{r}$ is a commutative family. Since $f(x_i) = f(x_i')$ the two subspaces, f(T) and f(span$\{x_i'\}$) in L_0, have the same dimension.

For each i, we have

$$\begin{aligned}
[x_i', x'] &= \Big[\sum_{\varphi \in U} \frac{1}{\varphi(x')} y_{\varphi}^i, x' \Big] \\
&= y_i^0 + \sum_{\text{all } \varphi} y_{\varphi}^i - \sum_{\varphi \in U} y_{\varphi}^i \\
&= y_i^0 + \sum_{\varphi \in U'} y_{\varphi}^i
\end{aligned}$$

and, from the definition of U', we get

$$[x',[x_i',x']] = 0.$$

We also have

$$[x_i',[x_i',x']] = [x_i,[x_i',x']] = \sum_{\varphi \in U'} \varphi(x_i) y_\varphi^i$$

which is zero by (5.14′).

The two computations show that, for each i, the element

$$z_i = [x_i', x']$$

commutes with x_i' and with x'. If, for some i, $z_i \neq 0$, then we have a copy of the Heisenberg Lie algebra contained in L. But, if, for all i, $z_i = 0$, then the commutative family $\{x_1', \cdots, x_r', x'\}$ satisfies

$$\dim f(\text{span}\{x_1', \cdots, x_r', x'\}) > \dim f(T),$$

and this cannot happen if T is chosen such that $\dim f(T)$ is maximal. It follows that, in this case, the split property (i) from the conclusion of Theorem 5.1 is satisfied.

We finally show that the two possibilities listed in Theorem 5.1 are mutually exclusive. We show that if (5.10) is split exact, then L cannot contain a Heisenberg Lie subalgebra. Indeed, if (5.10) splits, then L contains an abelian subalgebra T such that $L = T + L_1$. We shall show that if u_1,u_2 are elements in L such that $z = [u_1,u_2]$

commutes with both u_1 and u_2, then $z = 0$. Write

$$u_i = x_i + y_i\,, \quad x_i \in T, \quad y_i \in L_1, \quad i = 1,2,$$

and

$$y_i = y_i^0 + \sum_{\varphi} y_i^{\varphi} \tag{5.18}$$

with $y_i^0 \in L_1(0)$ and $y_i^{\varphi} \in L_1(\varphi)$ where we use the root–space decomposition of L_1 listed in (5.12) above.

We then have

$$\begin{aligned} z = [u_1,u_2] &= [y_1,x_2] + [x_1,y_2] \\ &= \sum_{\varphi} (\varphi(x_1)y_2^{\varphi} - \varphi(x_2)y_1^{\varphi}) \end{aligned} \tag{5.19}$$

and

$$\begin{aligned} [u_i, z] &= [x_i, z] \\ &= \sum_{\varphi} \varphi(x_i)(\varphi(x_1)y_2^{\varphi} - \varphi(x_2)u_1^{\varphi}) \\ &= 0. \end{aligned}$$

It follows that

$$\varphi(x_1)y_2^{\varphi} - \varphi(x_2)y_1^{\varphi} = 0 \quad \text{for all } \varphi,$$

and, by substitution into (5.19), we conclude that $z = 0$.

<u>**Proof of (c)**</u> (Theorem 5.1 continued). We must show that if x_1,x_2,z, $z = [x_1,x_2]$,

$x_i \in L$, generate a copy of the Heisenberg Lie algebra, then z is in the center of L.

Part (d) of the theorem amounts to the assertion that the two vectors $\{f(x_i) : i = 1,2\}$ are linearly independent in L_0 where $f : L \longrightarrow L_0$ is the canonical homomorphism.

For part (c), let x be an arbitrary element in L. Since $[L,L] \subset L_1$, each of the three elements z and $[x,x_i]$, $i = 1,2$, has a decomposition according to the formula (5.12) for diagonalizing the adjoint representation of L on L_1:

$$[x,x_i] = y_0^i + \sum_{\varphi} y_{\varphi}^i, \quad i = 1,2,$$

and

$$z = y_0^3 + \sum_{\varphi} y_{\varphi}^3 \tag{5.20}$$

where $y_0^i \in L_1(0)$, and $y_{\varphi}^i \in L_1(\varphi)$, $i = 1,1,3$, when φ runs over the set of nonzero eights.

It follows that

$$\begin{aligned}[x,z] &= [x,[x_1,x_2]] \\ &= [[x,x_1], x_2] + [x_1,[x,x_2]] \\ &= \sum_{\varphi} (\varphi(x_1)y_{\varphi}^2 - \varphi(x_2)y_{\varphi}^1)\end{aligned}$$

and we claim that each term in the (latter) sum is zero.

We also have

$$[x,z] = \sum_{\varphi} \varphi(x)y_{\varphi}^3$$

so it follows that

$$\varphi(x_1)y_\varphi^2 - \varphi(x_2)y_\varphi^1 = \varphi(x)y_\varphi^3 \tag{5.21}$$

Finally we have

$$0 = [x_i, z] = \sum_\varphi \varphi(x_i)y_\varphi^3$$

which implies the identity

$$\varphi(x_i)y_\varphi^3 = 0, \quad i = 1,2, \text{ all } \varphi. \tag{5.22}$$

Let

$$a_\varphi := \varphi(x_1)y_\varphi^2 - \varphi(x_2)y_\varphi^1.$$

If $\varphi(x) = 0$ or $y_\varphi^3 = 0$, then (5.21) shows that $a_\varphi = 0$. But if $y_\varphi^3 \neq 0$, then $\varphi(x_i) = 0$, $i = 1,2$, by (5.22), and then clearly $a_\varphi = 0$. It follows that $a_\varphi = 0$ for all φ as asserted. This shows that $[x, z] = 0$ for all $x \in L$, so z is in the center.

Turning now to (d), suppose first that $f(x_2) = 0$ where $\{x_1,x_2, z\}$ is a Heisenberg system. Then $x_2 \in L_1$ and it decomposes according to (5.12) as

$$x_2 = y_0 + \sum_\varphi y_\varphi.$$

$y_0 \in L_1(0)$, $y_\varphi \in L_1(\varphi)$, summation over the nonzero roots.

Therefore

$$z = [x_1,x_2] = \sum_\varphi \varphi(x_1)y_\varphi$$

and

$$0 = [x_1, z] = \sum_\varphi \varphi(x_1)^2 y_\varphi.$$

It follows that $\varphi(x_1)^2 y_\varphi = 0$ for all φ which implies $\varphi(x_1)y_\varphi = 0$, and $z = 0$; a contradiction.

If $\{f(x_i) : i = 1,2\}$ is not linearly independent, we may assume, by symmetry, that, for some $\lambda \in \mathbb{C}$, $f(x_2) - \lambda f(x_1) = 0$. But then $\{x_1, x_2 - \lambda x_1, z\}$ is a Heisenberg system, precisely as above, with the second element $x_1 - \lambda x_1$ in L_1. The earlier argument applies, and the proof of (d) is completed.

Proof of Corollary 5.2. Since the short exact sequence

$$0 \longrightarrow L_1 \longrightarrow L \longrightarrow L_0 \longrightarrow 0 \tag{5.23}$$

is assumed split, it follows from Theorem 5.1 that L does not contain a copy of the Heisenberg Lie algebra. But [L,L] is nilpotent since L is assumed solvable, and it is well known that a non–abelian nilpotent Lie algebra must contain a copy of the Heisenberg. So [L,L] must be abelian as asserted.

If conversely [L,L] is abelian, then L cannot contain a copy of the Heisenberg, and it follows from Theorem 5.1 that (5.23) splits.

Proof of Corollary 5.3. This result is a corollary of the proof of Theorem 5.1 rather than the theorem itself. The construction of T was carried out in the proof of Theorem 5.1(a), and part (a) follows. If f(T) is a proper subspace of L_0 we saw how to pick a basis $\{f(x_i) : i = 1,\cdots,d\}$ for f(T), and $x' \in L$, such that $f(x') \notin f(T)$, but $[x', x_i] \in \mathcal{Z}(L)$, and $[x_i, x_j] = 0$ for all $i,j = 1,\cdots,d$. It follows from this that the subspace U_1 defined by $U_1 = \operatorname{span}_i\{x_i, x'\}$, satisfies the two conditions, $[U_1,U_1] \subset \mathcal{Z}(L)$ and

$$\dim f(U_1) = \dim f(T) + 1.$$

We now proceed by induction: If $f(U_1)$ is a proper subspace of L_0 we proceed as before to construct a subspace U_2 satisfying $[U_2, U_2] \subset \mathscr{Z}(L)$, $f(U_1) \subset f(U_2)$, and $\dim f(U_2) = \dim f(U_1) + 1$.

In the inductive step, we use that, for every subspace U of L satisfying $[U, U] \subset \mathscr{Z}(L)$, the adjoint representation of U, acting on L, is abelian. Indeed, if $u, u' \in U$, and $y \in L$, then

$$[\mathrm{ad} u, \mathrm{ad}\, u'](y) = \mathrm{ad}[u, u'](y) = [[u, u'], y] = 0$$

since $[u, u'] \in \mathscr{Z}(L)$.

The proof of parts (b) and (c) of the corollary is completed.

<u>Proof of Corollary 5.4</u>. We consider a finite–dimensional Lie subalgebra L of $E_A(M)$ which is assumed real with respect to a compatible involution $*$ on $E_A(M)$. This means that the two algebras L_1 and L_0, defined by $L_0 = f(L)$ and $L_1 = \ker(f|_L)$ are also real as Lie algebras, and we have a short exact sequence

$$0 \longrightarrow L_1 \longrightarrow L \longrightarrow L_0 \longrightarrow 0 \tag{5.23}$$

in the category of *real* Lie algebras.

It is assumed that L_0 is abelian, and that the adjoint representation of L on L_1 is completely diagonalizable with generalized eigenspaces in the complexification $L_1^{\mathbb{C}}$. It follows from this that

$$L_1 = L_1(0) \oplus [L, L_1] \tag{5.24}$$

where $L_1(0)$ denotes the centralizer of L in L_1. Moreover $[L, L_1]$ has even real dimension, and is the direct sum of two–dimensional minimal ideals.

By assumption, we have

$$L_1^{\mathbb{C}} = L_1^{\mathbb{C}}(0) + \sum_{\varphi}^{\oplus} L_1^{\mathbb{C}}(\varphi) \tag{5.25}$$

where

$$L_1^{\mathbb{C}}(\varphi) = \{y \in L_1^{\mathbb{C}} : (\mathrm{ad}x)(y) = i\varphi(x)y,\ x \in L\} \tag{5.26}$$

and φ is a real valued linear functional on L. We may write the element y in the form

$$y = u + iv, \qquad u,v \in L_1$$

where $u = \frac{y-y^*}{2}$ and $v = \frac{y+y^*}{2i}$. It then follows that

$$[x, u] = -\varphi(x)v$$

and (5.27)

$$[x, v] = \varphi(x)u.$$

It follows that the elements u and v span a two–dimensional minimal ideal in L_1, and that [L, L] is the direct sum of the two–dimensional ideals corresponding to the nonzero weights.

The Lie algebra ad L, acting on $L_1^{\mathbb{C}}$,, is compact, so, if $(\mathrm{ad}x)^k(y) = 0$ for $x \in L$, $y \in L_1$ and some positive integer k, it follows from the compactness of ad L that $(\mathrm{ad}x)(y) = 0$, i.e., that k may be taken to be one. This concludes the proof of the assertions about the decomposition (5.24) above.

Returning now to the abelian subspace T of L chosen such that f(T) has maximal dimension in L_0.

Let

$$\Phi^{\mathbf{C}}(0) = \{y \in L^{\mathbf{C}} : \forall x \in T \ \exists k \ \text{s.t.} \ (\text{ad } x)^k(y) = 0\}.$$

It can be shown that, if $y \in \Phi^{\mathbf{C}}(0)$, then we may take $k = 2$. The argument from the proof of Theorem 5.1 shows that

$$L_0^{\mathbf{C}} \simeq \Phi^{\mathbf{C}}(0)/(\Phi^{\mathbf{C}}(0) \cap L_1^{\mathbf{C}}).$$

If the dimension of f(T) is smaller than that of L_0, it follows that $\Phi^{\mathbf{C}}(0)$ contains an element whose real part is not in $T \cap L_1$. But $\Phi^{\mathbf{C}}(0)^* = \Phi^{\mathbf{C}}(0)$ so the real part must also be in $\Phi^{\mathbf{C}}(0)$, and the proof now proceeds exactly as in the proof above for Theorem 5.1.

§6. INFINITE DIMENSIONAL LIE ALGEBRAS

The results on extensions of Lie algebras (from the previous section) are now applied to the infinite–dimensional Lie algebra of derivations, $\mathrm{Der}(\mathfrak{A}^\infty,\mathfrak{A}^\infty)$ of the noncommutative tori. In the special case when the bicharacter ρ is assumed to satisfy a certain generic (diophantine) condition, we get a complete analysis of the smooth structures, and their representations. For further details, we also refer to [Jo 5] and [BEGJ].

Let ρ be a given antisymmetric nondegenerate bicharacter on $\Gamma = \mathbb{Z}^d$, and let $\mathfrak{A}$ be the corresponding simple C^*–algebra with generators $\{u(\xi) : \xi \in \Gamma\}$ satisfying the commutation relation (3.1). Recall that $\mathfrak{A}$ is the C^*–norm completion of the algebra $\mathfrak{A}_0$ spanned by the $u(\xi)$'s. Finally, let $\mathfrak{A}^\infty$ be the dense algebra of smooth elements in $\mathfrak{A}$ for the canonical ergodic action of $\mathbb{T}^d$ which is defined in (2.4) an (4.1). We also recall that the space of all derivations

$$\delta : \mathfrak{A}_0 \longrightarrow \mathfrak{A} \tag{6.1}$$

carries a positive definite inner product, which was defined in (3.2). The corresponding Hilbert space was denoted $\mathscr{H}$, and it consists of module derivations from $\mathfrak{A}_0$ into $L^2(\mathfrak{A},\tau)$ where $L^2(\mathfrak{A},\tau)$ is the trace Hilbert space with Hilbert norm defined from the trace τ on $\mathfrak{A}$, and containing $\mathfrak{A}$ as a dense linear subspace.

We have an abelian Lie algebra of derivations A given as follows,

$$A = \{\delta_x : x \in \mathbb{C}^d\}, \tag{6.2}$$

and

$$\delta_x(u(\xi)) = \sqrt{-1}\,(x\cdot\xi)u(\xi), \quad \text{for } \xi \in \Gamma. \tag{6.3}$$

Let M denote the smooth approximately inner derivations. By virtue of Theorem

3.4, we have

$$M = \mathrm{Der}(\mathfrak{A}^{\infty}) \cap (\mathscr{H} \ominus A) \tag{6.4}$$

If M_0 denotes the finite inner derivations, then

$$M_0 = \mathrm{Der}(\mathfrak{A}_0) \cap (\mathscr{H} \ominus A) \tag{6.5}$$

We saw in Corollary 4.1 that both M and M_0 are Lie ideals in $\mathrm{Der}(\mathfrak{A}^{\infty})$, resp., $\mathrm{Der}(\mathfrak{A}_0)$ when equipped with the commutator bracket. We define a representation of A as follows,

$$F_x(\delta) = [\delta_x, \delta] = \delta_x \delta - \delta\delta_x, \quad \text{for all } x \in \mathbb{C}^d, \text{ and } \delta \in M. \tag{6.6}$$

The decomposition theorem (Theorem 3.4 above) may now be summarized as follows:

$$E_A(M) \simeq \mathrm{Der}(\mathfrak{A}^{\infty}) \tag{6.7}$$

and

$$E_A(M_0) \simeq \mathrm{Der}(\mathfrak{A}_0) \tag{6.8}$$

where the isomorphism is defined (from left to right) by,

$$(a,m) \longrightarrow a+m, \quad a \in A, \ m \in M$$

and

$$(a,m_0) \longrightarrow a+m_0, \quad a \in A, \ m_0 \in M_0,$$

respectively.

Define M_1 as the Lie algebra of all smooth and *inner* derivations. If the bicharacter ρ is further assumed to satisfy the generic diophantine condition (4.9), then we have the Lie isomorphism

$$E_A(M_1) \simeq \mathrm{Der}(\mathfrak{A}^{\infty})$$

given by

$$(a,m_1) \longrightarrow a+m_1, \quad a \in A, \quad m_1 \in M_1.$$

In the present application, $A = \{\delta_x : x \in \mathbb{C}^d\}$ is abelian, and, it follows, for any extension,

$$0 \longrightarrow L_1 \longrightarrow L \longrightarrow L_0 \longrightarrow 0$$

with $L_0 = f(L)$ and $L_1 = \mathrm{Ker}(f|_L)$, that L_0 is abelian. We shall consider below the question of when the adjoint representation of L on the ideal L_1 is completely diagonalizable.

Since A is abelian, we have

$$[E_A(M), E_A(M)] \subset M \tag{6.9}$$

where M is identified with an ideal in $E_A(M)$.

<u>**Theorem 6.1**</u>. *Let* L *be a finite-dimensional Lie subalgebra of* $E_A(M)$ *and assume that*

$$L \subseteq E_A(M_1)$$

where M_1 *denotes the inner smooth derivations.*

Assume further that

$$L = \{\delta^* : \delta \in L\}$$

where

$$\delta^*(a) := -\delta(a^*)^*, \quad a \in \mathfrak{A}^\infty. \tag{6.10}$$

Then it follows that the adjoint representation of L *is completely diagonalizable on* L_1, *and as a consequence the results from Section* 5 *apply.*

__Proof__. This theorem is, in fact, a corollary to a result about representations of Lie algebras which we include as 6.2 and 6.3 below. This material is from [BEGJ(a)] and [JM, Ch. 9, Appendix G].

__Perturbations of abelian operator Lie algebras.__

Recall that a real Lie algebra $\mathfrak{L}$ of operators on a domain D in a given Banach space $\mathcal{X}$ is said to be exponentiable if there is a strongly continuous representation $\rho : G_{\mathfrak{L}} \longrightarrow \mathcal{B}(\mathcal{X})$ where $G_{\mathfrak{L}}$ is the simply connected Lie group with Lie algebra isomorphic to $\mathfrak{L}$, such that

(1) $D \subseteq \mathcal{X}^\infty(\rho)$,

(2) $d\rho(X)\big|_D = X\big|_D \quad (X \in \mathfrak{L})$, and

(3) D is a core for $d\rho(X) \quad (X \in \mathfrak{L})$.

In general, the correspondence between representations of Lie groups on $\mathcal{X}$ and representations of Lie algebras is inexact, because a representation of a Lie algebra may fail to exponentiate. However, we shall single out a class of Lie algebras of smooth operators which do exponentiate to smooth representations of the corresponding simply connected Lie groups. Let α be a representation of a group G with Lie algebra $\mathfrak{G}$.

A Lie algebra $\mathfrak{P}$ of bounded operators on $\mathcal{X}$ is said to be a perturbation class for $\mathfrak{L}_0 = d\alpha(\mathfrak{G})$ if

(1) the C^∞–vectors for α, $\mathcal{X}^\infty(\alpha)$ is invariant under the operators in $\mathfrak{P}$, and

(2) $[\mathfrak{L}_0,\mathfrak{P}] \subseteq \mathfrak{P}$.

$\mathfrak{P}$ is permitted to be infinite–dimensional. For example,let $(\mathfrak{A},G,\alpha)$ be a C^*–dynamical system and set

$$\mathfrak{P} = \{ad(h) : h \in \mathfrak{A}^{\infty}(\alpha), \text{ h skew adjoint}\}. \tag{6.11}$$

For $\delta \in d\alpha(\mathfrak{G})$ and $ad(h) \in \mathfrak{P}$, $[\delta,ad(h)] = ad(\delta(h))$, so $\mathfrak{P}$ is a perturbation class for $\mathfrak{L}_0$.

Note that $\mathfrak{L}_0 + \mathfrak{P}$ is a Lie subalgebra of $End(\mathcal{X}^{\infty})$. We will consider finite dimensional Lie subalgebras of $\mathfrak{L}_0 + \mathfrak{P}$; such Lie algebras were called "semi–direct product perturbations of $\mathfrak{L}_0$" in [JM], Chapter 9.

Proposition 6.2. *Let $\mathfrak{P}$ be a perturbation class for $\mathfrak{L}_0 = d\alpha(\mathfrak{G})$, and let $\mathfrak{L}$ be a finite dimensional Lie subalgebra of $\mathfrak{L}_0 + \mathfrak{P}$. Then $\mathfrak{L}$ exponentiates to a smooth representation of the simply connected Lie group L with Lie algebra isomorphic to $\mathfrak{L}$.*

Proof. That $\mathfrak{L}$ exponentiates to a continuous representation ρ follows at once from [JM], Theorem 9.9. Because $d\rho(X)$ extends X $(X \in \mathfrak{L})$, $\mathcal{X}^{\infty}(\alpha)$ is invariant under $d\rho(X)$ and hence $\mathcal{X}^{\infty}(\alpha) \subseteq \mathcal{X}^{\infty}(\rho)$, where $\mathcal{X}^{\infty}(\alpha)$ and $\mathcal{X}^{\infty}(\rho)$ denote the respective spaces of C^{∞}–vectors.

We next show that for each $\mathcal{X}^{\infty}(\alpha)$ is invariant under ρ_h $(h \in G_{\mathfrak{L}})$. For this, it is enough to show that $\mathcal{X}^{\infty}(\alpha)$ is invariant under $\exp(X+P)$ for $X \in \mathfrak{L}_0$ and $P \in \mathfrak{P}$.

We assert that for each $n \in \mathbb{N}$, P maps $\mathcal{X}^{\infty}(\alpha)$, the space of C^n–elements for the action α, into itself, that P is bounded with respect to the norm $\| \|_n$ of $\mathcal{X}^n(\alpha)$, and, finally, that $\exp(X+P)$ leaves $\mathcal{X}^n(\alpha)$ invariant. Let $Y_1,\cdots,Y_d$ be a basis of $\mathfrak{L}_0$. Then $Y_iP = PY_i + [Y_i,P]$, as operators on $\mathcal{X}^{\infty}(\alpha)$, ad $[Y_i,P] \in \mathfrak{P} \subseteq B(\mathcal{X})$. Hence

$$\|Y_i Pa\| \leq \|P\|\,\|Y_i a\| + \|[Y_i,P]\|\,\|a\| \quad (a \in \mathcal{X}^\infty(\alpha)).$$

This shows that P is bounded with respect to the norm

$$\|a\|_1 = \max\{\|Y_i a\| + \|a\| : 1 \leq i \leq d\},$$

and it follows also that for $a \in \mathcal{X}^1(\alpha)$, $Pa \in \bigcap_{1 \leq i \leq d} D(Y_i) = \mathcal{X}^1(\alpha)$.

The one–parameter group of operators, $t \longmapsto \exp tX$, restricts to a strongly continuous group on $\mathcal{X}^1(\alpha)$. Indeed, if $a \in \mathcal{X}^1(\alpha)$ and $Y \in \mathfrak{L}_0$, then

$$\lim_{t \to 0} \|d\alpha(Y)(\alpha(\exp(tX)a - a)\| = \lim_{t \to 0} \|\alpha(\exp(tX)d\alpha(\mathrm{Ad}(\exp(-tX)(Y))a - d\alpha(Y)a\|,$$

which is zero due to the uniform boundedness of $\{\alpha(\exp(tX) : |t| \leq 1\}$ and the continuity of the adjoint representation of G on the finite–dimensional space $\mathfrak{L}_0$. Denote the infinitesimal generator of this restricted group by X_1; if a is in the domain of X_1 then also $a \in D(X)$ and $X_1(a) = X(a)$, since

$$\|t^{-1}(e^{tX}a-a) - Xa\| \leq \|t^{-1}(e^{tX}a-a) - X_1 a\|_1.$$

Since $P_1 := P|_{\mathcal{X}_1(\alpha)}$ is a bounded operator in the Banach space $\mathcal{X}^1(\alpha)$, it follows from Phillips' perturbation theorem that $X_1 + P_1$ is also the infinitesimal generator of a strongly continuous one–parameter group on $\mathcal{X}^1(\alpha)$, and it is easy to see that this group agrees with $\exp t(X+P)$ on $\mathcal{X}^1(\alpha)$, because for $a \in D(X_1)$,

$$\frac{d}{dt} e^{-t(X+P)} e^{t(X_1+P_1)} a = 0.$$

The invariance of $\mathscr{X}^1(\alpha)$ under $\exp t(X+P)$ is immediate from this.

This establishes the case $n = 1$ of our assertion. The general case $(n > 1)$ follows at once by induction, as the space $\mathscr{X}^{n+1}(\alpha)$ is the space of C^1–vectors or the action of G on the Banach space $\mathscr{X}^n(\alpha)$.

Since $\mathscr{X}^n(\alpha)$ is invariant under $\exp(X+P)$ for all n, so also is $\mathscr{X}^\infty(\alpha)$. Hence ρ is a smooth representation.

The following theorem, a restatement of results from [JM], Appendix G, is a useful tool for analyzing Lie algbras of operators which exponentiate to uniformly bounded Lie group representations, i.e., representations whose image in $\mathscr{B}(\mathscr{X})$ is norm bounded. The theorem generalizes a result of Singer [Si] for unitary representations; see also [KS], [SvN], and [Sak 6].

Given a real Lie algebra $\mathfrak{L}$ of operators in $\mathscr{X}$, we let $\mathfrak{L}_b$ denote the Lie algebra of bounded elements in $\mathfrak{L}$, and $\mathfrak{L}^{\mathbb{C}}$ the complexification of $\mathfrak{L}$, which may be identified with the complex span of $\mathfrak{L}$ in the operators on $\mathscr{X}$.

<u>**Theorem 6.3**</u> ([JM]); The Generalized Singer Theorem). *Let $\mathfrak{L}$ be a Lie algebra of operators in a Banach space which exponentiates to a uniformly bounded Lie group representation. Then*

(a) *$\mathfrak{L}_b$ is an ideal in $\mathfrak{L}$.*

(b) *For all $\xi \in \mathfrak{L}$, $\operatorname{ad} \xi\big|_{\mathfrak{L}_b^{\mathbb{C}}}$ is diagonalizable, with purely imaginary eigenvalues.*

(c) *Let $\mathfrak{L} = \mathfrak{S} + \mathfrak{R}$ be a Levi decomposition of $\mathfrak{L}$ into the solvable radical ideal $\mathfrak{R}$ and a semisimple subalgebra $\mathfrak{S}$. Then $\mathfrak{S}_b$ and $\mathfrak{R}_b$ commute, and $\mathfrak{L}_b = \mathfrak{S}_b + \mathfrak{R}_b$. In other words, $\mathfrak{L}_b$ is the direct sum of the commuting ideals $\mathfrak{S}_b$ and $\mathfrak{R}_b$. Furthermore, $\mathfrak{S}_b$ is compact and $\mathfrak{R}_b$ is abelian.* #

To be able to use Theorem 6.3, we must make sure that our exponentiable Lie algebras generate uniformly bounded Lie group representations. There is no problem with this in the case of Lie algebras of $*$–derivations in C^*–algebras, the application of primary interest to us, since these generate representations by $*$–automorphisms, which are isometric.

Recall that $\mathrm{Der}(\mathfrak{A}^{\infty})$ carries an involution which is defined as follows:

$$\delta^*(a) := -\delta(a^*)^*, \qquad a \in \mathfrak{A}^{\infty}. \tag{6.12}$$

For $\delta \in \mathrm{Der}(\mathfrak{A}^{\infty})$, define further

$$\mathrm{Re}\ \delta = \frac{\delta - \delta^*}{2},$$

$$\mathrm{Im}\ \delta = \frac{\delta + \delta^*}{2i},$$

and

$$L' = \{\mathrm{Re}\ \delta : \delta \in L\},$$

$$L'' = \{\mathrm{Im}\ \delta : \delta \in L\}.$$

We are, in fact, applying Proposition 6.2 to the two Lie algebras L' and L'', and not to L directly.

An additional argument is needed in the case the given Lie subalgebra L has $\mathbb{C}$ as ground field. In that case, L does not exponentiate to a group of diffeomorphisms. But we note that L has a decomposition of the form

$$L = L' + iL'' \tag{6.13}$$

where L′ and L″ are real Lie algebras which do exponentiate to groups of automorphisms, and, in particular, to uniformly bounded representations.

Theorem 6.3 then applies to the component Lie algebras L′ and L″ from (6.13) which are both real Lie algebras, see below:

A subalgebra L of $\mathrm{Der}(\mathfrak{A}^\infty)$ is said to be real if

$$\delta^* = -\delta \quad \text{for all} \quad \delta \in L.$$

If L is real, we conclude from Corollary 4.1′ that the inner product (δ_1, δ_2) is real valued for all $\delta_1, \delta_2 \in L$.

Note that the two Lie algebras L′ and L″ in the decomposition (6.13) above are both real.

Since $E_A(M_0) \simeq \mathrm{Der}(\mathfrak{A}_0)$, and $E_A(M_1) \simeq \mathrm{Der}(\mathfrak{A}^\infty)$, when the bicharacter ρ is assumed to satisfy the generic diophantine condition, it follows that the conclusions of Theorem 5.1, and Corollaries 5.2 thru 5.4 apply to any finite–dimensional subalgebras of $\mathrm{Der}(\mathfrak{A}_0)$, or of $\mathrm{Der}(\mathfrak{A}^\infty)$ in the generic case, cf. (4.9).

§7. DIFFERENTIAL GEOMETRY OF THE CUNTZ ALGEBRAS

In this section, we include some considerations concerning Lie algebras of derivations acting on the simple Cuntz–algebras $\mathcal{O}_n$, $n = 2,3,\cdots$. We recall that the C^*–algebra $\mathcal{O}_n$ is simple [Cu 1], and generated (with C^*–norm completion) by isometries $s_1,\cdots,s_n$ and adjoints satisfying the relations,

$$s_i^* s_j = D_{ij}\,1, \quad \sum_{i=1}^{n} s_i s_i^* = 1 \tag{7.1}$$

where D_{ij} denotes the Kronecker delta–function, and 1 refers to the unit–element in $\mathcal{O}_n$.

We show that the Lie algebraic techniques from Section 5, and the Hilbert space inner product from Section 3 apply to a situation which is quite different from the one for the noncommutative tori, described in Sections 2–3. We show that known results [BEvGJ] as well as new results about decompositions of derivations in $\mathcal{O}_n$ may be derived as Hilbert space decompositions relative to our new inner product on the *derivations* of $\mathcal{O}_n$.

There is a canonical action of the noncompact reductive Lie group $U(n,1)$ on $\mathcal{O}_n$ which was found by Voiculescu in [Voi], and further analyzed in [C–E] and [Jo 8]. The group $U(n,1)$ contains as a subgroup $U(n)$, the group of unitary n by n complex matrices, $g = (g_{ij}) \in U(n)$, and the action of the subgroup α_g is given on the generators s_i as follows,

$$\alpha_g(s_i) = \sum_{j=1}^{n} g_{ji}\, s_j, \tag{7.2}$$

and there is an extension of this action to an automorphic action (also denoted by α),

$$\alpha : U(n,1) \longrightarrow \mathrm{Aut}(\mathcal{O}_n). \tag{7.3}$$

It is convenient to work instead with the corresponding infinitesimal action of the Lie algebra of $U(n,1)$. This Lie algebra will be denoted $\mathfrak{g}(n,1)$. Let $\mathcal{P}_n$ be the dense polynomial subalgebra, $\mathcal{P}_n \subset \mathcal{O}_n$ generated by the s_i's and their adjoints. Then we have a Lie algebra representation, $X \longrightarrow \delta_X$, of $\mathfrak{g}(n,1)$ into the Lie algebra $\mathrm{Der}(\mathcal{P}_n)$ of all derivations of $\mathcal{P}_n$. Specifically,

$$\delta : \mathfrak{g}(n,1) \longrightarrow \mathrm{Der}(\mathcal{P}_n),$$

and

$$\delta_X : \mathcal{P}_n \longrightarrow \mathcal{P}_n,$$

satisfying

$$\delta_X(ab) = \delta_X(a)b + a\delta_X(b)$$

for $X \in \mathfrak{g}(n,1)$ and all $a,b \in \mathcal{P}_n$. Since the δ–action is obtained as the infinitesimal action of α, we also have

$$\delta_X(a^*) = \delta_X(a)^*, \quad a \in \mathcal{P}_n;$$

in other words, $\{\delta_X : X \in \mathfrak{g}(n,1)\}$ is a *real* Lie algebra of derivations.

The diagonal matrices in $U(n)$, with z down the main diagonal $(z \in \mathbf{T}^1)$, act on $\mathcal{P}_n$ as follows,

$$\alpha_z(s_i) = zs_i, \quad z \in \mathbf{T}^1 \approx U(1), \quad i = 1,\cdots,n, \tag{7.4}$$

and this is called the *gauge–action.*

A mapping in $\mathcal{O}_n$, or $\mathcal{P}_n$, is said to be gauge–invariant if it commutes with $\{\alpha_z : z \in \mathbf{T}^1\}$; and elements a in $\mathcal{O}_n$ are called gauge–invariant if

$$\alpha_z(a) = a, \quad z \in \mathbf{T}^1. \tag{7.5}$$

The subalgebra $\mathcal{O}_n'$ of all gauge–invariant elements in $\mathcal{O}_n$ is known to be isomorphic to the UHF–algebra $\bigotimes_1^\infty M_n$ where M_n denotes the n by n complex matrices. In fact, we may identify

$$\{s_{i_1} \cdots s_{i_r} s_{j_r}^* \cdots s_{j_1}^* : 1 \le i_1, \cdots, i, j_1, \cdots, j_r \le n\}$$

in $\mathcal{O}_n'$ with canonical matrix–units

$$e_{i_1 j_1} \otimes \cdots \otimes e_{i_r j_r} \tag{7.6}$$

in $\bigotimes_1^r M_n \subset \bigotimes_1^\infty M_n$ if $\{e_{ij} : 1 \le i,j \le n\}$ denote the familiar canonical matrix units for n by n matrices.

Let

$$\epsilon := \int_{\mathbf{T}^1} \alpha_z \, dz \tag{7.7}$$

denote the canonical conditional expectation of $\mathcal{O}_n$ onto the subalgebra $\mathcal{O}_n'$ of all gauge–invariant elements in $\mathcal{O}_n$. Let τ be the trace on $\mathcal{O}_n' \simeq \bigotimes_1^\infty M_n$. Then the state ω on $\mathcal{O}_n$, defined by

$$\omega := \tau \circ \epsilon \tag{7.8}$$

is faithful. We call it *the canonical state* on $\mathcal{O}_n$.

Let δ_1 and δ_2 be arbitrary derivations from $\mathcal{P}_n$ into $\mathcal{O}_n$, and define the inner product (δ_1,δ_2) as follows,

$$(\delta_1,\delta_2) = \sum_{i,j} \omega(s_i\delta_1(s_i)^*\delta_2(s_j)s_j^*). \tag{7.9}$$

We leave to the reader the verification that this is indeed a Hilbert space inner product. (See also Lemma 7.2 below.)

We shall need some technical observations:

Lemma 7.1. *Let δ_1 and δ_1 be derivations from $\mathcal{P}_n$ into $\mathcal{O}_n$, and assume that one of the two is gauge–invariant.*

Then the inner product is

$$(\delta_1,\delta_2) = \sum_{i} \omega(\delta_2(s_i)\delta_1(s_i)^*). \tag{7.10}$$

Proof. Compute!

The lemma shows that, while the definition of the inner product (7.9) involves a double summation, $1 \leq i,j \leq n$, this simplifies in the event one of the two derivations is assumed to be gauge–invariant: Formula (7.10) has just a single summation index!

Lemma 7.2. *There is a one–to–one correspondence between the space of **all** derivations* $\mathrm{Der}(\mathcal{P}_n,\mathcal{O}_n)$ *and the elements in $\mathcal{O}_n$ given as follows:*

(i) *If $L \in \mathcal{O}_n$ there is a unique derivation δ determined by the two formulas:*

$$\delta(s_i) = Ls_i\,, \quad \delta(s_i^*) = -s_i^*L. \tag{7.11}$$

(ii) *If, conversely,* $\delta \in \mathrm{Der}(\mathcal{P}_n, \mathcal{O}_n)$ *is given, then the element* L *defined by*

$$L = \sum_i \delta(s_i)s_i^* \tag{7.12}$$

satisfies the conditions in (i).

(iii) *The isomorphism* $\delta \longrightarrow L_\delta$ *carries the involution on* $\mathrm{Der}(\mathcal{P}_n, \mathcal{O}_n)$, $\delta \longrightarrow \delta^*$ *with* $\delta^*(a) := -\delta(a^*)^*$, $a \in \mathcal{P}_n$, *onto the* C^**-involution of* $\mathcal{O}_n$, *i.e.,* $L \longrightarrow L^*$.

(iv) *Let* $\sigma : \mathcal{O}_n \longrightarrow \mathcal{O}_n$ be the shift–endomorphism given by

$$\sigma(a) = \sum_i s_i a\, s_i^*, \quad a \in \mathcal{O}_n. \tag{7.13}$$

Then a derivation δ *is inner if and only if the corresponding element* L_δ *is in* $(I-\sigma)(\mathcal{O}_n)$; *and* δ *is approximately inner on* $\mathcal{P}_n$ *iff* L_δ *is in the norm–closure (relative to the* C^**–norm) of the set.*

(v) *A derivation* δ *leaves invariant the polynomial subalgebra* $\mathcal{P}_n$ *generated by the* s_i*'s and* s_i^**'s iff the corresponding element* L_δ *is in* $\mathcal{P}_n$.

Proof. In the special case when δ is assumed to satisfy

$$\delta(a^*) = \delta(a)^*, \quad a \in \mathcal{P}_n, \tag{7.14}$$

or equivalently $\delta^* = -\delta$, the proof is contained in [C–E] and [BEvGJ]. So we will add only comments about the modification needed for the case when δ is general and not assumed to satisfy (7.14). In this case, the formula for $\delta(s_i^*)$ does not follow from the one for $\delta(s_i)$. In fact, it is not even clear that δ is well defined. This is why, in part (i) above, both $\delta(s_i)$ and $\delta(s_i^*)$ must be defined! If they are defined as in (7.11), then it follows that the two relations for $\mathcal{O}_n$, $s_i^* s_j = D_{ij} 1$ and $\sum_i s_i s_i^* = 1$, are both consistent with (7.11), and therefore a derivation is well defined when L is an *arbitrarily given* element in $\mathcal{O}_n$. It passes to the quotient by the ideal on the two relations.

It remains to show that, if δ is a given derivation, and L is defined by (7.12), then the two formulas (7.11) hold.

We compute:

$$L s_k = \sum_i \delta(s_i) s_i^* s_k = \sum_i \delta(s_i) D_{ik} = \delta(s_k),$$

and

$$\begin{aligned} s_k^* L &= \sum_i s_k^* \delta(s_i) s_i^* \\ &= -\sum_i \delta(s_k^*) s_i s_i^* \\ &= -\delta(s_k^*) \sum_i s_i s_i^* \\ &= -\delta(s_k^*), \qquad 1 \leq k \leq n, \end{aligned}$$

where Leibniz' rule is applied in the second computation.

The other points of the lemma are simple computations and are left to the reader.

__Lemma 7.3__. *Let* δ_1 *and* δ_2 *be two derivations, and let* L_1 *and* L_2 *be the respective elements in* $\mathcal{O}_n$, *given by formula* (7.12). *Then the inner product* (δ_1, δ_2) *is given by the formula*

$$(\delta_1, \delta_2) = \omega(L_1^* L_2). \tag{7.15}$$

__Proof__. Compute!

§8. COMMUTATORS AND COCYCLES

Let $H_n = \mathbb{C}^n$ be the n–dimensional complex Hilbert space with inner product denoted $\langle\cdot,\cdot\rangle$. Then the Lie algebra $\mathfrak{g}(n,1)$ may be expressed as $(n+1)$ by $(n+1)$ complex matrices,

$$X = \begin{bmatrix} x_0 & \langle v,\cdot\rangle \\ v & x \end{bmatrix} \tag{8.1}$$

where $v \in H_n$, $x_0 \in i\mathbb{R}^1$, and $x = -x^* \in M_n$. If $x = (x_{ij})$ is the matrix entry representation in M_n, we define

$$sxs^* := \sum_{i,j} x_{ij}\, s_i s_j^* \tag{8.2}$$

and similarly,

$$s(v) := \sum_i v_i s_i \tag{8.3}$$

where $v \in \mathbb{C}^n$ is expressed as $v = (v_1,\cdots,v_n)$.

<u>Lemma 8.1</u>. *There is a Lie algebra representation,*

$$X \longrightarrow \delta_X,$$

of $\mathfrak{g}(n,1)$ *into the commutator Lie algebra* $\mathrm{Der}(\mathcal{P}_n,\mathcal{P}_n)$ *when* δ_X *is the derivation determined by*

$$\delta_X(s_i) = L(X)s_i, \quad 1 \le i \le n \tag{8.4}$$

and, for $X = (v,x) \in \mathfrak{g}(n,1)$,

$$L(X) := x_0 1 + s(v) - s(v)^* + sxs^*. \tag{8.5}$$

We have

$$\delta_X(a^*) = \delta_X(a)^*, \quad a \in \mathcal{P}_n;$$

and the element $L([X,Y])$ *corresponding to the derivation*

$$\delta_{[X,Y]} := [\delta_X, \delta_Y]$$

is given by

$$(L[X,Y]) = [L(Y),L(X)] + \delta_X(L(Y)) - \delta_Y(L(X)),$$

where $[X,Y]$ *is the matrix commutator in* M_{n+1}.

Proof. Compute, or refer to [BEvGJ].

Lemma 8.2. *The approximately inner derivations in* $\mathrm{Der}(\mathcal{P}_n)$ *form a* $\mathfrak{g}(n,1)$*–module when the action of* $\mathfrak{g}(n,1)$ *on* $\mathrm{Der}(\mathcal{P}_n)$ *is defined by*

$$X,\delta \longrightarrow [\delta_X, \delta] \tag{8.6}$$

for $X \in \mathfrak{g}(n,1)$ *and* $\delta \in \mathrm{Der}(\mathcal{P}_n)$.

Proof. It is enough to show that the commutator $[\delta_X,\delta] = \delta_X\delta - \delta\delta_X$ is approximately inner for every approximately inner $\delta \in \mathrm{Der}(\mathcal{P}_n)$.

Let L be the element associated to δ, and let L′ be the element associated to $[\delta_X,\delta]$ according to formulas (7.11) and (7.12) of Lemma 7.2.

We then have

$$L' = [L,L(X)] + \delta_X(L) - \delta(L(X)) \tag{8.7}$$

as can be seen by a computation.

Let L_N, $N = 1,2,\cdots$ be a sequence in $(I-\sigma)(\mathcal{P}_n)$ such that $\lim_{N\to\infty}\|L-L_N\| = 0$, and let δ_N be the derivation given by (cf. Lemma 7.2):

$$\delta_N(s_i) = L_N s_i, \quad \delta_N(s_i^*) = -s_i^* L_N.$$

Then we have,

$$\lim_{N\to\infty}[L_N,L(X)] = [L,L(X)]$$

and

$$\lim_{N\to\infty}\delta_N(L(X)) = \delta(L(X))$$

where, in both cases, the limit exists relative to the C^*–norm of $\mathcal{O}_n$.

We note that we may use Lemma 8.1 to complete the proof by showing first the limit relation,

$$\lim_{N\to\infty} \delta_X(L_N) = \delta_X(L)$$

for the second term on the right hand side in formula (8.7) above (for the element L' associated to the derivation commutator).

The remaining step follows, in turn, from the following:

Sublemma 8.3. *Let σ be the shift in the polynomial subalgebra $\mathcal{P}_n$, and let $\mathcal{R}_n(\sigma)$ be the range of $I-\sigma$, i.e., $\mathcal{R}_n(\sigma) := \{H-\sigma(H) : H \in \mathcal{P}_n\}$. Let $\{L_r' : r = 1,2,\cdots\}$ be a sequence in $\mathcal{R}_n(\sigma)$, and let $L \in \mathcal{P}_n$ be given such that*

$$\lim_{r\to\infty} ||L'_r - L|| = 0.$$

Then there is a sequence $\{L_r\}$ *in* $\mathcal{R}_n(\sigma)$ *such that*

$$\lim_{r\to\infty} ||L_r - L|| = 0$$

and

$$\lim_{r\to\infty} ||\delta_X(L_r) - \delta_X(L)|| = 0$$

for all $X \in \mathfrak{g}(n,1)$.

Proof. Consider the set J of all finite multi–indices $\alpha = (\alpha_1, \alpha_2, \cdots, \alpha_r)$, where r is variable with entries $\alpha_i \in \{1,2,\cdots,n\}$. Define the length $|\alpha| := r$. Define

$$s(\alpha) := s_{\alpha_1} \cdots s_{\alpha_r},$$

and

$$s(\alpha,) := s(\alpha)s(\beta)^*, \quad \alpha,\beta \in J. \tag{8.8}$$

For $k = 1,2,\cdots,$ let

$$\mathcal{A}(k) := \text{span}\{s(\alpha,\beta) : |\alpha| \leq k, \ |\beta| \leq k\}. \tag{8.9}$$

We have $\bigcup_k \mathcal{A}(k) = \mathcal{P}_n$. (We shall drop the subscript n in the sequel.) Recall that the group U(n,1) acts on $\mathcal{O}_n$, and the restriction to the subgroup U(n) leaves $\mathcal{P}$ invariant. In fact, it follows form the definition of the restricted action, cf. (7.2), that each $\mathcal{A}(k)$ is invariant, for U(n) (but not for U(n,1)). By integration of group characters for U(n) against the action, we obtain a bounded spectral projection π_k of $\mathcal{O}_n$ onto $\mathcal{A}(k)$ for each $k = 1,2,\cdots$. Since $L \in \mathcal{P}$, there is a k such that $\pi_k(L) = L$. It follows that

$$\lim_{r\to\infty} \pi_k(L_r') = \pi_k(L) = L.$$

Now k is fixed, and we consider the following system of linear equations,

$$\pi_k(L_r') = H_r - \sigma(H_r), \quad r = 1,2,\cdots, \tag{8.10}$$

where the elements H_r are unknown. This is a problem in finite–dimensional linear algebra, and the Fredholm alternative applies. (Details below.)

In solving the system (8.10), we shall make use of the following simple observations (which may be checked by inspection):

(i) We have

$$\psi(\pi_k(L_r')) = 0$$

for every continuous linear functional ψ on $\mathcal{O}_n$ which is invariant under π_k, and is assumed to satisfy $\psi(\mathcal{R}_n(\sigma)) = 0$.

(ii) Recall that the mapping, $a \longrightarrow a - \sigma(a)$, has one–dimensional kernel. (This can be checked, using the known simplicity of $\mathcal{O}_n$, cf. [Cu 1–2].)

(iii) By [BEvGJ] it is possible to choose an inner product such that σ is contractive. Let σ^* denote the adjoint mapping relative to this chosen inner product. Then a satisfies $\sigma(a) = a$ iff $\sigma^*(a) = a$ holds.

It follows that a solution $H_r \in \mathcal{A}(k)$ exists for each $r = 1,2,\cdots$. We choose H_r such that $\omega(H_r) = 0$.

Now define

$$L_r := H_r - \sigma(H_r) = \pi_k(L'_r).$$

We noted that $\lim_{r\to\infty} L_r = L$.

We claim that δ_X maps $\mathscr{A}(k)$ into $\mathscr{A}(k+1)$ which is also finite–dimensional. This follows from the following formula for $\delta_X(s_i)$ where $X = (v,x_0,x) \in \mathfrak{g}(n,1)$ is given by $v \in \mathbb{C}^n$, $v = (v_1,\cdots,v_n)$, $x_0 \in \sqrt{-1}\,\mathbb{R}$, $x = -x^* \in M_n$:

$$\delta_X(s_i) = x_0 s_i + s(v)s_i - \bar{v}_i + s(x^{(i)}) \tag{8.11}$$

where $x^{(i)}$ denotes the i–th column of the matrix x.

Since we are now in finite dimensions, δ_X is continuous from $\mathscr{A}(k)$ into $\mathscr{A}(k+1)$ and

$$\pi_{k+1} \circ \delta_X \circ \pi_k = \delta_X \circ \pi_k.$$

It follows that

$$\lim_{r\to\infty} \delta_X(L_r) = \delta_X(L)$$

as asserted in the conclusion of the sublemma.

__Corollary 8.4__. *Let* M *be the Lie algebra of all approximately inner derivations in* $\mathrm{Der}(\mathscr{P}_n)$, *let* $A = \{\delta_X : X \in \mathfrak{g}(n,1) \otimes \mathbb{C}$, *and let* $E_A(M)$ *be the universal extension of* A *by* M. *Then the mapping*

$$(a,m) \longrightarrow a+m : E_A(M) \longrightarrow \mathrm{Der}(\mathscr{P}_n) \tag{8.12}$$

is an embedding of the Lie algebra $E_A(M)$ *into* $\mathrm{Der}(\mathscr{P}_n)$.

Remark 8.5. (i) We conjecture that all approximately inner derivations in

$\text{Der}(\mathscr{P}_n)$ are in fact inner, but we have only been able to prove this for the

gauge–invariant approximately inner derivations, cf. [BEvGJ].

(ii) The representation of A on M, implicit in the construction of $E_A(M)$, is the

adjoint representation, given by

$$F_X(\delta) := \text{ad } \delta_X(\delta) = \delta_X\delta - \delta\delta_X \text{ for all } X \in \mathfrak{g}(n,1) \text{ and } \delta \in M, \qquad (8.13)$$

and it follows from Lemma 8.2 that this is indeed a representation.

(iii) It is not known whether or not the mapping in (8.12) is *onto*. The mapping

being onto is equivalent to every derivation $\delta : \mathscr{P}_n \longrightarrow \mathscr{P}_n$ having a decomposition of the

form

$$\delta = \delta_X + \tilde\delta \qquad (8.14)$$

with δ_X in the complexification of $\mathfrak{g}(n,1)$, and $\tilde\delta$ approximately inner, and it is not

(yet) known if such a decomposition always exists. But if it does, it is unique (implicit in

the corollary!).

The mapping in (8.12) is a Lie homomorphism by virtue of the definition of the Lie

bracket in $E_A(M)$:

$$[(a,m),(a',m')] = ([a,a'], F_a(m') - F_{a'}(m) + [m,m']).$$

To show that (8.12) is one–to–one, we check that if $a+m = 0$, $a \in A$, $m \in M$, then

$a = 0 = m$. This is the uniqueness assertion in the decomposition (8.14) mentioned above.

We shall need the Cuntz–states from [Cu 2] and [BEvGJ], and recall here (next

section), briefly, the definition.

The Cuntz–states is a family of states, denoted $\mathscr{C}$ below, with the following two properties:

(i) For $\forall\ \delta \in A_{\mathbb{C}}\backslash\{0\}$, $\exists\ \psi \in \mathscr{C}$ s.t. $\psi(L_\delta) \neq 0$

and

(ii) $\psi(L_\delta) = 0$ for all approximately inner derivations δ.

The uniqueness assertion needed in (8.12) is immediate from this.

§9. THE CUNTZ STATES

Let u, v, w denote vectors in $\mathbb{C}^n$, with coordinates $u = (u_1,\cdots,u_n)$, etc., let $e_i = (0,\cdots,1,0,\cdots,0)$ with 1 on the i–th place, and let $\langle\cdot,\cdot\rangle$ be the inner product in $\mathbb{C}^n$, $\langle u,v\rangle = \sum_i \bar{u}_i v_i$. The states ψ_w are parametrized by unit–vectors w, $\|w\|^2 = \langle w,w\rangle = 1$, and determined by

$$\psi_w(s(u)) = \langle w,u\rangle = s(w)^* s(v) \tag{9.1}$$

for $s(u) = \sum_i u_i s_i$, $u \in \mathbb{C}^n$, and, in particular, $s_i = s(e_i)$, $i = 1,2,\cdots,n$. It follows from, e.g., [BEvGJ, Proposition 2.1] that ψ_w is a state; and that it is shift–invariant, i.e.,

$$\psi_w(\sigma(a)) = \psi_w(a)\ , \quad a \in \mathcal{O}_n, \quad w \text{ as above,}$$

and

$$\psi_w(s_i a) = \psi_w(s_i)\psi_w(a)\ , \quad a \in \mathcal{O}_n.$$

It follows further that

$$\psi_w(s(u_1)\cdots s(u_r)s(v_s)^*\cdots s(v_1)^*) = \prod_i \langle w,u_i\rangle \prod_j \langle v_j,w\rangle$$
$$\text{for all } u_1,\cdots,u_r,\ v_1,\cdots,v_s \in \mathbb{C}^n,\ r,s = 1,2,\cdots. \tag{9.2}$$

Let $Z = X + iY \in \mathfrak{g}(n,1) \otimes \mathbb{C}$, $X,Y \in \mathfrak{g}(n,1)$, and let

$$\delta_Z = \delta_X + i\delta_Y$$

with corresponding

$$L_Z = L_X + iL_Y.$$

We then have

$$\delta_Z(s_j) = L_Z\, s_j \quad \text{and} \quad \delta_Z(s_j^*) = -s_j^* L_Z \quad \text{for } j = 1,\cdots,n. \tag{9.3}$$

Moreover the element L_Z is uniquely determined, from δ_Z, by the two equations (8.3), and vice versa.

It follows that $A^{\mathbb{C}} := \mathfrak{g}(n,1) \otimes \mathbb{C}$ is parametrized by $\mathbb{C} \times M_n \times \mathbb{C}^n$. Let $(z,x,v) \in \mathbb{C} \times M_n \times \mathbb{C}^n$, $z \in \mathbb{C}$, $x \in M_n$, $v \in \mathbb{C}^n$, be coordinates for the matrix $Z = X + iY$. Then

$$\psi_w(L_Z) = z + 2i\,\mathrm{Im}\langle w,v\rangle + \langle w,x^T w\rangle \tag{9.4}$$

where x^T denotes the transposed matrix $x^T = (x_{ji})$ for $x = (x_{ij})$. Since $\psi_w(L) = 0$ for all unit–vectors w, and all elements L in the norm–closure of $\mathcal{R}_n(\sigma)$, it follows from (9.4) that δ_Z is not approximately inner unless $Z = 0$ in $\mathfrak{g}(n,1) \otimes \mathbb{C}$. This concludes the uniqueness proof for the decomposition (8.14) which was the remaining item in the proof of the corollary.

Remark 9.1. The states ψ_w are very useful, cf. [BEvGJ], and the canonical state $\omega := \tau \circ \epsilon$ may be expressed in terms of the ψ_w's as follows. The ψ_w's are indexed by $w \in \mathbb{C}^n$, $\|w\| = 1$, i.e., $\sum_i \overline{w}_i w_i = 1$. Let S_n be the unit–vectors in $\mathbb{C}^n$, and let dw be the probability measure on S_n which arises from the Haar measure on $U(n)$ when S_n is realized as the compact symmetric space $S_n \approx U(n)/U(n-1)$, cf. [Hel, Ch. VII]. We then have the formula,

$$\omega = \int_{S_n} \psi_w \, dw \tag{9.5}$$

which can be checked by computation, and use of Proposition 2.1 in [BEvGJ].

Let π denote the GNS–representation from the canonical state ω, and let π_w denote the GNS–representation from the Cuntz–state ψ_w, $w \in S_n$. Then the following direct integral formula

$$\pi = \int_{S_n}^{\oplus} \pi_w \, dw$$

follows immediately from (9.5).

The formula is useful for a further detailed analysis of the derivations with $\mathcal{P}_n$ as domain of definition. (See Section 12 below.) Let δ_1, $\delta_2 \in \mathrm{Der}(\mathcal{P}_n, \mathcal{O}_n)$ and let (δ_1, δ_2) denote the inner product from formula (7.15) above. Recall, if

$$L_k = \sum_i \delta_k(s_i) s_i^*, \quad k = 1,2,$$

then

$$(\delta_1, \delta_2) = \omega(L_1^* L_2).$$

But $\mathrm{Der}(\mathcal{P}_n, \mathcal{O}_n)$ also carries pseudo–inner products, indexed by $w \in S_n$, and defined by

$$(\delta_1, \delta_2)_w := \psi_w(L_1^* L_2),$$

although $(\cdot,\cdot)$ is not strictly positive definite. We have

$$(\delta_1, \delta_2) = \int_{S_n} (\delta_1, \delta_2)_w \, dw. \tag{9.5'}$$

Occasionally, we shall need to know the individual components ψ_w in the decomposition (9.5), or, similarly, in (9.5′). See Corollary 11.3 below.

As another example, we used, in the proof of Corollary 8.4, the formula

$$\psi_{w}(L_{Z}) = z + 2i\ \mathrm{Im}\langle w,v\rangle + \langle w,x^{T}w\rangle \tag{9.6}$$

rather than just

$$\omega(L_{Z}) = z.$$

§10. THE ORTHOGONAL COMPLEMENT OF THE TANGENT SPACE

We are now ready to give the orthogonality relations for the inner product on the derivations of $\mathcal{O}_n$ which are analogous to the orthogonality relations on the derivations of the noncommutative tori of Section 3 (Lemma 3.1) above.

We saw that the complex Lie algebra $A_{\mathbb{C}} := \mathfrak{g}(n,1) \otimes \mathbb{C}$ is coordinatized by $\mathbb{C} \times M_n \times \mathbb{C}^n$, in the form $X = (x_0,x,u)$, $x_0 \in \mathbb{C}$, $x \in M_n$, $u \in \mathbb{C}^n$, and similarly for $Y = (y_0,y,v)$, $y \in \mathbb{C}$, $y \in M_n$, $v \in \mathbb{C}^n$.

__Proposition 10.1__. (i) *For derivations* δ_X *and* δ_Y *in the complex Lie algebra* $A_{\mathbb{C}}$, *we have*

$$(\delta_X,\delta_Y) = \bar{x}_0 y_0 + (1+1/n)\langle u,v\rangle + \operatorname{trace}(x^* y). \tag{10.1}$$

(ii) *In general, the two Lie algebras* $A_{\mathbb{C}}$ *and* M *are not orthogonal.*

(iii) *If* $\delta_X \in A_{\mathbb{C}}$ *is of the form* $X = (x_0,x,u)$ *with* $x_0 \in \mathbb{C}$, $x \in M_n$, $u \in \mathbb{C}^n$, *then*

$$(\delta,\delta_X) = \tau(sx\ \delta(s)^*) + x_0\ \tau(L_\delta^*) \tag{10.2}$$

for all gauge-invariant δ *in* $\operatorname{Der}(\mathcal{P}_n,\mathcal{O}_n)$.

(iv) *If* δ *in* $\operatorname{Der}(\mathcal{P}_n,\mathcal{O}_n)$ *is gauge-invariant and* $\tau(L_\delta) = 0$, *then* $\delta \perp A$ *iff* $\tau(s_i^* L_\delta s_j) = 0$ for all $i \neq j$, $1 \leq i, j \leq n$.

(v) *Let* $L' \in \mathcal{O}_n'$ *(gauge-invariant),* $\tau(L') = 0$, *let* α *be a multi-index, let* $L := s_\alpha L' s_\alpha^*$, *and let* δ *be the unique derivation satisfying,*

$$\delta(s_i) = Ls_i, \quad \delta(s_i^*) = -s_i^* L.$$

Then $\delta \in \mathcal{H} \ominus \mathrm{A}$ *although* δ *is generally not approximately inner.*

Corollary 10.2. *A gauge–invariant* $\delta \in \mathrm{Der}(\mathcal{P}_n, \mathcal{O}_n)$ *is in the orthogonal complement of* A *iff* $\tau(\delta(s_i)s_j^*) = 0$, *all* $1 \le i,j \le n$, *and* $\tau(L_\delta) = 0$.

Proofs. We do only part (iv) of Proposition 10.1 since the other parts are straightforward computations.

Let $\delta \in \mathrm{Der}(\mathcal{P}_n, \mathcal{O}_n)$ be given, gauge–invariant, and let $L = \sum_i \sigma(s_i)s_i^*$. Then

$$\begin{aligned}(\delta,\delta_X) &= \omega(L^*L_X) \\ &= \tau(L^*\epsilon(L_X)) \\ &= \tau(L^*(x_0 1 + sxs^*)) \\ &= x_0\tau(L^*) + \tau(L^*sxs^*) \\ &= x_0\overline{\tau(L)} - \tau(\delta(s^*)^*xs^*) \\ &= x_0\overline{\tau(L)} + \tau(sx\delta(s)^*)\end{aligned}$$

as asserted.

We have used the short–hand notation $sx\delta(s)^*$ which is defined by

$$sx\delta(s)^* := \sum_{ij} x_{ij}s_i\delta(s_j)^*. \tag{10.3}$$

Since

$$\tau(s_i\delta(s_j)^*) = \overline{\tau(\delta(s_j)s_i^*)},$$

the corollary follows.

(iv) We now show that the inner product must vanish if $\tau(s_i^*L_\delta s_j) = 0$ for all $i \neq j$, or if δ has the form specified in part (v) of the proposition. We have the identification,

$$s(\alpha,\beta) \longleftrightarrow e_{\alpha_1\beta_1} \otimes e_{\alpha_2\beta_2} \otimes \cdots \tag{10.4}$$

valid for multi–indices $|\alpha| = |\beta| = r$ where the tensors represent elementary matrix units in

$$\bigotimes_1^r M_n \subset \bigotimes_1^\infty M_n \simeq \mathcal{O}'_n. \tag{10.5}$$

Using the identification, we show that both of the terms from (iv) must vanish when the derivation δ is as specified. Clearly, $\tau(L) = 0$ if $L \in \mathcal{R}_n(\sigma)$, or in the norm closure, or as specified in (v). So we must check that

$$\tau(L\, s_i s_j^*) = 0 \qquad \text{for all} \quad i,j.$$

By linearity in L, and by continuity, it is enough to consider terms of the form $\tau(s(\alpha,\beta)s_i s_j^*)$ where $|\alpha| = |\beta|$, and $\alpha_1 = \beta_1$. Let $|\alpha| = |\beta| = r$, and let $\alpha_1 = \beta_1 = 1$ (for specificity).

We have

$$\begin{aligned}
\tau(s_{\alpha_1} \cdots s_{\alpha_r} s_{\beta_r}^* \cdots s_{\beta_1}^* s_i s_j^*) &= \tau((e_{\alpha_1\beta_1} \otimes e_{\alpha_2\beta_2} \otimes \cdots)(e_{ij} \otimes 1)) \\
&= \tau((e_{11}\, e_{ij}) \otimes e_{\alpha_2\beta_2} \otimes \cdots) \\
&= D(1,i)\tau(e_{1j} \otimes e_{\alpha_2\beta_2} \otimes \cdots) \\
&= D(1,i)D(1,j)\tau(e_{\alpha_2\beta_2} \otimes \cdots) \\
&= D(1,i)D(1,j)\tau(s_1^* L s_1) = 0.
\end{aligned}$$

The weighted sum of the terms vanishes identically since $= D(1,i)D(1,j) \neq 0$ only if $i = 1 = j$, and, in this case,

$$\tau(s_\alpha s_\beta^* s_i s_i^*) = \tau(s_\alpha s_\beta^* s_1 s_1^*) = \tau(s(\alpha,\beta)).$$

It follows that

$$\begin{aligned}\tau(L\, s_i s_j^*) &= \sum_{\alpha\beta} L_{\alpha\beta}\, \tau(s(\alpha,\beta) s_i s_j^*) \\ &= \sum_{\alpha\beta} L_{\alpha\beta}\, \tau(s(\alpha,\beta)) \\ &= \tau\Big[\sum_{\alpha\beta} L_{\alpha\beta}\, s(\alpha,\beta)\Big] \\ &= \tau(L) \\ &= 0\end{aligned}$$

for all i,j as asserted. We used the assumption $\tau(L) = 0$.

When the arguments above are combined we get the following:

Corollary 10.3. *Let* $\delta \in \mathrm{Der}(\mathcal{P}_n, \mathcal{O}_n)$ *and let* $L = \sum_i \delta(s_i) s_i^*$. *Then* $\delta \perp A$ *if and only if*

(i) $$\omega(L) = 0;$$

(ii) $$\tau(s_i^*\, \epsilon(L) s_j) = 0, \ \forall\, i \neq j;$$

and

(iii) $$\omega(s_i L) = \omega(s_i^* L), \ \forall\, i.$$

Proof. Compute the terms in the inner product $(\delta_X, \delta) = \omega(L_X^* L)$ for $X \in \mathfrak{g}(n,1) \otimes \mathbb{C}$.

Example 10.4. *The orthogonal complement to* A *contains non-approximately inner derivations: Let*

$$L = s_1(s_1s_1^* - \sigma(s_1s_1^*)), \tag{10.6}$$

and let δ *be the unique derivation determined by*

$$\delta(s_i) = Ls_i, \qquad \delta(s_i^*) = -s_i^*L.$$

Then $\delta \perp A$ *but* δ *is not approximately inner.*

<u>Proof</u>. Show (by computation) that L satisfied the conditions (i)–(iii) in Corollary 10.3. We conclude that δ is in the orthogonal complement. But it is known to be non–approximately inner, cf. [BEvGJ, §2].

The example contrasts the corresponding situation for the noncommutative tori where we showed that the orthogonal complement to the canonical Lie algebra coincides with the approximately inner derivations.

<u>Corollary 10.5</u>. *Let the involution on* $\mathrm{Der}(\mathcal{P}_n, \mathcal{O}_n)$ *be given by*

$$\delta^*(a) = -\delta(a^*)^*, \quad a \in \mathcal{P}_n.$$

Then $(\delta_1, \delta_2) = (\delta_2^*, \delta_1^*)$ *if at least one of the two derivations is assumed gauge–invariant.*

<u>Proof</u>. Left to the reader.

§11. THE ORTHOGONAL COMPLEMENT OF THE APPROXIMATELY INNER DERIVATIONS IN $\mathcal{O}_n$

In this section, we further analyze the two types of approximately inner derivations in $\mathcal{O}_n$:

(i) Approximation with respect to the C^*–norm on $\mathcal{O}_n$, pointwise on $\mathcal{P}_n$. A derivation δ satisfies this condition iff there is a sequence (h_p) in $\mathcal{O}_n$ such that

$$\lim_{p\to\infty} \| h_p - \sigma(h_p) - L_\delta \| = 0$$

where $\|\cdot\|$ is the C^*–norm, and $L_\delta = \sum_i \delta(s_i)s_i^*$.

(ii) Approximation with a sequence in $\{ad(h) : h \in \mathcal{O}_n\}$ relative to the Hilbert-norm on the derivations.

While (i) and (ii) are equivalent for the noncommutative tori, they are not for $\mathcal{O}_n$.

We have characterized (in Section 10) the orthogonal complement of the canonical Lie algebra $A \simeq u(n,1) \otimes \mathbb{C}$ in $\mathrm{Der}(\mathcal{P}_n,\mathcal{O}_n)$ where $\mathcal{P}_n$ is the dense polynomial subalgebra of $\mathcal{O}_n$. The analogy to the case of the noncommutative tori is broken in that A is *not* contained in the orthogonal complement of the approximately inner derivations.

We now find this complement. Let M_1 be the set of all approximately inner elements in $\mathrm{Der}(\mathcal{P}_n,\mathcal{O}_n)$. Recall $\delta \in M_1$ iff L_δ is in the C^*–norm–closure of

$$\mathcal{R}_n(\sigma) = (I-\sigma)(\mathcal{P}_n). \qquad (11.1)$$

Theorem 11.1. *The orthogonal complement* $\mathrm{Der}(\mathcal{P}_n,\mathcal{O}_n) \ominus M_1$ *(relative to the inner product on the derivations) is one–dimensional and consists of scalar multiples of the* *generator of the gauge–action.*

Proof. Let

$$s(\alpha,\beta) = s_\alpha s_\beta^*, \quad \alpha,\beta \in J.$$

The gauge–action $\{\alpha_z : z \in \mathbf{T}^1\}$ satisfies

$$\alpha_z(s(\alpha,\beta)) = z^{|\alpha|-|\beta|}\, s(\alpha,\beta)\,, \quad z \in \mathbf{T}^1.$$

If δ_g denotes the corresponding infinitesimal generator (the generalized number operator), then

$$\delta_g(s(\alpha,\beta)) = \sqrt{-1}\,(|\alpha|-|\beta|)\;\; s(\alpha,\beta)$$

and

$$L_g := \sum_i \delta_g(s_i)s_i^* = \sqrt{-1}. \tag{11.2}$$

We showed that a given δ in $\mathrm{Der}(\mathcal{P}_n,\mathcal{O}_n)$ is orthogonal to M_1 iff it is orthogonal to the inner derivations. Let $L = L_\delta = \sum_i \delta(s_i)s_i^*$. Let $b \in \mathcal{O}_n$, and let $\mathrm{ad}(b)$ be the corresponding inner derivation.

Then the formula for the Hilbert space–inner product yields

$$(\delta,\, \mathrm{ad}(b)) = \omega(L^*(b-\sigma(b))). \tag{11.3}$$

Let π_ω be the GNS–representation of $\mathcal{O}_n$ associated to the state ω, and let $\mathcal{L}^2_\omega := \mathcal{L}^2(\mathcal{O}_n,\omega)$ denote the representation space of π_ω with cyclic vector $1_\omega \in \mathcal{L}^2_\omega$. Since $\omega \circ \sigma = \omega$, it follows that σ implements an isometry T in $\mathcal{L}^2_\omega$ which is determined by

$$T(\pi_\omega(a)\, 1_\omega) = \pi_\omega(\sigma(a))1_\omega\,, \quad a \in \mathcal{O}_n, \tag{11.4}$$

and satisfies the covariance relation,

$$T^*\pi_\omega(\sigma(a))T = \pi_\omega(a)\,, \quad a \in \omega_n.$$

Since ω is faithful, there is a linear embedding,

$$\varphi_\omega : \mathcal{O}_n \longrightarrow \mathcal{L}^2_\omega$$

with dense range, given by,

$$\varphi_\omega(a) := \pi_\omega(a)\, 1_\omega.$$

(It identifies $\mathcal{O}_n$ with a dense linear subspace of $\mathcal{L}^2_\omega$.) We have

$$\omega(a) = (1_\omega, \pi_\omega(a)1_\omega)_{\mathcal{L}^2_\omega} = (\varphi_\omega(1), \varphi_\omega(a))_{\mathcal{L}^2_\omega}\,, \quad a \in \mathcal{O}_n.$$

It follows that

$$(\delta,\mathrm{ad}(b)) = (\varphi_\omega(L), \varphi_\omega(b-\sigma(b)))_{\mathcal{L}^2_\omega} = (\varphi_\omega(L), \varphi_\omega(b)-T\varphi_\omega(b))_{\mathcal{L}^2_\omega}.$$

We are now in a position to apply elementary geometry of isometries (on the isometry T). We have:

$$\begin{aligned}\delta \in M_1^{\perp} &\Longleftrightarrow \varphi_{\omega}(L) \in \mathcal{R}(I-T)^{\perp}\\ &\Longleftrightarrow T^*(\varphi_{\omega}(L)) = \varphi_{\omega}(L)\\ &\Longleftrightarrow T(\varphi_{\omega}(L)) = \varphi_{\omega}(L)\\ &\Longleftrightarrow \sigma(L) = L\\ &\Longleftrightarrow L \in \{\lambda 1 : \lambda \in \mathbb{C}\}\\ &\Longleftrightarrow \delta \in \{\lambda\delta_g : \lambda \in \mathbb{C}\}.\end{aligned}$$

This concludes the proof.

Definition 11.2. *We say that* $\delta \in \mathrm{Der}(\mathcal{P}_n, \mathcal{O}_n)$ *is* ***approximately inner in mean*** *if there is a sequence* $\{h_N : N = 1,2,\cdots\}$ *in* $\mathcal{O}_n$ *such that*

$$\lim_{N\to\infty} \|\delta - \mathrm{ad}(h_N)\|_{\mathcal{H}} = 0 \tag{11.5}$$

where the norm is defined by the inner product on derivations; equivalently,

$$\lim_{N\to\infty} \omega((h_N - \sigma(h_N) - L_{\delta})^*(h_N - \sigma(h_N) - L_{\delta})) = 0$$

Corollary 11.3. *Let* $\delta \in \mathrm{Der}(\mathcal{P}_n, \mathcal{O}_n)$. *Then* δ *is approximately inner in mean iff* $\psi_w(L_\delta) = 0$ *for all* $w \in S_n$ *were* $L_\delta = \sum_i \delta(s_i)s_i^*$.

Proof. The inner product on derivations gives rise to a Hilbert space $\mathcal{H}$ of (module–) derivations, and the set of inner derivations M_0 forms a subspace of $\mathcal{H}$. The orthogonal complements below will be relative to this Hilbert space $\mathcal{H}$.

We have:

$$M_0^{\perp} = M_1^{\perp} = \{\lambda\delta_g : \lambda \in \mathbb{C}\}$$

and

$$\begin{aligned} M_0^{\perp\perp} &= \text{(Hilbert–norm closure)} \\ &= \{\delta : L_\delta \in N(I-T^*)^{\perp}\} \\ &= \{\delta : (\xi, L_\delta) = 0 \quad \text{for all } \xi \in \mathscr{L}_2(\omega) \text{ s.t. } T\xi = \xi\}. \end{aligned}$$

Since T is isometric on $\mathscr{L}^2(\omega)$, a vector $\xi \in \mathscr{L}^2(\omega)$ satisfies $T\xi = \xi$ iff $T^*\xi = \xi$. Using the detailed properties of the Cuntz–states ψ_w, $w \in S_n$, and formula (9.5), we show that solutions to $T\xi = \xi$ may be expressed as

$$\xi = \int_{S_n}^{\oplus} f(w) 1_w \, dw$$

relative to the direct integral decomposition induced from (9.5), where $f \in \mathscr{L}^2(S_n)$, and 1_w denotes the cyclic vector for the representation of ψ_w. Since

$$(\xi, L_\delta) = \int_{S_n} \overline{f(w)} \, \psi_w(L_\delta) dw,$$

it follows that $\delta \in \{\text{inner}\}^{\perp\perp}$ iff $\psi_w(L_\delta) = 0$ for all $w \in S_n$ as asserted.

Definition 11.4. A derivation $\delta \in \text{Der}(\mathscr{P}_n, \mathcal{O}_n)$ is said to be *singular* if the element $L_\delta := \sum_i \delta(s_i)s_i^*$ satisfies $\omega(L_\delta) = 0$, but $\psi_w(L_\delta) \neq 0$ for some $w \in S_n$.

Note that the derivations (other than the gauge–generator) in the Lie algebra of the U(n,1)–action are singular.

§12. A DECOMPOSITION THEOREM FOR THE DERIVATIONS IN $\mathcal{O}_n$

Corollary 12.1. (i) *Every* δ *in* $\mathrm{Der}(\mathcal{P}_n, \mathcal{O}_n)$ *decomposes uniquely as follows:*

$$\delta = \lambda\delta_g + \mathcal{Y} \tag{12.1}$$

where $\lambda \in \mathbb{C}$, δ_g *is the generator for the gauge–action, and* $\mathcal{Y}$ *is singular or approximately inner in mean.*

(ii) *For every mean approximately inner,* $\delta \in \mathrm{Der}(\mathcal{P}_n, \mathcal{O}_n)$ *such that the commutator* $\delta' := [\delta_g, \delta]$ *is defined,* δ' *is automatically approximately inner in mean.*

(iii) *The gauge–invariant elements in* $\mathrm{Der}(\mathcal{P}_n, \mathcal{O}_n)$ *which are approximately inner in mean form an ideal.*

Remarks 12.2. We conjecture that the assumption of gauge–invariance in (iii) may be omitted. By (ii), the conjecture is equivalent to the assertion that the derivations which are approximately inner in mean are closed under taking commutator bracket. We have only been able to prove this under the assumption of gauge–invariance.

A result in [BEvGJ] states that gauge–invariant approximately inner elements in $\mathrm{Der}(\mathcal{P}_n, \mathcal{P}_n)$ are automatically inner. We note that this conclusion fails for $\delta \in \mathrm{Der}(\mathcal{P}_n, \mathcal{P}_n)$ if δ is only assumed approximately inner *in mean.*

As an example, take $L = s_1(s_2^* - \sigma(s_2^*))$, and define δ by,

$$\delta(s_i) = Ls_i\,, \quad \delta(s_i^*) = -s_i^* L.$$

Then δ is not approximately inner. But $\psi_w(L) = \overline{w}_1(w_2 - w_2) = 0$ for all $w \in S_n$, so δ is approximately inner in mean.

Proof. Just use the Hilbert space decomposition relative to

$$\mathrm{Der}(\mathcal{P}_n,\mathcal{O}_n) \subset \mathcal{H} = \{T\xi = \xi\} \oplus \{\text{mean approx. inner}\}. \tag{12.2}$$

The corollary

(i) follows since

$$\{T\xi = \xi,\delta_g\}^{\perp} = M_0^{\perp\perp} = (\text{Hilbert space norm–closure of } M_0).$$

(ii) Suppose $\delta_1 \in \mathrm{Der}(\mathcal{P}_n,\mathcal{O}_n)$, $\delta_2 \in \overline{M_0}$ (Hilbert norm–closure), and $[\delta_1,\delta_2] = \delta_1\delta_2 - \delta_2\delta_1 \in \mathrm{Der}(\mathcal{P}_n,\mathcal{O}_n)$. We must show that $[\delta_1,\delta_2] \in \overline{M_0}$. Let $L = \sum_i [\delta_1,\delta_2](s_i)s_i^*$. By Corollary 11.3, it is enough to show that $\psi_w(L) = 0$ for all w.

We now turn to the details:

First note that the commutator $\delta' = [\delta_g,\delta]$ is in $\overline{M_0}$ for all $\delta \in \mathrm{Der}(\mathcal{P}_n,\mathcal{O}_n)$ such that the commutator is well defined on $\mathcal{P}_n$. Let δ be such, and let $L = \sum_i \delta(s_i)s_i^*$ and $L_g = (\sqrt{-1})1$. Then the element

$$L' = [L,L_g] + \delta_g(L) - \delta(L_g)$$

satisfies

$$[\delta_g,\delta](s_i) = L's_i\,, \quad 1 \leq i \leq n,$$

and we must show that, for each ψ_w, $\psi_w(L') = 0$. But $L' = \delta_g(L)$ since L_g is a scalar multiple of 1, and $\psi_w(\delta_g(L)) = 0$ since each $\psi_w(L) = 0$. It follows that $[\delta_g,\delta] \in \overline{M_0}$ as asserted.

(An important example of a Lie algebra $\mathcal{L}$, contained in $\mathrm{Der}(\mathcal{P}_n,\mathcal{O}_n)$ where the commutator is well defined for any pair of elements in $\mathscr{L}$, is,

$$L = \mathrm{Der}(\mathcal{O}_n^\infty, \mathcal{O}_n^\infty)$$

where $\mathcal{O}_n^\infty$ denotes the dense subalgebra consisting of all C^∞–elements for the canonical U(n,1)–action on $\mathcal{O}_n$.)

(iii) Let $\delta_1, \delta_2 \in \overline{M_0}$ and assume that both are gauge–invariant. Let

$$L_k = \sum_i \delta_k(s_i)s_i^*, \quad k = 1,2,$$

and

$$L := [L_2, L_1] - \delta_1(L_2) - \delta_2(L_1).$$

Since $\epsilon(L) = L$, we have $\omega(L) = \tau(L) = \tau(\delta_1(L_2)) - \tau(\delta_2(L_1))$. But the two derivations δ_1 and δ_2 are now restricted to the invariant algebra $\mathscr{P}_n' \subset \mathcal{O}_n'$ and $\mathcal{O}_n' \simeq \bigotimes_1^\infty M_n$. Furthermore, we have $\tau \circ \delta_k = 0$ on $\mathscr{P}_n'$ by a well known result on derivations in UHF–algebras, cf. [PSa 2] or [B–R, §3.2.5].

It follows that $\psi_w(L) = 0$ for all w which implies $[\delta_1, \delta_2] \in \overline{M_0}$ as asserted.

§13. BIBLIOGRAPHICAL AND CONCLUDING COMMENTS

Let $\mathfrak{A}$ be a C^*–algebra, and let $\mathfrak{A}_0$ be a specified dense subalgebra of $\mathfrak{A}$. Let $\mathscr{D} = \mathrm{Der}(\mathfrak{A}_0,\mathfrak{A})$ be the derivations which are defined on $\mathfrak{A}_0$ and mapping into $\mathfrak{A}$. An *energy form* on $\mathscr{D}$ is defined to be a sesquilinear, strictly positive definite, form on $\mathscr{D}$, and is denoted $(\cdot,\cdot)$. Let $E(\delta) = (\delta,\delta)$, $\delta \in \mathscr{D}$. In classical analysis, and in the study of harmonic maps [At], such forms have proven useful.

Let $\mathscr{B} \subset \mathscr{D}$ be a given subset which is assumed closed under complex scalar multiplication. We define a *singular point* δ relative to $\mathscr{B}$ by the following condition:

$$(D_\beta E)(\delta) = 0\,, \quad \beta \in \mathscr{B}, \tag{13.1}$$

where

$$(D_\beta E)(\delta) = \frac{d}{dt}\Big|_{t=0} E(\delta+t\beta).$$

It is immediate that δ is singular relative to $\mathscr{B}$ iff $(\delta,\beta) = 0$ for all $\beta \in \mathscr{B}$. We are thus led to consider the orthogonal complement $\mathscr{B}^\perp$ with respect to a given inner product, and a given perturbation class $\mathscr{B}$.

Sakai suggested [Sak 7] that the differential geometric features of simple C^*–algebras should arise from choosing $\mathscr{B}$ to consist of all the approximately inner derivations.

We defined an energy form for each of the following examples, reviewed below:

(i) $\mathfrak{A}$ = (the CAR–algebra over a given separable complex Hilbert space);

(ii) $\mathfrak{A}$ = (the simple C^*–algebra over a given nondegenerate antisymmetric bicharacter on an abelian discrete group Γ of rank d, a noncommutative d–torus);

(iii) $\mathcal{O}_n$ = (the Cuntz–algebra).

In each case, the above–mentioned subalgebra $\mathfrak{A}_0$ is defined in terms of a specific finite set of generators.

If $\mathcal{B} \subset \mathcal{D}$ is chosen to consist of the approximately inner derivations, we showed, in the three cases:

(i) $\mathcal{B}^{\perp} = 0$;

(ii) $\mathcal{B}^{\perp} =$ (the d–dimensional abelian Lie algebra of the canonical ergodic action of $\hat{\Gamma}$ on $\mathfrak{A}$.)

(iii) $\mathcal{B}^{\perp} =$ (scalar multiples of the infinitesimal generator of the gauge–action by $\mathbf{T}^1$).

Note that (i) follows from a special case of Sakai's theorem [Sak 5], and at (ii) can be shown (see Section 4 above) to imply the decomposition theorem in [BEJ]; while, finally, (iii) leads to an analogous decomposition theorem for the derivations in $\mathcal{O}_n$. In fact, there are several such decomposition results for $\mathcal{O}_n$, as indicated in Section 7 above, and sketched below.

First, we are led to consider two distinct notions of approximately inner derivations; one defined relative to pointwise convergence in the C^*–norm, and the other defined relative to the Hilbert norm associated to the given energy quadratic form. The second notion is called "approximately inner *in mean*". We showed that the two notions are equivalent only in cases (i) an (ii) (the CAR–algebra, and the noncommutatiave tori), but not for (iii) (the algebra $\mathcal{O}_n$).

Secondly, there is a variety of Lie algebras yielding distinct decomposition results.

Approximately inner derivations were studied in mathematical physics [HHW], and it was shown in [PSa 2] that approximately inner derivations have β–KMS states for all values of (inverse temperature) β provided the underlying C^*–algebra has trace. They have ground states (and ceiling states) in general, also when there is no trace. Strictly speaking, the Powers–Sakai result is for one–parameter groups of automorphisms, but the result also holds for derivations when the domain is appropriately specified. We refer to [BR, vol. II, Theorem 5.3.15] for details on this last point.

For $\mathcal{O}_n$ (case (iii)), there are derivations (e.g., generators for the U(n,1)–action) which are known [C–E] *not* to have ground states, or ceiling states. It follows that the Powers–Sakai theorem [PSa 2] cannot be strengthened in this setting, i.e., (iii):

A derivation which generates an $\mathbb{R}$–action and is approximately inner *in mean* (i.e., the weaker condition) does generally not have ground states, and also not β–KMS states. (The gauge–generator has a unique β–KMS state for one and only one value of β, [Cu 2].)

There is a number of distinct Lie algebras $\mathfrak{g}$ in $\mathrm{Der}(\mathcal{P}_n,\mathcal{O}_n)$ with a differential geometric significance (viewing $\mathcal{O}_n$ as noncommutative hyperbolic space) consisting of *non*approximating inner derivations. For some of them, we have decomposition results. We now list the cases:

(i) Let $\mathfrak{g}_U$ be the Lie algebra obtained by differentiating the canonical U(n)–action on $\mathcal{O}_n$. It is a compact Lie algebra consisting of real (i.e., $\delta^* = -\delta$) derivations,

$$\mathfrak{g}_U = \{\delta : L_\delta = x_0 1 + sxs^*\} \tag{13.2}$$

where

$$x_0 \in \sqrt{-1}\,\mathbb{R} \quad \text{and} \quad x = -x^* \in M_n.$$

Let $\mathcal{O}_n^\infty(U(n))$ be the C^∞–elements for the U(n)–action on $\mathcal{O}_n$. Then a result of [BGoo] states that every $\delta \in \mathrm{Der}(\mathcal{O}_n^\infty(U(n)),\mathcal{O}_n)$ decomposes uniquely as:

$$\delta = \delta_x + \tilde{\delta} \tag{13.3}$$

where $\delta_x \in \mathfrak{g}_U \otimes \mathbb{C}$ and $\tilde{\delta}$ is inner.

(ii) Let $\mathfrak{g}_0$ be the one–dimensional Lie algebra spanned by the gauge–generator. Then we have, from Theorem 11.1 above, that $\{\text{inner}\}^\perp \cap \text{Der} = \mathfrak{g}_0$, and

$$\{\text{inner}\}^{\perp\perp} \cap \text{Der} = \{\delta : \psi_w(L_\delta) = 0, \ \forall \ w \in S_n\} \tag{13.4}$$
$$= (\text{all mean approximately inner derivations}).$$

The next results (which we state below) are related to Corollary 10.3 above, and they will be contained in a sequel paper [Jo 11] by the author.

(iii) Let $\mathfrak{g}_U$ be the compact Lie algebra (13.2) defined in (i) above. Let

$$\mathfrak{g}_U(m) = \{\delta_x : L_x \in \sigma^m(x_0 1 + sxs^*)\}$$

where $m = 0,1,2,\cdots$, $x_0 \in \sqrt{-1}\,\mathbb{R}$, and $x = -x^* \in M_n$. Then $\{\mathfrak{g}_U(m)\}$ is a family of mutually isomorphic and mutually commuting Lie algebras. As a result

$$\mathfrak{g}_U(\infty) := \text{span}_\mathbb{R}\{\mathfrak{g}_U(m) : m = 0,1,\cdots\}$$

is a Lie algebra. It consists of nonapproximately inner derivations. We show in [Jo 11] that

$$\mathfrak{g}_U(\infty)^\perp = \{\delta : \epsilon(L_\delta) = 0\} \tag{13.5}$$

where $\epsilon(L) = \int_{\mathbb{T}^1} \alpha_z(L)dz.$

Note that when the action of $\mathfrak{g}_U(\infty)$ is restricted to the gauge–invariant elements

$\mathcal{O}'_n \simeq \bigotimes_1^\infty M_n$, this is just the infinite tensor product (possibly different actions on different tensor slots) action.

Let $T \subset U$ be maximal abelian, and let $\mathfrak{g}_T$ be the corresponding abelian Lie subalgebra of $\mathfrak{g}_U$. Now apply iterated shifts to $\mathfrak{g}_T$ to construct an abelian Lie subalgebra $\mathfrak{g}_T^\infty := \text{span}_{\mathbb{R}}\{\mathfrak{g}_T(m) : m = 0,1,\cdots\}$ of $\mathfrak{g}_U^\infty$. Then we show in [Jo 11] that $\mathfrak{g}_T^\infty$ is a *maximal abelian* Lie subalgebra of $\text{Der}(\mathcal{P}_n, \mathcal{P}_n)$, and also of $\text{Der}(\mathcal{O}_n^\infty, \mathcal{O}_n^\infty)$.

(iv) Let $\mathfrak{g}_{SU}$ be the Lie algebra defined by the natural $SU(n)$–action on $\mathcal{O}_n$, viewing $SU(n)$ as a subgroup of $U(n)$,

$$\mathfrak{g}_{SU} = \{\delta : L_\delta = sxs^*, x \in su(n)\}$$

where $su(n) = \{x : x = -x^* \in M_n, \text{trace}(x) = 0\}$ is the matrix Lie algebra of the group $SU(n)$. Let

$$\mathfrak{g}_{SU}(m) = \{\delta_x : L_x \in \sigma^m\{sxs^* : x \in su(n)\}\}.$$

We show in [Jo 11] that $\{\mathfrak{g}_{SU}(m) : m = 0,1,\cdots\}$ is a family of mutually isomorphic, mutually *orthogonal*, and mutually commuting Lie algebras. If

$$\mathfrak{g}_{SU}(\infty) := \text{span}_{\mathbb{R}}\{\mathfrak{g}_{SU}(m) : m = 0,1,\cdots\},$$

then

$$\mathfrak{g}_{SU}(\infty)^\perp = \{\delta : \epsilon(L_\delta) \in \mathbb{C}1\}. \tag{13.6}$$

(v) Let $\mathfrak{g}(n,1)$ be the derivation Lie algebra obtained by differentiating the noncompact hyperbolic $U(n,1)$–action on $\mathcal{O}_n$,

$$\mathfrak{g}(n,1) = \{\delta : L_\delta = x_0 1 + s(v) - s(v)^* + sxs^*\}$$

where $x_0 \in \sqrt{-1}\,\mathbb{R}$, $v \in \mathbb{C}^n$, and $x = -x^* \in M_n$.

Let

$$\mathfrak{g}(n,1)^m := \{\delta_x^{(m)} : L(\delta_x^{(m)}) = \sigma^m(L(\delta_x)),\ \delta_x \in \mathfrak{g}(n,1)\}.$$

Then we show in [Jo 11] that $\{\mathfrak{g}(n,1)^m : m = 0,1,\cdots\}$ is a family of mutually isomorphic Lie algebras, and that

$$\mathfrak{g}(n,1)^\infty := \mathrm{span}_{\mathbb{R}}\{\mathfrak{g}(n,1)^m : m = 0,1,\cdots\}$$

satisfies

$$(\mathfrak{g}(n,1)^\infty)^\perp = \{\delta : L_\delta \text{ satisfies (a)–(b) below}\} \tag{13.7}$$

where

(a)
$$\epsilon(L) = 0, \tag{13.7a}$$

(b)
$$\omega(\sigma^m(s_i)L) = \omega(\sigma^m(s_i^*)L) \text{ for all } m \text{ and } i. \tag{13.7b}$$

The derivation δ which is determined by $L_\delta = L$ from formula (10.6) in Example 10.4, is an example of a non–inner derivation in

$$(\mathfrak{g}(n,1)^\infty)^\perp \cap \mathrm{Der}(\mathcal{P}_n,\mathcal{P}_n).$$

(It is approximately inner in mean since clearly $\psi_w(L) = 0$ for all the Cuntz–states ψ_w, $w \in S_n$.)

The following decomposition theorem follows from this:

Let $\mathrm{Der}(\mathcal{P}_n,\mathcal{O}_n)$ be decomposed relative to the gauge–action,

$$\delta = \sum_{k\in\mathbb{Z}}^{\oplus} \delta(k)\,, \qquad \delta(k) = \int_{\mathbf{T}^1} \bar{z}^k\, \alpha_z \circ \delta \circ \alpha_z^{-1}\, dz.$$

Since α_z preserves the inner product on $\mathrm{Der}(\mathscr{P}_n, \mathcal{O}_n)$, this is an orthogonal decomposition corresponding to the decomposition of L_δ as follows:

$$L = \sum_{k\in\mathbb{Z}}^{\oplus} L(k)\,, \qquad L(k) = \int_{T^1} \bar{z}^k\, \alpha_z(L)dz$$

where $L = \sum_{i=1}^{n} \delta(s_i)s_i^*$.

Let

$$\mathscr{D}^\alpha(k) = \{\delta : \alpha_z(L_\delta) = z^k L_\delta,\ z \in T^1\}.$$

Then

$$\sum_k^{\oplus} \{\mathscr{D}^\alpha(k) : |k| > 1\} \subset (\mathfrak{g}(n,1)^\infty)^\perp,$$

and every δ in $\mathscr{D}^\alpha(1) + \mathscr{D}^\alpha(-1)$ decomposes uniquely as a sum of two linearly independent terms as follows:

$$\delta = \delta_x + \gamma \tag{13.8}$$

where $\delta_x \in \mathfrak{g}(n,1)^\infty \otimes \mathbf{C}$ and γ is approximately inner.

(Note that the uniqueness in this decomposition (13.8) follows from the observation in Section 9 to the effect that $\mathfrak{g}(n,1)^\infty \backslash \{0\}$ consists of nonapproximately inner derivations. Formula (9.4) is crucial here. The two terms in (13.8) are just linearly independent but generally not orthogonal.)

(vi) Some of the specific results quoted in (i)–(v) are based on the following lemma on the shift which may be of independent interest:

Lemma. (a) *Let* $\delta \in \mathrm{Der}(\mathcal{P}_n, \mathcal{O}_n)$, $L = \sum_i \delta(s_i)s_i^*$, *and* $m \in \{0,1,2,\cdots\}$; *and let* $\delta^{(m)}$ *be the (unique) derivation determined by*

$$\delta^{(m)}(s_i) = \sigma^m(L)s_i, \quad \delta^{(m)}(s_i^*) = -s_i^* \sigma^m(L).$$

Then we have the commutation relation,

$$\sigma^m \circ \delta = \delta^{(m)} \circ \sigma^m. \tag{13.9}$$

(b) *The mapping,* $\delta \longrightarrow \delta^{(m)}$, *is a* 1–1 *Lie homomorphism, and it is isometric in the Hilbert norm from the inner product on* $\mathrm{Der}(\mathcal{P}_n, \mathcal{O}_n)$. *It maps integrable Lie algebras (cf.* [JM] *and* [GJ 2]) *onto integrable Lie algebras.*

REFERENCES

[Am] W. Ambrose, Spectral resolution of groups of unitary operators, Duke Math. J. **11**(1944), 589–595.

[Ar] H. Araki, Bogoliubov automorphisms and Fock representations of canonical anticommutation relations, in Operator Algebras and Mathematical Physics, Contemporary Mathematics, **62**, Amer. Math. Soc., Providence, RI, 1986.

[Arv] W.B. Arveson, On groups of automorphisms of operator algebras, J. Funct. Anal. **15**(1974), 217–243.

[At] M.F. Atiyah & R. Bott, The Yang–Mills equations over Riemann surfaces, Philos. Trans. Roy. Soc. London, **308A**(1982), 523–615.

[Bat] C.J.K. Batty, Derivations on compact spaces, Proc. London Math. Soc. (3) **42**(1981), 299–330.

[Bla] B. Blackadar, K–theory for Operator Algebras, Springer–Verlag, New York, 1986.

[Bou] N. Bourbaki, Groupes et algebres de Lie, Ch. 1, Hermann, Paris, 1960; Ch. 2 et 3, Hermann, Paris, 1972; Ch. 4, 5 et 6, Hermann, Paris, 1974; Ch. 7 et 8, Hermann, Paris, 1975; Ch. 9, Masson, Paris, 1982.

[Bra] O. Bratteli, Derivations, Dissipations, and Group Actions on C^*–algebras, Springer–Verlag, Heidelberg, New York, 1987 (LNM 1229).

[BEE] O. Bratteli, G.A. Elliott & D.E. Evans, Localilty and differential operators on C^*–algebras, J. Diff. Equations, **64**(1986), 221–273.

[BER] O. Bratteli, G.A. Elliott & D.W. Robinson, The characterizatio of differential operators by locality: Classical flows, Comp. Math. **58**(1986), 279–319.

[BGoo] O. Bratteli & F.M. Goodman, Derivations tangential to compact group actions: Spectral conditions in the weak closure, Canad. J. Math. **37**(1985), 160–192.

[BEGJ(a)] O. Bratteli, G.A. Elliott, F.M. Goodman & P.E.T. Jorgensen, Smooth Lie group actions on noncommutative tori, Preprint 1986. To appear in "Nonlinearity".

[BEGJ(b)] __________________, On Lie algebras of operators, Preprint 1987. To appear in J. Funct. Anal.

[BEvGJ] O. Bratteli, D.E. Evans, F.M. Goodman & P.E.T. Jorgensen, A dichotomy for derivations on O_n, Publ. of the RIMS, Kyoto U. **22**(1986), 103–117.

[BEJ] O. Bratteli, G.A. Elliott & P.E.T. Jorgensen, Decomposition of unbounded derivations into invariant and approximately inner parts, J. Reine Angew. Math. (Crelle's J.) **346**(1985), 247–289.

[BKJR] O. Bratteli, A. Kishimoto, P.E.T. Jorgensen & D.W. Robinson, A C^*-algebraic Shoenberg theorem, Ann. Inst. Fourier (Grenoble) **33**(1984), 155–187.

[BJ 1] O. Bratteli & P.E.T. Jorgensen, Unbounded $*$–derivations and infinitesimal generators on operator algebras, in Proceedings of the AMS Summer Institute on Operator Algebras, Kingston, Ontario, 1980. Symposia in Pure Math. (38) **2**(1982), 353–365.

[BJ 2] __________, Unbounded derivations tangential to compact groups of automorphisms, J. Func. Anal. **48**(1982), 107–133.

[BJ 3] ________________, Derivations commuting with abelian gauge actions on lattice systems, Commun. Math. Phys. **87**(1982), 353–364.

[BR] O. Bratteli & D.W. Robinson, Operator Algebras and Quantum Statistical Mechanics, I and II, Springer–Verlag, New York, 1979/81 (vol. I in new ed. 1987).

[CE] A.L. Carey & D.E. Evans, On an automorphic action of $U(n,1)$ on O_n, J. Func. Anal **70**(1987), 90–110.

[Ca] P. Cartier, Quantum mechanical commutation relations and theta functions, Proc. Symp. Pure Math. **9**(1966), 361–383.

[Cob] L. Coburn, The C^*–algebra generated by an isometry, Bull. Amer. Math. Soc. **73**(1967), 722–726.

[Con 1] A. Connes, Sur la théorie non–commutative de l'integration, Lect. Notes in Math. 725, 19–143, Springer–Verlag, NY, 1979.

[Con 2] ________________, A survey of foliations and operator algebras, in: Operator Algebras and Applications, Proc. Symp. Pure Math. (38) vol. 1, Amer. Math. Soc., Providence, RI, 1980, 521–632.

[Con 3] ________________, C^*–algèbres et géométrie différentielle, C.R. Acad. Sci. Paris **290**(1980), 599–604.

[Con 4] ________________, Spectral sequences and homology of currents for operator algebras, Tagungsbericht 42/81, Oberwolfach 1981, 4–5.

[Con 5] ________________, Cohomologie cyclique et foncteurs Ext^n, C.R. Acad. Sci. Sér. A, Paris **296**(1983), 953–958.

[Con 6] ________________, Cyclic cohomology and the transverse fundamental class of a foliation, Preprint, I.H.E.S. M/84/7 (1984).

[Con 7] ________________, Non–commutatiave differential geometri, I. The Chern character in K–homology: II. De Rham homology and non–commutative algebra, Publ. Math. IHES **62**(1985), 257–360.

[CR] A. Connes & M.A. Rieffel, Yang–Mills for noncommutative two–tori, in: "Operator Algebras and Mathematical Physics," Contemporary Math. vol **62** (eds. P.E.T. Jorgensen & P.S. Muhly), Amer. Math. Soc., Providence, RI, 1986.

[Cu 1] J. Cuntz, Simple C^*–algebras generated by isometries, Commun. Math. Phys. **57**(1977), 173–185.

[Cu 2] ________________, Automorphisms of certain simple C^*–algebras, in: Quantum Fields—Algebras, Processes, L. Streit, ed., Springer–Verlag, Wien–New York, 1980, 187–196.

[CEℓGJ] J. Cuntz, G.A. Elliott, F.M. Goodman & P.E.T. Jorgensen, On the classification of non–commutative tori. II, C.R. Math. Rep. Acad. Sci. Canada **7**(1985), 189–194.

[Dix] J. Dixmier, Les algèbres d'operateurs dans l'espace hilbertien, Gauthier-Villars, Paris, 1969.

[DM] J. Dixmier & P. Malliavin, Factorisations de functions et de vecteurs indefiniment differentiables, Bull. Sci. Math. 2e sér. **102**(1978), 305–330.

[Eℓℓ] G.A. Elliott, On the K–theory of the C^*–algebra generated by a projective representation of a torsion–free discrete abelian group, Operator Algebras and Group Representation, I, New York, 1983, 157–184.

[Fr] I.B. Frenkel, Two constructions of affine Lie algebra representations and boson–fermion correspondence in quantum field theory, J. Funct. Anal. **44**(1981), 259–327.

[Ga] L. Gårding, Note on continuous representations of Lie groupos, Proc. Nat. Acad. Sci. USA **33**(1947), 331–332.

[Ge] I.M. Gelfand, On one–parametrical groups of operators in a normed space, Dokl. Akad. Nauk SSSR **25**(1939), 713–718 (Russian).

[GP] I.M. Gelfand & V.Ya. Ponomarev, Remarks on the classification of a pair of commuting linear transformation in a finite–dimensional space, Funkcional. Anal. i Prilozen. **3**(1969), 81–82.

[GJ 1] F. Goodman & P.E.T. Jorgensen, Unbounded derivations commuting with compact group actions, Comm. Math. Phys. **82**(1981), 399–405.

[GJ 2] ________________, Lie algbras of unbounded derivations, J. Funct. Anal. **52**(1983), 369–384.

[HHW] R. Haag, N.M. Hugenholtz & M. Winnink, On the equilibrium states in quantum statistical mechanics, Commun. Math. Phys. **5**(1967), 215–236.

[HS] M. Hausner & J. Schwartz, Lie Groups; Lie Algebras, Gordon and Breach, New York–London–Paris, 1968, 229.

[Hel] S. Helgason, Differential Geometry, Lie Groups, and Symmetric Spaces, Academic Press, New York, 1978.

[Hoc] G. Hochschild, The Structure of Lie Groups, Holden–Day, San Francisco, 1965.

[Ja] J. Jacobson, Lie Algebras, Interscience, New York, 1962.

[Jo 1] P.E.T. Jorgensen, Trace states and KMS states for approximately inner dynamical one–parameter groups of ∗–automorphisms, Commun. Math. Phys **53**(1977), 135–142.

[Jo 2] ————————, Ergodic properties of one–parameter automorphism groups on operator algebras, J. Math. Anal. Appl. **87**(1982), 354–372.

[Jo 3] ————————, Extensions of unbounded ∗–derivations in UHF C^*-algebras, J. Funct. Anal. **45**(1982), 341–356.

[Jo 4] ————————, The integrability problem for infinite–dimensional representations of finite–dimensional Lie algebras, Expositiones Math. **4**(1983), 289–306.

[Jo 5] ————————, A structure theorem for Lie algebras of unbounded derivations in C^*–algebras & Appendix, Compositio Math. **52**(1984), 85–98.

[Jo 6] ————————, New results on unbounded derivations and ergodic groups of automorphisms, Expositiones Math. **2**(1984), 3–24.

[Jo 7] ————————, Unitary dilations and the C^*–algebra O_2, Israel J. Math. (8) **20**(1986).

[Jo 8] ——————, Ergodicity of the automorphic action of U(1,H) on $C^*(\ell(H))$, Abstracts Amer. Math. Soc. **5**(1984), 212.

[Jo 9] ——————, Book review of "Derivations, Dissipations and Group Actions on C^*–algebras" (O. Bratteli) for the Amer. Math. Soc. Bull. **17**(1987), 202–209.

[Jo 10] ——————, "Operators and Representation Theory," North Holland Publ. Co., Amsterdam 1988.

[Jo 11] ——————, Infinite–dimensional Lie algebras of derivations acting on the Cuntz–algebra O_n, (in preparation).

[JM] P.E.T. Jorgensen & R.T. Moore, "Operator Commutation Relations," D. Reidel Publ. Co., Mathematics and Its Applications, Dordrecht–Boston-Lancaster, 1984.

[JPr] P.E.T. Jorgensen & G.L. Price, Extending quasi–free derivations on the CAR–algebra, J. Operator Theory **16**(1986), 147–155.

[Kad 1] R.V. Kadison, Derivations on operator algebras, Ann. Math. **83**(1966), 280–293.

[Kad 2] —————— (ed.), Operator algebras and applications, Proc. Symposia Pre Math., vol. **38**, parts 1 and 2, Amer. Math. Soc. Providence, RI, 1982.

[KR] R.V. Kadison & J.R. Ringrose, Fundamentals of the Theory of Operator Algebras, Vols. I and II, Academic Press, 1983/86.

[KS] R.V. Kadison & I.M. Singer, Some remarks on representations of connected groups, Proc. Nat. Acad. Sci. USA **38**(1952), 419–423.

[Kap] I. Kaplansky, Modules over operator algebras, Amer. J. Math. **75**(1953), 839–859.

[Kas] G.G. Kasparov, The operator K–functor and extensions of C^*–algebras, Math. USSR–Izv. **44**(1981), 513–572.

[Katn] Y. Katznelson, An Introduction to Harmonic Analysis, Dover Publ., Inc., New York, 1976.

[KiRo] A. Kishimoto & D.W. Robinson, Derivations, dynamical systems and spectral restrictions, Math. Scand. **56**(1985), 159–168.

[KuS] R.A. Kunze & E.M. Stein, Uniformly bounded representations, III. Intertwining Operators for the Principal Series on Semisimple Groups, Amer. J. Math. **89**(1967), 385–442.

[La] S. Lang, Introduction to Differentiable Manifolds, Interscience, New York, 1962.

[Las] R.K. Lashof, Lie algebras of locally compact groups, Pacific J. Math. **7**(1957), 1145–1162.

[Lo] R. Longo, Automatic relative boundedness of derivations in C^*–algebras, J. Functional Anal. **34**(1979), 21–28.

[Ma] G.W. Mackey, Induced representations of locally compact groups and applications, in: Functional Analysis and Related Fields, F.E. Browder, ed., Springer–Verlag, Berlin–Heidelberg, 1970, 132–171.

[MZ] D. Montgomery & L. Zippin, Topological Transformation Groups, Interscience Publ. Inc., New York, 1955.

[Ne] E. Nelson, Topics in Dynamics, I: Flows, Math. Notes, Princeton Univ. Press, 1969.

[OPT] D. Olesen, G.K. Pedersen & M. Takesaki, Ergodic actions of compact abelian groups, J. Operator Theory, **3**(1980), 237–269.

[Ped] G.K. Pedersen, C^*–algebras and Their Automorphism Groups, Academic Press, London–New York–San Francisco, 1979.

[PiPo] M. Pimsner & S. Popa, The Ext groups of some C^*–algebras considered by J. Cuntz, Rev. Roumaine Math. Pures Appl. **23**(1978), 1069–1076.

[PiVo 1] M. Pimsner & D. Voiculescu, Imbedding the irrational rotation C^*–algebra into an AF–algebra, J. Operator Theory **4**(1980), 201–210.

[PiVo 2] ——————, Exact sequences for K–groups and Ext–groups of certain cross–product C^*–algebras, J. Operator Theory **4**(1980), 93–118.

[Po] N.S. Poulsen, On C^∞–vectors and intertwining bilinear forms for representations of Lie groups, J. Funct. Anal. **9**(1972), 87–120.

[PSa 1] R.T. Powers & S. Sakai, Unbounded derivations in operator algebras, J. Funct. Anal. **19**(1975), 91–95.

[PSa 2] ——————, Existence of ground states and KMS states for approximately inner dynamics, Comm. Math. Phys. **39**(1975), 273–288.

[Rie 1] M.A. Rieffel, C^*–algebras associated with irrational rotations, Pacific J. Math. **93**(1981), 415, 429.

[Rie 2] ——————, Critical points of Yang–Mills for non–commutative two–tori, Preprint 1988.

[Rob] D.W. Robinson, Differential operators on C^*–algebras, in: Operator Algebras and Mathematical Physics, P.E.T. Jorgensen & P. Muhly, eds., Contemporary Mathematics, vol. **62**, Amer. Math. Soc., Providence, 1986.

[Ru] W. Rudin, Fourier Analysis on Groups, Interscience Publ., 1967.

[Sak 1] S. Sakai, On a conjecture of Kaplansky, Tohoku Math. J. **12**(1960), 31–33.

[Sak 2] ————————, Derivations of W^*–algebras, Ann. Math. **83**(1966), 273–279.

[Sak 3] ————————, Derivations of simple C^*–algebras, J. Funct. Anal. **2**(1968), 202–206.

[Sak 4] ————————, C^*–algebras and W^*–algebras, Springer–Verlag, Berlin-Heidelberg–New York, 1971.

[Sak 5] ————————, On one–parameter groups of $*$–automorphisms on operator algebras and the corresponding unbounded derivations, Amer. J. Math. **98**(1976), 427–440.

[Sak 6] ————————, Theory of unbounded derivations on C^*–algebras, Lecture Notes, Copenhagen Univ. and Univ. of Newcastle upon Tyne, 1977.

[Sak 7] ————————, Developments in the theory of unbounded derivations in C^*–algebras, Proc. of Symposia in Pure Mathematics II, **38**(1982), 309–331.

[Sch] W.M. Schmidt, Diophantine Approximation, Lecture Notes in Mathematics, **785**, Berlin–Heidelberg–New York, 1980.

[Se] I.E. Segal, Irreducible representations of operator algebras, Bull. Amer. Math. Soc. **53**(1947), 73–88.

[SvN] I.E. Segal & J. von Neumann, A theorem on unitary representations of semisimple Lie groups, Ann. of Math. **52**(1950), 509–517.

[SH] Ja.G. Sinai & A.Ja. Helmskii, A description of differentiations in algebras of the type of local observables of spin systsems, Funk. Anal. Prilo. **6**(1973), 343–344 (English translation).

[Si] I.M. Singer, Uniformly continuous representations of Lie groups, Ann. Math. (2) **56**(1952), 242–247.

[Sla] J. Slawny, On factor representations and the C^*–algebra of canonical commutation relations, Comm. Math. Phys. **24**(1971), 151–170.

[St] M.H. Stone, Linear Transformations in Hilbert Space, Colloq. Publ. **15**, Amer. Math. Soc., Providence, RI, 1932.

[Ta] H. Takai, On a problem of Sakai in unbounded derivations, J. Funct. Anal. **43**(1981), 202–208.

[Tom] J. Tomiyama, The Theory of Closed Derivations in the Algebra of Continuous Functions on the unit interval, Lecture Notes, Tsing Hua Univ., 1983.

[Voi 1] D. Voiculescu, Symmetries of some reduced free product C^*–algebras, Preprint, INCREST, 1983, in: Springer Lecture Notes in Math. vol. **1132**, 556–588.

[Voi 2] ——————, Dual algebraic structures on operator algebras related to free products, J. Operator Theory **17**(1987), 85–98.

[Wal] P. Walters, An Introduction to Ergodic Theory, GTM, Springer–Verlag, New York, 1982.

[We] A. Weil, L'integration dans les groupes topologiques et ses applications, Hermann, Paris, A.S.I., 1940.

[Wit] E. Witten, Noncommutative geometry and string field theory, Preprint, 1985, Princeton Univ. Press.

[Yang] C.–T. Yang, Hilbert's fifth problem and related problems on transformation groups, in: Proc. Symposia Pure Math. **28**(1976), Amer. Math. Soc., Providence, RI.

[Yo] K. Yosida, Functional Analysis (3rd edn.), Springer–Verlag, New York, 1971.

INDEX: SUBDIVISION OF GENERAL REFERENCES

1. Differential Geometry and Derivations.
   (a) Commutative: [At], [Bra], [Heℓ], [La], [MZ], [Tom].
   (b) Noncommutative: [Ar], [Arv], [Bat], [BEE], [BJ 1], [BR], [Con 2–7], [CEℓGJ], [Jo 9–10], [Ka 1], Sa 2–7].

2. General Theory.
   (a) Operator Algebras: [BR], [KR], [Ped].
   (b) Lie Groups: [HS], [Heℓ], [Hoc].
   (c) Lie Algebras: [Bou], [Ja].
   (d) K–theory: [Bℓa], [Con 7], [Eℓℓ], [Kas], [PiPo], [PiVo 2].
   (e) Ergodic Theory: Commutative: [Ne], [Waℓ]; Noncommutative: [Arv], [BKJR], [CE], [Jo 2&8], [OPT], [Ped], [SH].

3. Quantum Mechancis and C^*–algebras: [Ar], [BR], [Ca], [HHW], [Jo 1], [Sℓa], [SH], [Wit].

4. Harmonic Analysis:
   (a) Commutative: [Am], [Katn], [Ru].
   (b) Noncommutative: [Con 1–2], [Las], [We], [Yang].

5. Representation Theory.
   (a) Groups: [DM], [Gå], [JM], [Ku S], [Ma], [Po].
   (b) Algebras: [Dix], [Fr], [Kap], [PiVo], [Se].
   (c) Integration of Lie Algebra Representations: [BEGJ], [Ge], [GJ 2], [Jo 4], [JM].

Derivations in Commutative C*-Algebras

H. KUROSE

1. Introduction and Preliminaries

Prof. S. Sakai began the systematic study of unbounded *-derivations in C^*-algebras after his work on bounded *-derivations. For the theory of the unbounded *-derivations, he posed many questions in his lecture notes and his talks ([S1, S2]). In the case of commutative C^*-algebras, several authors have developed the theory by trying to solve his problems ([Ba], [G]). In consequence, roughly speaking, now we may say that the structure of closed *-derivations has been almost clarified when the underlying space of the commutative C^*-algebra is of 0- or 1-dimension, though a few problems have been left unsolved. Furthermore the structure of *-derivations commuting with group actions has rapidly become clear in the last decade. In this note, we shall not mention these structures, using as our references [Bra] and [T].

In Section 2, we shall discuss closed *-derivations in higher dimensional spaces. We will fix our notations for later use. Let Ω be a compact Hausdorff space, $C(\Omega)$ the C^*-algebras of all continuous $\mathbb{C}$-valued functions on Ω and δ a closed *-derivation in $C(\Omega)$. The partial derivatives and the generators of continuous flows are familiar examples of closed *-derivations. Here we will give the fundamental properties of closed *-derivations on compact spaces.

(1) The domain $\mathcal{D}(\delta)$ of δ is a regular Banach algebra which admits the C^1-functional calculus. Furthermore, $\mathcal{D}(\delta)$ is of type C.

(2) δ is a local operator, that is, for each open subset U of Ω, we get $\delta(f) = 0$ on U whenever $f = 0$ on U and $f \in \mathcal{D}(\delta)$.

As in (2), if we get $\delta(f) = 0$ on S whenever $f = 0$ on S and $f \in \mathcal{D}(\delta)$, the subset S of Ω is said to be self-determining for δ. In this case we can consider the restriction δ_S of δ to S by

$$\delta_S(f|_S) = \delta(f)|_S$$

and

$$\mathcal{D}(\delta_S) = \{f\big|_S; f \in \mathcal{D}(\delta)\}.$$

Each open subset of Ω is self-determining for δ and it is known that the restriction of δ to the closure of the open subset is always closable.

2. Closed *-Derivations on Higher Dimensional Spaces

All known examples of closed *-derivations on compact spaces are given by *-derivations on one-dimensional spaces. A closed *-derivation δ is said to have the one-dimensional structure at a point ω in Ω iff there exists a topological embedding ϕ of $[-1,1]$ into Ω such that $\phi(0) = \omega$, $\phi([-1,1])$ is self-determining for δ and $\delta_{\phi([-1,1])}$ is closable. Set $A_\delta = \{x \in \Omega;\ \delta(f)(x) \neq 0 \text{ for some } f \in \mathcal{D}(\delta)\}$, then the following conjecture seems to be true.

Conjecture. Every closed *-derivation on compact spaces has one-dimensional structure, that is, there exists a dense subset of A_δ at each point of which the derivation has one-dimensional structure.

To solve this conjecture, first we have to seek one-dimensional subsets of Ω which are self-determining for δ. The example of the partial derivative indicates that we may consider the kernel of the given derivation. In fact we can get the desired one-dimensional subsets for the partial derivative $\partial/\partial x$ by taking stationary subsets for the kernel of $\partial/\partial x$, where the stationary subsets for a family of functions are defined by the maximal subsets on each of which every function in the family is constant. For our purpose, the kernel of the given *-derivation has to be sufficiently large. Here, we encounter the first difficulty; there is a simple example of a closed *-derivation on $[0,1] \times [0,1]$ that has a zero kernel. But Nishio proved a "local kernel property."

Theorem ([Ni]). *Let δ be a closed *-derivation on $[0,1] \times [0,1]$ with* Range $\delta = C([0,1] \times [0,1])$. *Then, for every open subset U of $[0,1] \times [0,1]$, there exists a connected open set V in U such that the kernel of $\delta_{\bar{V}}$ is non-zero.*

The above theorem indicates that, for the study of δ in the theorem, we may consider the closure of $\delta_{\bar{V}}$. It is important to study the following:

Problem. Prove the local kernel property for general closed *-derivations on compact spaces without use of the range condition.

Now we assume that the kernel of a closed *-derivation δ is sufficiently large and that we can find one-dimensional subsets as the stationary subsets for the kernel of δ. In restricting δ to these one-dimensional subsets, we encounter the second difficulty. We have

Example ([K]). There exists a closed *-derivation δ on $[0,1]\times[0,1]$ such that the kernel of δ is equal to $\mathbf{1}\otimes C([0,1])$ and the subset $[0,1]\times\{0\}$ is not self-determining for δ.

But we can overcome the second difficulty as follows.

Theorem ([K]). *Let δ be a closed *-derivation on a compact Hausdorff space Ω and set* Kernel $\delta = C(\Delta)$ *for some compact Hausdorff space Δ. For each $\sigma \in \Delta$, we denote the maximum self-determining subset in $\phi^{-1}(\{\sigma\})$ by S_σ, where ϕ is the canonical continuous mapping from Ω onto Δ. Then we have*

(i) $\bigcup_{\sigma\in\Delta} S_\sigma$ *is dense in* Ω;

(ii) δ_{S_σ} *is closed for each* $\sigma \in \Delta$.

In particular, if δ is a closed *-derivation on $[0,1]\times\Delta$ such that Kernel $\delta \supset \mathbf{1}\otimes C(\Delta)$, as a corollary of the above theorem, we can give an affirmative answer to the conjecture for δ.

To solve the conjecture for a general closed *-derivation δ, it may help to prove that the local kernel of δ is sufficiently large. For this purpose, sheaf theory seems to be useful.

REFERENCES

[Ba] C.J.K. Batty, *Derivations on compact spaces,* Proc. London Math. Soc. **42** (1981), 299–330.

[Bra] O. Bratteli, *Dissipations and Group Actions on C^*-Algebras,* Springer Lecture Notes in Math., **1229** (1986).

[G] F.M. Goodman, *Closed derivations in commutative C^*-algebras,* J. Funct. Anal. **39** (1980), 308–346.

[K] H. Kurose, *Closed derivations on compact spaces,* J. London Math. Soc. **34** (1986), 524–533.

[Ni] K. Nishio, *A local kernel property of closed derivations on $C(I\times I)$,* Proc. Amer. Math. Soc. **95** (1985), 573–576.

[S1] S. Sakai, *Theory of Unbounded Derivations in C^*-Algebras,* Lecture Notes, Copenhagen Univ. and Univ. of Newcastle upon Tyne, 1977.

[S2] S. Sakai, *Development in the theory of unbounded derivations in C^*-algebras,* Symposia in Pure Mathematics Vol. 38, Providence, R.I., 1982, 309–331.

[T] J. Tomiyama, *The Theory of Closed Derivations in the Algebra of Continuous Functions on the Unit Interval,* Lecture Notes, Tsing Hua Univ., 1983.

H. Kurose
Department of Applied Mathematics
Fukuoka University
Fukuoka 814-01
Japan

Representation of Quantum Groups

TETSUYA MASUDA, KATSUHISA MIMACHI,

YOSHIOMI NAKAGAMI, MASATOSHI NOUMI AND KIMIO UENO

The concept of quantum groups is important for the study of the quantum Yang–Baxter equations, Drinfeld [2], Jimbo [4], Manin [5] and others. On the other hand, Woronowicz [10] introduced the concept of compact matrix pseudogroups through the study of the dual object of groups. As pointed out by Rosso in [8], these two concepts are related to each other as quantum Lie algebras and quantum Lie groups. In this talk we want to indicate that the ideas of Kac algebras studied by Takesaki [9] and Enock and Schwartz [3] et al. are helpful for the study of quantum groups. As a result we can give a geometric interpretation for a q-analogue of a certain class of special functions, which has been a long standing problem of q-analogues.

1. Hopf Algebra of Quantum Group

For the moment, we fix a commutative field $\mathcal{C}$.

We introduce the quantum group $SL_q(2)$ for $q \in \mathcal{C}, q \neq 0$ as the one corresponding to the Hopf algebra $A = A(SL_q(2))$ defined as follows.

As a $\mathcal{C}$-algebra, A is generated by four elements x, u, v and y with the relation:

$$(1.1) \qquad \begin{cases} ux = qxu, vx = qxv, yu = quy, yv = qvy, \\ vu = uv \quad \text{and} \quad xy - q^{-1}uv = yx - quv = 1. \end{cases}$$

The coproduct $\Gamma: A \mapsto A \otimes_{\mathcal{C}} A$ and the counit $\varepsilon: A \mapsto \mathcal{C}$ of A are the unique $\mathcal{C}$-algebra homomorphisms satisfying

$$(1.2) \quad \Gamma\begin{pmatrix} x & u \\ v & y \end{pmatrix} = \begin{pmatrix} x & u \\ v & y \end{pmatrix} \otimes \begin{pmatrix} x & u \\ v & y \end{pmatrix} \quad \text{and} \quad \begin{pmatrix} x & u \\ v & y \end{pmatrix} = \begin{pmatrix} 1 & 0 \\ 0 & 1 \end{pmatrix}.$$

This is an abbreviated notation for $\Gamma(x) = x \otimes x + u \otimes v$, $\Gamma(u) = x \otimes u + u \otimes y$, etc. The antipode $\kappa: A \mapsto A$ of A is an $\mathcal{C}$-algebra antihomomorphism determined by the condition

(1.3)
$$\kappa\begin{pmatrix} x & u \\ v & y \end{pmatrix} = \begin{pmatrix} y & -qu \\ -q^{-1}v & x \end{pmatrix}.$$

In connection with the coproduct Γ, the antipode κ satisfies

(1.4)
$$\varepsilon \circ \kappa = \kappa \quad \text{and} \quad \Gamma \circ \kappa = \sigma \circ (\kappa \otimes \kappa) \circ \Gamma,$$

where $\sigma: A \otimes_{\mathcal{C}} A \mapsto A \otimes_{\mathcal{C}} A$ is the "flip" automorphism defined by $\sigma(a \otimes b) = b \otimes a$ for $a, b \in A$. Note that $\kappa^2 \neq 1$ if $q^2 \neq 1$.

In what follows, we fix an automorphism $\lambda \mapsto \overline{\lambda}$ of order 2 of the coefficient field $\mathcal{C}$ and suppose that $\overline{q} = q$; $\overline{\lambda}$ will be called the conjugation of λ. Then, there exists a unique involution: $a \in A \mapsto a^* \in A$ satisfying

(1.5)
$$\begin{pmatrix} x^* & u^* \\ v^* & y^* \end{pmatrix} = \begin{pmatrix} y & -q^{-1}v \\ -qu & x \end{pmatrix}.$$

One can immediately show that

Lemma 1.1. (i) $a^{**} = a$ *and* $\kappa(\kappa(a)^*)^* = a$ *for all* $a \in A$. (ii) $\Gamma(a^*) = \sum_i b_i^* \otimes c_i^*$ *if* $a \in A$ *and* $\Gamma(a) = \sum_i b_i \otimes c_i$.

When $\mathcal{C} = \mathbb{C}$ and $\lambda \mapsto \overline{\lambda}$ is the complex conjugation (hence $q \in \mathbb{R}$), the above *-operation can be considered as specializing the *real form* $SU_q(2)$ of $SL_q(2;\mathbb{C})$ corresponding to $SU(2)$ in the classical case. In fact, if $q > 0$, the Hopf algebra $A = A(SL_q(2))$ endowed with the *-operation is identified with the compact matrix pseudogroup $S_\mu U(2)$ of Woronowicz [10] by the correspondence

(1.6)
$$\mu = q^{-1} \quad \text{and} \quad \alpha = x, \gamma = v, \alpha^* = y, -\mu\gamma^* = u \text{ if } q \geq 1,$$

and by

(1.7)
$$\mu = q \quad \text{and} \quad \alpha = y, \gamma = u, \alpha^* = x, -\mu\gamma^* = v \text{ if } 0 < q \leq 1.$$

2. Revised Kac Algebras of Quantum Group

Let $\mathcal{A}$ be the universal enveloping C^*-algebra of $A = A(G)$ with $\mathcal{C} = \mathbb{C}$. Let ψ be the normalized Haar measure, i.e.

$$(id \otimes \psi)(\Gamma(a)) = \psi(a)1, \quad a \in \mathcal{A}.$$

Let H be the Hilbert space obtained by completing $\mathcal{A}$ with respect to the inner product $\langle a, b\rangle = \psi(b^* a)$. This H plays the role of the L^2 space for the quantum group G.

The (left) regular representation of a quantum group is given by the Kac–Takesaki operator W on a Hilbert space $H \otimes H$ defined by

$$W(a \otimes b) = \Gamma(b)(a \otimes 1), \qquad a, b \in \mathcal{A}. \tag{2.1}$$

It is a unitary operator implementing the coproduct Γ:

$$\Gamma(a) = W(1 \otimes a)W^*, \qquad a \in \mathcal{A}. \tag{2.2}$$

Now, we consider the quantum group $G = SU_q(2)$. For each spin $\ell \in (1/2)\mathbb{N}$, we fix a basis $\{\xi_i^{(\ell)}; i \in I_\ell\}$ as follows:

$$\xi_i^{(\ell)} = x^{\ell+i} v^{\ell-i}, \qquad i \in I_\ell, \tag{2.3}$$

where $I_\ell = \{-\ell, -\ell+1, \ldots, \ell\}$. We denote by $w_{i,j}^{(\ell)}$ $(i, j \in I_\ell)$ the matrix elements obtained by

$$\Gamma(\xi_i^{(\ell)}) = \sum_{j \in I_\ell} w_{i,j}^{(\ell)} \otimes \xi_j^{(\ell)}, \qquad j \in I_\ell. \tag{2.4}$$

Then the coassociativity of the coproduct Γ implies

$$\Gamma(w_{i,j}^{(\ell)}) = \sum_{k \in I_\ell} w_{i,k}^{(\ell)} \otimes w_{k,j}^{(\ell)}, \qquad i, j \in I_\ell. \tag{2.5}$$

Here, we identify the $(2\ell+1) \times (2\ell+1)$ matrix $(w_{i,j}^{(\ell)})$ with a representation $w^{(\ell)}$ of G on the vector space H_ℓ spanned by $\{w_{i,j}^{(\ell)} : i, j \in I_\ell\}$. Hence $w^{(\ell)}$ is an invertible operator on $H \otimes H_\ell$ satisfying

$$\{(\sigma \otimes id)(1 \otimes w^{(\ell)})\}(1 \otimes w^{(\ell)}) = (\Gamma \otimes id)(w^{(\ell)}).$$

The Peter–Weyl theorem is then obtained as follows:

Theorem 2.1. *Let $W^{(\ell)}$ be the restriction of the Kac–Takesaki operator W to $H \otimes H_\ell$. Then*

(i) *$W^{(\ell)}$ is a representation of G similar to $w^{(\ell)}$;*

(ii) *$W^{(\ell)}$ is $2\ell + 1$ copies of irreducible representations;*

and

(iii) $\{W, H \otimes H\} = \sum^{\oplus} \{W^{(\ell)}, H \otimes H_\ell\}$.

Next, we consider the L^1 space for the quantum group G. The dual space $\mathcal{A}^*$ turns out to be an algebra with respect to the (convolution) product

$$\varphi_1 * \varphi_2 = (\varphi_1 \otimes \varphi_2) \circ \Gamma. \tag{2.6}$$

Let $\mathcal{A}_\sim$ be the subalgebra generated by $\{\omega_{a,b}: a, b \in A\}$, where $\omega_{a,b}(c) = \psi(b^* ca)$. Then $\mathcal{A}_\sim$ is invariant under the involution: $\varphi \mapsto \varphi^{\hat{\#}}$ defined by

$$\varphi^{\hat{\#}}(a) = \overline{\varphi(\kappa(a)^*)}, \qquad a \in A. \tag{2.7}$$

By virtue of the symmetry $\psi = \psi \circ \kappa$ of the Haar measure we find that if $\varphi = \omega_{a,b}$, then $\varphi^{\hat{\#}} = \omega_{\kappa(a)^*, \kappa(b^*)}$. Denote the closure of $\mathcal{A}_\sim$ by $\mathcal{A}_*$. Then this $\mathcal{A}_*$ will play the role of L^1 space for G.

Let $\mathcal{D}_\psi$ be the set of all $\omega \in \mathcal{A}_*$ such that

$$|\omega(a^*)| \leq \mu \|a\|_\psi, \qquad a \in \mathcal{A} \tag{2.8}$$

for some $\mu > 0$, where the norm $\| \ \|_\psi$ is given by $\|a\|_\psi^2 = \langle a, a \rangle$. The space $\mathcal{D}_\psi$ is considered to be the intersection of the L^1 and the L^2 spaces for G. In this case, each element ω in $\mathcal{D}_\psi$ is identified with an element of H such that

$$\langle \omega, a \rangle = \omega(a^*), \qquad \omega \in \mathcal{D}_\psi, \ a \in \mathcal{A}. \tag{2.9}$$

Theorem 2.2. *The algebra $\mathcal{A}_\sim$ is a left Hilbert algebra with respect to the product* (2.6) *and the involution* (2.7).

We now define the Fourier transform $\lambda(\varphi)$ of $\varphi \in \mathcal{A}_*$. It is a bounded operator on H given by

$$\lambda(\varphi)\omega = \varphi * \omega, \qquad \omega \in \mathcal{A}_\sim. \tag{2.10}$$

The left representation $\{\pi_\ell, H\}$ of the left Hilbert algebra $\mathcal{A}_\sim$ is nothing but the Fourier transform: $\pi_\ell(\omega) = \lambda(\omega), \omega \in \mathcal{A}_\sim$, and $\pi_\ell(\mathcal{A}_\sim)'' = \lambda(\mathcal{A}_*)''$. It is easy to see that the Fourier transform $\lambda(\varphi)$, $\varphi \in \mathcal{A}_*$ is of the form

$$\lambda(\varphi)a = ((\varphi \circ \kappa) \otimes id)(\Gamma(a)), \qquad a \in \mathcal{A}. \tag{2.11}$$

Before going into the definition of the dual quantum group, we introduce the algebra $\hat{\mathcal{A}}_\ell$, which is the set of left multiplication operators L_a, $a \in M_{2\ell+1}(\mathbb{C})$ on H_ℓ defined by

$$L_a(w_{i,j}^{(\ell)}) = a(w_{i,j}^{(\ell)}).$$

Then $\{\hat{\mathcal{A}}_\ell, H_\ell\}$ is isomorphic to $\{M_{2\ell+1}(\mathbb{C}) \otimes \mathbb{C}1,\ V_\ell^L \otimes V_\ell^R\}$.

Let $\hat{\mathcal{A}}$ be the von Neumann algebra generated by $\{\lambda(\varphi): \varphi \in \mathcal{A}_*\}$. Then the algebra $\{\hat{\mathcal{A}}, H\}$ is the weak closure $\sum_{\ell\in(1/2)\mathbb{N}}^{\oplus}\{\hat{\mathcal{A}}_\ell, H_\ell\}$ of the algebraic direct sum of $\hat{\mathcal{A}}_\ell$'s.

Using the unitary operator $\hat{W} = \sigma(W^*)$, and the adjoint F of the $\#$-operator S on H: $Sa = a^*$, $a \in \mathcal{A}$, we set

$$\hat{\Gamma}(a) = \hat{W}(1 \otimes a)\hat{W}^*, \qquad a \in \hat{\mathcal{A}}, \tag{2.12}$$

$$\hat{\kappa}(b) = (Fb^*F)^-, \qquad b \in \hat{A}, \tag{2.13}$$

where $\hat{A} = \{\lambda(\omega): \omega \in A_\sim\}$. It is straightforward to verify that $\hat{\Gamma}$ and $\hat{\kappa}$ are the coproduct and the antipode on $\hat{\mathcal{A}}$. The canonical weight $\hat{\psi}$ for the left Hilbert algebra $\mathcal{A}_\sim$:

$$\hat{\psi}(\lambda(\omega)^*\lambda(\omega)) = \langle\omega, \omega\rangle, \qquad \omega \in \mathcal{A}_\sim, \tag{2.14}$$

is the Plancherel measure on $\hat{\mathcal{A}}$. It is easy to see that $\hat{\psi}$ is a Haar measure on $\hat{\mathcal{A}}$. Consequently, $\{\hat{\mathcal{A}}, \hat{\Gamma}, \hat{\kappa}, \hat{\psi}\}$ is considered as the dual quantum group. The counit $\hat{\varepsilon}$ is given by

$$\hat{\varepsilon}(\lambda(\varphi)) = \varphi(1), \qquad \varphi \in \mathcal{A}_*. \tag{2.15}$$

Now, let $\hat{S}$ denote the $\#$-operator: $\varphi \mapsto \varphi^\#$ for $\mathcal{A}_\sim$ and $\hat{F}$ its adjoint. Since

$$\langle\varphi^\#, a\rangle = \hat{\varphi^\#}(a^*) = \langle\kappa(a^*), \varphi\rangle, \qquad \varphi \in \mathcal{A}_\sim,$$

it follows that $a \in \mathrm{Dom}(\hat{F})$ and $\hat{F}a = \kappa(a^*)$. By symmetry of the Haar meausre ψ, we find that

$$\langle\hat{F}b, a\rangle = \psi(a^*\kappa(b^*)) = \psi(b^*\kappa(a)^*) = \langle\kappa(a)^*, b\rangle$$

for any $a, b \in A$. Since A is a core for $\hat{F}$, it follows that $a \in \mathrm{Dom}(\hat{S})$ and $\hat{S}a = \kappa(a^*)$. Thus we have the formulae: $\hat{S}a = \kappa(a)^*$ and $\hat{F}a = \kappa(a^*)$. Hence the modular operator $\hat{\Delta} = \hat{F}\hat{S}$ on H satisfies

$$\hat{\Delta}a = \kappa^2(a), \qquad a \in A. \tag{2.16}$$

It is worthwhile emphasizing the following fact: The square κ^2 of the antipode κ is realized as the value of the "modular" automorphism implemented by $\hat{\Delta}$ at $-i$, i.e., $\kappa^2(a) = \hat{\Delta}a\hat{\Delta}^{-1}$ for $a \in A$.

Bearing the above facts in mind, we have the following theorem:

Theorem 2.3. *The Plancherel measure $\hat{\psi}$ on $\hat{\mathcal{A}}$ satisfies on each $\hat{\mathcal{A}}_\ell$,*

$$(\hat{\psi}\,|_{\hat{\mathcal{A}}_\ell})(a) = \mathrm{Tr}(L_{h_\ell} a), \qquad a \in \hat{\mathcal{A}}_\ell,$$

where

$$h_\ell = \begin{pmatrix} q^{2\ell} & & & & \\ & q^{2\ell-2} & & & \\ & & \ddots & & \\ & & & q^{-2\ell+2} & \\ & & & & q^{-2\ell} \end{pmatrix} \in \hat{\mathcal{A}}_\ell.$$

This theorem asserts that the Plancherel measure is given by a q-analogue of the trace, which corresponds to the fact that the Haar measure ψ on $\mathcal{A}$ is written by the Jackson integral in q-analogue, [10, 6].

The roles in $\hat{\mathcal{A}}$ induced from the coproduct Γ and the antipode κ on $\mathcal{A}$ are given by the formulae:

$$\lambda(\omega_1)\lambda(\omega_2) = \lambda(\omega_1 * \omega_2) = \lambda((\omega_1 \otimes \omega_2) \circ \Gamma)$$

$$\lambda(\omega^*)^* = \lambda(\omega \circ \kappa).$$

We turn our discussion to the dual quantum group $\{\hat{\mathcal{A}}, \hat{\Gamma}, \hat{\kappa}, \hat{\psi}\}$ in place of $\{\mathcal{A}, \Gamma, \kappa, \psi\}$. Repeating the above argument in a similar way, we can define the inverse Fourier transform $\hat{\lambda}$, which is a mapping from the predual $\hat{\mathcal{A}}_*$ of $\hat{\mathcal{A}}$ to the von Neumann algebra $\hat{\hat{\mathcal{A}}}$ of the second dual quantum group. Then we obtain the formula:

$$\hat{\lambda}(\theta) = \kappa(\lambda_*(\theta)), \qquad \theta \in \hat{\mathcal{A}}_*, \tag{2.17}$$

where λ_* is the restriction $\lambda^*|_{\hat{\mathcal{A}}_*}$ of the dual mapping $\lambda^*: \hat{\mathcal{A}}^* \mapsto (\mathcal{A}_*)^*$.

Theorem 2.4. *If $\mathcal{A}$ is represented by the left multiplication on H, then it will be a weakly dense *-subalgebra of $\hat{\hat{\mathcal{A}}}$. Moreover, the coproduct $\hat{\hat{\Gamma}}$, the antipode $\hat{\hat{\kappa}}$ and the Haar measure $\hat{\hat{\psi}}$ for $\hat{\hat{\mathcal{A}}}$ will agree with Γ, κ and ψ on $\mathcal{A}$.*

Finally, we discuss the relation between the dual quantum group and the Hopf algebra structure of the universal enveloping algebra $U_q(sl(2,\mathbb{C}))$. To begin with we extend the domain of the Fourier transform λ to a larger class $\mathcal{A}^\sim$ ($\subset \mathcal{A}^*$) of functionals satisfying a growth condition of Payley–Wiener type. Let $\{\hat{\mathcal{A}}_{alg}, H_{alg}\}$ be the algebraic direct sum of $\{\hat{\mathcal{A}}_\ell, H_\ell\}$'s. Then H_{alg} is dense in H. For each $\varphi \in A^*$ we denote the $(2\ell+1) \times (2\ell+1)$

matrix $(\varphi \circ \kappa^{-1}(w_{i,j}^{(\ell)}))$ by $\lambda^\ell(\varphi)$. While the direct sum $\sum_\ell^\oplus \lambda^\ell(\varphi)$ has a meaning on a dense subspace H_{alg}, the closability is not assured. Let $\mathcal{A}^\sim$ be the set of $\varphi \in A^*$ such that

(i) $\sum_\ell^\oplus \lambda^\ell(\varphi)$ and $\sum_\ell^\oplus \lambda^\ell(\varphi)^*$ are closable and the adjoint of the former is the closure of the latter; and

(ii) if $\varphi \in \mathcal{A}^\sim$, then $\varphi \circ \kappa^{-1}$, $\varphi^* \in \mathcal{A}^\sim$.

Denote the closure of the first direct sum in (i) by $\lambda(\varphi)$ for $\varphi \in \mathcal{A}^\sim$. Then we can extend the coproduct, the antipode and the counit for these operators.

Theorem 2.5. *If $\varphi_1, \varphi_2 \in A^*$ are characters and if $\chi \in A^*$ satisfies the twisted derivation property:*

$$\chi(ab) = \chi(a)\varphi_1(b) + \varphi_2(a)\chi(b), \qquad a, b \in A,$$

then

(i) *the coproduct $\hat{\Gamma}$ satisfies*

$$\hat{\Gamma}(\lambda(\chi \circ \kappa)) = \lambda(\chi \circ \kappa) \otimes \lambda(\varphi_1 \circ \kappa) + \lambda(\varphi_2 \circ \kappa) \otimes \lambda(\chi \circ \kappa),$$

$$\hat{\Gamma}(\lambda(\varphi_j \circ \kappa)) = \lambda(\varphi_j \circ \kappa) \otimes \lambda(\varphi_j \circ \kappa);$$

(ii) *the antipode $\hat{\kappa}$ satisfies*

$$\hat{\kappa}(\lambda(\omega)) = \lambda(\omega \circ \kappa); \quad \textit{and}$$

(iii) *the counit $\hat{\varepsilon}$ satisfies*

$$\hat{\varepsilon}(\lambda(\varphi_j \circ \kappa)) = 1 \quad \textit{and} \quad \hat{\varepsilon}(\lambda(\chi \circ \kappa)) = 0.$$

When we choose the values of functionals $\varphi_1, \varphi_2, \chi_0, \chi_1, \chi_2$ at $\begin{pmatrix} x & u \\ v & y \end{pmatrix}$ as the following: with $\alpha = q^{1/2}$,

$$\begin{pmatrix} \alpha^{-1} & 0 \\ 0 & \alpha \end{pmatrix}, \begin{pmatrix} \alpha & 0 \\ 0 & \alpha^{-1} \end{pmatrix}, \begin{pmatrix} 0 & \alpha^{-1} \\ 0 & 0 \end{pmatrix}, \begin{pmatrix} \alpha & 0 \\ 0 & -\alpha \end{pmatrix}, \begin{pmatrix} 0 & 0 \\ \alpha & 0 \end{pmatrix},$$

then the operators $\Pi_i = \lambda(\varphi_i \circ \kappa)$ and $\nabla_j = \lambda(\chi_j \circ \kappa)$ correspond to the generators of $U_q(sl(2, \mathbb{C}))$ used in [4] in such a way that

$$\Pi_1 = \hat{k}^{-1}, \Pi_2 = \hat{k}, \nabla_0 = \nabla_e, \nabla_2 = \nabla_f.$$

Then these operators are affiliated with $\hat{\mathcal{A}}$. Conversely, if $\nabla_j = U_j|\nabla_j|$ is the polar decomposition of ∇_j, then Π_j, U_j and the spectral projections of $|\nabla_j|$ generates $\hat{\mathcal{A}}$.

3. Little q-Jacobi Polynomials and Casimir Operator

In this section, we shall express the matrix elements of the representations and give an orthogonality relation for the little q-Jacobi polynomials. Before stating our theorem, we give a definition of the little q-Jacobi polynomial as follows:

$$P_n^{(\alpha,\beta)}(z;q) = \sum_{r\geq 0} \frac{(q^{-n};q)_r(q^{d+\beta+n+1};q)_r}{(q;q)_r(q^{d+1};q)_r}(qz)^r, \tag{3.1}$$

where

$$(a;q)_n = (1-a)(1-aq)\dots(1-aq^{n-1}).$$

The little q-Jacobi polynomial is a q-analogue of the Jacobi polynomial, and plays a crucial role in our argument.

Theorem 3.1. *The matrix elements $w_{i,j}^{(\ell)}$ of the spin ℓ representation are expressed in terms of the little q-Jacobi polynomials in $\zeta = -q^{-1}uv$:*

$$x^{-i-j}v^{i-j}q^{(\ell+j)(j-i)}\begin{bmatrix}\ell+i\\ \ell+j\end{bmatrix}_{q^2} P_{\ell+j}^{(i-j,-i-j)}(\zeta;q^2), \tag{I}$$

$$\text{if} \quad i+j\leq 0,\ j\leq i,$$

$$x^{-i-j}u^{j-i}q^{(\ell+i)(i-j)}\begin{bmatrix}\ell-i\\ \ell-j\end{bmatrix}_{q^2} P_{\ell+i}^{(j-i,-i-j)}(\zeta;q^2), \tag{II}$$

$$\text{if} \quad i+j\leq 0,\ i\leq j,$$

$$q^{(j-i)(j-\ell)}\begin{bmatrix}\ell-i\\ \ell-j\end{bmatrix}_{q^2} P_{\ell-j}^{(j-i,i+j)}(\zeta;q^2)u^{j-i}y^{i+j}, \tag{III}$$

$$\text{if} \quad 0\leq i+j,\ i\leq j,$$

$$q^{(i-j)(i-\ell)}\begin{bmatrix}\ell+i\\ \ell+j\end{bmatrix}_{q^2} P_{\ell-i}^{(i-j,i+j)}(\zeta;q^2)v^{i-j}y^{i+j}, \tag{IV}$$

$$if \quad 0 \leq i+j, \ j \leq i.$$

where

$$\begin{bmatrix} n \\ k \end{bmatrix}_q = \frac{(q;q)_n}{(q;q)_k (q;q)_{n-k}} \ \ if \ \ 0 \leq k \leq n, \ \ and \ \begin{bmatrix} n \\ k \end{bmatrix}_q = 0 \ otherwise.$$

The Casimir element C of $U_q(sl(2))$ is given by

$$C = \frac{qk^2 + q^{-1}k^{-2} - 2}{(q - q^{-1})^2} + fe, \tag{3.2}$$

(see, for example, [4]), which is represented as the Casimir operator on $A(G)$ in terms of the twisted derivations ∇_e, ∇_f, and $\hat{k}$,

$$\hat{C} = \frac{q\hat{k}^2 + q^{-1}\hat{k}^{-2} - 2}{(q - q^{-1})^2} + \nabla_e \cdot \nabla_f. \tag{3.3}$$

Proposition 3.2. *The algebra $A(G)$ has the eigenspace decomposition, $A(G) = \oplus_{\ell \in (1/2) I\!N} H_\ell$, with respect to the Casimir operator $\hat{C}$. The eigenvalue of $\hat{C}$ on H_ℓ is $\frac{q^{2\ell+1} + q^{-2\ell-1} - 2}{(q-q^{-1})^2}$.*

Let T_{q^2} be the operator defined by $T_{q^2} f(\zeta) = f(q^2\zeta)$ for $f(\zeta) \in \mathbb{C}[\zeta]$. Then we have the following proposition.

Proposition 3.3. *The little q-Jacobi polynomial $P_n^{(\alpha,\beta)}(\zeta; q^2)$ satisfies the following q-difference equation*

(3.4)

$$\begin{aligned} \{q^{\alpha+\beta}(q^{-2\beta} - q^2\zeta)T_{q^2} + q(q^{2n+\alpha+\beta+1} + q^{-2n-\alpha-\beta-1})\zeta \\ - q^{-\alpha-\beta}(1 + q^{2\alpha}) + q^{-\alpha-\beta}(1 - \zeta)T_{q^2}^{-1}\} P_n^{(\alpha,\beta)}(\zeta; q^2) = 0. \end{aligned}$$

Finally, we give the Rodrigues formula for the little q-Jacobi polynomials. Let $D_{q^2} f(\zeta) = \{f(q^2\zeta) - f(\zeta)\}/(q^2\zeta - \zeta)$.

Proposition 3.4. *For any $n, \alpha, \beta \in I\!N$, we have*

$$\begin{aligned} P_n^{(\alpha,\beta)} = q^{n(2\alpha+n-1)} \frac{(1-q^2)^n}{(q^{2\alpha+2}; q^2)_n} \zeta^{-\alpha} (q^2\zeta; q^2)_\beta^{-1} \\ \times (T_{q^2}^{-1} D_{q^2})^n \{\zeta^{\alpha+n} (q^2\zeta; q^2)_{\beta+n}\}. \end{aligned} \tag{3.5}$$

REFERENCES

[1] Andrews, G.E. and R. Askey, *Enumeration of partitions. The role of Eulerian series and the q-orthogonal polynomials, Higher Combinatorics,* edited by M. Aigner, 2–26, Reidel, Dordrecht, Holland, 1977.

[2] Drinfel'd, V.G., *Quantum groups,* Proc. of International Congress of Math., Berkeley, California, USA, 1986, 798–820.

[3] Enock, M. and J.M. Schwartz, *Une dualité dans algèbres de von Neumann,* Bull. Soc. Math. France Suppl. Mémoire, **44** (1975), 1–144.

[4] Jimbo, M., *A q-difference analogue of $U(g)$ and Yang–Baxter equation,* Lett. in Math. Phys. **10** (1985), 63–69.

[5] Manin, Y.I., *Some remarks on Koszul algebras and quantum groups,* Ann. de l'Inst. Fourier, 1987 (Colloque en l'honeur de J.-L. Koszul).

[6] Masuda, T., K. Mimachi, Y. Nakagami, M. Noumi and K. Ueno, *Representations of quantum groups and a q-analogue of polynomials,* C.R. Acad. Sc. Paris, **307** (1988), 559–564.

[7] Poldeš, P., *Quantum spheres,* Lett. in Math. Phys. **14** (1987), 193–202.

[8] Rosso, M., *Comparison des groupes $SU(2)$ quantiques de Drinfeld et de Woronowicz,* C.R. Acad. Sc. Paris **304** (1987), 323–326.

[9] Takesaki, M., *Duality and von Neumann algebras,* Lecture Notes in Math. **247** (1972), 665–785, Springer-Verlag, Berlin.

[10] Woronowicz, S.L., *Twisted $SU(2)$ group. An example of a noncommutative differential calculus,* Publ. RIMS, Kyoto Univ. **23** (1987), 117–181.

[11] Woronowicz, S.L., *Compact matrix pseudogroups,* Commun. Math. Phys. **111** (1987), 613–665.

T. Masuda
Institute of Mathematics, University of Tsukuba
1, Tennoudai, Tsukuba, 305 Japan

K. Mimachi
Department of Mathematics, Nagoya University
Furou-cho, Chikusa-ku, Nagoya, 464 Japan

Y. Nakagami
Department of Mathematics, Yokohama City University
22-2, Seto, Kanazawa-ku, Yokohama, 236 Japan

M. Noumi
Department of Mathematics, Sophia University
7, Kioi-cho, Chiyoda-ku, Tokyo, 102 Japan

K. Ueno
Department of Mathematics, Waseda University
3, Ohkubo, Shinjuku-ku, Tokyo, 160 Japan

Automorphism Groups and Covariant Irreducible Representations

AKITAKA KISHIMOTO

1. Introduction

Given a C^*-dynamical system (A, G, α), I would like to consider the problem of analyzing $(\hat{A}, G, \alpha^*)$, where A is a C^*-algebra with its dual $\hat{A}$, G is a locally compact group, and α is a continuous action of G on A by automorphisms with α^* being the transposed action on $\hat{A}$. In other words, I would like to interpret the non-commutative system (A, G, α) in terms of the commutative-like system $(\hat{A}, G, \alpha^*)$. As this is still too general a problem, my main concern will be with 'type I orbits' (which will be defined soon) in more restricted situations; especially with covariant irreducible representations. The latter was also the subject of my talk at the previous US-Japan seminar (cf. [10]).

There are now a few results in this direction; most of them were obtained in collaboration with O. Bratteli, G.A. Elliott, D.E. Evans, and D.W. Robinson ([2,3,5,6,11,12]). (However the framework given above is not necessarily shared with the others.) In this talk I will present some results in [11,12] with proof. Before going into details, I have to give a few definitions.

The orbit type is defined as follows ([11]). For a $\pi \in \hat{A}$ regarded as an irreducible representation on some Hilbert space, say H_π, one constructs a representation $\tilde{\pi}$ of A by using the direct integral

$$\tilde{\pi} = \int_G^\oplus \pi \circ \alpha_t \, dt$$

on $L^2(G, H_\pi)$, and defines the type of the orbit through π under α^* to be the type of $\tilde{\pi}(A)''$ as a von Neumann algebra. Since it is shown that $\tilde{\pi}(A)''$ is homogeneous, the orbit type is, in particular, of type I, II, or III. If A is commutative, then of course all orbits are of type I. If A is not of type I, G is not discrete, and α is 'non-trivial,' there will be non-type I orbits, and in some cases (like an ergodic compact action on a simple C^*-algebra) there will be no type I orbits. The easiest orbits to analyze seem to be of type I (as always).

The type I orbits can be characterized as follows:

1.1 Proposition [11,14]. *Suppose that A is separable and G has a countable basis. Let $\pi \in \hat{A}$ and define $G_\pi = \{t \in G;\ \pi \circ \alpha_t = \pi\}$. Then the following conditions are equivalent:*

(i) *$\tilde{\pi}(A)''$ is of type* I.

(ii) *G_π is closed and $\tilde{\pi}(A)'' \cap \tilde{\pi}(A)' = L^\infty(G_\pi \setminus G) \otimes \mathbf{C}1$.*

When G is abelian (as we will assume from now on), we will denote by $\Delta(\pi, \alpha)$ the spectrum of (the weak extension of) α restricted to the center of $\tilde{\pi}(A)''$. Thus, if $\{\alpha_t^*\pi;\ t \in G\}$ is of type I, it follows that $\Delta(\pi, \alpha) = G_\pi^\perp$ $(\subset \hat{G})$.

The type I orbits we will treat are in fact more restrictive; they are G_π-covariant in the sense that there is a unitary representation of G_π on H_π that implements $\alpha \mid G_\pi$.

The Connes spectrum $\Gamma(\alpha)$ of α is defined as a closed subgroup of the dual group $\hat{G}$ of G as follows:

$$\Gamma(\alpha) = \bigcap_B \mathrm{Sp}(\alpha \mid B)$$

where B runs over the set of non-zero α-invariant hereditary C^*-subalgebras of A, and $\mathrm{Sp}(\alpha \mid B)$ denotes the Arveson spectrum of the restriction of α to B. We now define another spectrum denoted by $\Gamma_2(\alpha)$ as follows: $p \in \hat{G}$ belongs to $\Gamma_2(\alpha)$ if for any non-zero $x \in A$, any compact neighbourhood U of p, and any $\varepsilon > 0$, there is an $a \in A$ such that $\mathrm{Sp}_\alpha(a) \subset U$, $\|a\| = 1$, and

$$\|x(a + a^*)x^*\| \geq (2 - \varepsilon)\|x\|^2.$$

Note that $p \in \Gamma(\alpha)$ if and only if for any non-zero $x \in A$ and any compact neighbourhood U of p, there are $t \in G$ and $a \in A$ such that $\mathrm{Sp}_\alpha(a) \subset U$ and

$$xa\alpha_t(x^*) \neq 0.$$

It thus follows that $\Gamma(\alpha) \supset \Gamma_2(\alpha)$.

We will now state the main result:

1.2. Theorem. *Let A be a separable prime C^*-algebra, G a locally compact abelian group with countable basis, and α a continuous action of*

G on A. Let H be a closed subgroup of G with $H \supset \Gamma(\alpha)^{\perp}$. Then the following conditions are equivalent:

(i) *There is a faithful α-covariant irreducible representation of A.*

(ii) *$\Gamma_2(\hat{\alpha}) = G$, where $\hat{\alpha}$ is the dual action of $\hat{G}$ on the crossed product $A \times_{\alpha} G$.*

(iii) *$\Gamma_2(\hat{\alpha} \mid H^{\perp}) = (H^{\perp})^{\wedge} \cong G/H$ and there is a faithful $\alpha \mid H$-covariant irreducible representation π of A such that $\Delta(\pi, \alpha) = H^{\perp}$.*

(iv) *$\Gamma_2(\hat{\alpha} \mid \Gamma(\alpha)) = \Gamma(\alpha)^{\wedge}$ and $\Gamma_2(\alpha) = \Gamma(\alpha)$.*
If $\Gamma(\alpha) = \hat{G}$ or $G = \mathbb{R}$ or $\mathbf{Z}^n$ or $\mathbf{T}^n$, the above conditions are equivalent to

(v) *$\Gamma_2(\alpha) = \Gamma(\alpha)$.*

Remark. In the above conditions, 'a faithful representation' can be replaced by 'a faithful family of representations.'

When $\Gamma(\alpha) = \hat{G}$, the above theorem is Theorem 2 in [12]. The rest is mostly Theorem 7 there. But the proof there was incomplete as it was based on Lemma 8 whose proof was wrong. Hence we will concentrate on this part of the theorem in the remainder after describing some techniques used in the proof.

2. Techniques Used

The proof of 1.2 uses some techniques which were obtained at different times. We will explain them here to some extent.

In general it is easy to treat the case where the group is compact or discrete. We first treat these two cases, and then compute Γ_2 (maybe both for the action and the dual action) using duality (for crossed products) to get a desired result.

(I) *Properly Outer Automorphisms.* We shall need some properties of properly outer (or non-properly outer) automorphisms (cf. [15,8]). For example:

2.1. Proposition. *Let B be a separable C^*-algebra and β be an automorphism of B. Suppose that β is universally weakly inner. Then there is a decreasing sequence $\{I_n\}$ of essential ideals of B such that there is a unitary u in $\overline{\cup M(I_n)}$ with $\beta = Ad\, u$ where $M(I_n)$ is the multiplier algebra of I_n and, since I_{n+1} is essential in I_n, $M(I_n)$ can be regarded as a subalgebra of $M(I_{n+1})$. (All ideals will be closed and two-sided.)*

(II) *Irreducible Representations.* We shall have to construct an irreducible representation with certain properties. This is done by constructing a pure state of that algebra or a crossed product with the corresponding properties. For example, the method for proving the following is often used.

2.2. Proposition. *Let B be a separable C^*-algebra and let $\{I_n\}$ be a decreasing sequence of essential ideals of B. Then there is an irreducible representation π of B such that $\pi \mid I_n \neq 0$ for all n.*

(III) *The Case G is Discrete.* If all α_t are properly outer for $t \in G\setminus\{1\}$, there is a method of constructing an irreducible representation π of A such that $\pi \circ \alpha_t$ is disjoint from π for $t \in G \setminus \{1\}$ (i.e., $\Delta(\pi, \alpha) = \hat{G}$) (cf. [7]).

(IV) *The Case G is Compact.* When $\Gamma(\alpha) = \hat{G}$ and there is a faithful α-covariant irreducible representation, one can apply the technique developed by Glimm for the analysis of non-type I C^*-algebras to produce an irreducible representation π of A such that $\Delta(\pi, \alpha) = \hat{G}$ (and to get much more) (cf. [5]).

(V) *Duality.* There is a well-known duality for crossed products. In this context the following is relevant (cf. [9,10,12]):

2.3. Proposition. *Let H be a closed subgroup of G and let π be an irreducible representation of A such that π is $\alpha \mid H$-covariant and $\Delta(\pi, \alpha) = H^{\perp}$. Then there is an irreducible representation ρ of $A \times_\alpha G$ (induced by π over G/H) such that ρ is $\hat{\alpha} \mid H^{\perp}$-covariant and $\Delta(\rho, \hat{\alpha}) = H$.*

(VI) *The Spectrum $\Gamma_2(\alpha)$.* There is a close relation between $\Gamma_2(\alpha)$ and certain type I orbits. We explain this in the next section.

3. The Spectrum Γ_2

Let A be a separable C^*-algebra, G a locally compact abelian group with a countable basis, and α a continuous action of G on A as before. We collect some properties of Γ_2.

3.1. Proposition [3]. *The following conditions are equivalent:*

(i) $p \in \Gamma_2(\alpha)$.

(ii) *There exists a central sequence $\{x_n\}$ in A such that $\|x_n\| = 1$, $\mathrm{Sp}_\alpha(x_n)$ shrinks to $\{p\}$, and*

$$\lim \|x_n a\| = \|a\| \quad \textit{for any} \quad a \in A,$$

where '$\mathrm{Sp}_\alpha(x_n)$ shrinks to $\{p\}$' means that for any neighbourhood U of p, $Sp_\alpha(x_n)$ is included in U for all $n \geq N$ for some N.

This will follow, by using Lemma 1.1 in [1], from:

3.2. Proposition [12]. *The following properties hold:*

(i) *For any faithful family F of irreducible representations of A, it follows that*

$$\Gamma_2(\alpha) \supset \bigcap_{\pi \in F} \Delta(\pi, \alpha).$$

(ii) *There is a faithful family F of irreducible representations of A such that*

$$\Gamma_2(\alpha) = \bigcap_{\pi \in F} \Delta(\pi, \alpha).$$

Especially $\Gamma_2(\alpha)$ is a closed subgroup of G.

3.3. Proposition. *Suppose that A is α-prime (i.e., for any non-zero α-invariant ideals I and J, $I \cap J \neq (0)$). Let H be the set of $t \in G$ such that for any non-zero ideal I of A, $I \cap \alpha_t(I) \neq 0$. Then H is a closed subgroup of G and $\Gamma_2(\alpha) \supset H^\perp$.*

Proof. It is obvious that H is a subgroup. Let $t \notin H$. There is a non-zero ideal I of A such that $I \cap \alpha_t(I) = 0$. Replacing I by $\cap \{\alpha_s(I): s \in U\}$ with a small neighborhood U of $1 \in G$ (which is still non-zero), we may assume that $I \cap \alpha_s(I) = 0$ for any s around t (thus H is closed).

Let S be the set of non-zero α-invariant ideals of A. There is a sequence $\{J_n\}$ in S such that for any $J \in S$ there is a J_n with $J_n \subset J$. Let π be an irreducible representation of A such that $\pi \mid I \cap J_n \neq 0$ for any n. (There exists such a π because $I \cap J_n$ is essential in I for any n.)

Define a representation $\tilde{\pi}$ of A by

$$\tilde{\pi} = \int_G^\oplus \pi \circ \alpha_t \, dt.$$

Then $\tilde{\pi}$ is faithful and the identity of $\tilde{\pi}(I)''$ belongs to the center of $\tilde{\pi}(A)''$ which can be regarded as a subalgebra of $L^\infty(G)$, and this identity is 1 around $1 \in G$ and 0 around t. Hence there is a $p \in \Delta(\pi, \alpha) \subset \Gamma_2(\alpha)$ such that $\langle t, p \rangle \neq 1$, i.e., $t \notin \Gamma_2(\alpha)^\perp$. Thus, $H \supset \Gamma_2(\alpha)^\perp$, concluding the proof as $\Gamma_2(\alpha)$ and H are closed subgroups.

3.4. Lemma. *Let (A, G, α) be as above. Let H be a closed subgroup of G. Then the following conditions are equivalent:*

(i) $\Gamma_2(\alpha) = \hat{G}$.

(ii) $\Gamma_2(\alpha) \supset H^\perp$ *and* $\Gamma_2(\alpha \mid H) = \hat{H}$.

Proof. It is immediate that (i) implies (ii). Suppose (ii). Then one can construct an irreducible representation π of A such that $\Delta(\pi, \alpha) \supset H^\perp$ and $\Delta(\pi, \alpha \mid H) = \hat{H}$ (see [11]). Since the center of $\tilde{\pi}(A)''$ (as a closed subalgebra of $L^\infty(G)$) is translation-invariant, it is of the form $L^\infty(G/N)$ for some closed subgroup N of G. Since $\Delta(\pi, \alpha) \supset H^\perp$, it follows that $H = (H^\perp)^\perp \supset N$. Thus, $\tilde{\pi}$ is equivalent to the direct integral of

$$\tilde{\pi}_H \circ \alpha_{f(s)} = \int_H^\oplus \pi \circ \alpha_{tf(s)} dt$$

over G/H where f is a measurable function of G/H into G such that $f(s)H = s$, $s \in G/H$. Since the center of $\pi_H(A)''$ is $L^\infty(H)$, this shows that $N = \{1\}$. Since there is a faithful family of such π, this concludes the proof.

One can compute Γ_2 directly for some dynamical systems, e.g., asymptotic abelian systems (cf. [4]) and some systems associated with a compact abelian ergodic action (cf. [13]).

4. Covariant Irreducible Representations

In this section we shall prove the following, which is a generalization of Proposition 3 in [12].

4.1. Proposition. *Let A be a separable C^*-algebra, G a locally compact abelian group with countable basis, and α an action of G on A. Suppose that $\Gamma_2(\alpha) = \hat{G}$, and that $I \cap \alpha_t(I) \neq 0$ for any non-zero ideal I of A and any $t \in G$. Then $\Gamma_2(\hat{\alpha}) = G$ for the dual action $\hat{\alpha}$ of G on the crossed product $A \times_\alpha G$ (and hence there is a faithful family of α-covariant irreducible representations of A).*

Let H be a closed subgroup of G. Since $\Gamma_2(\alpha) = \hat{G}$, it follows that $\Gamma_2(\alpha \mid H) = \hat{H}$ ($\cong \Gamma_2(\alpha)/H^\perp$). Hence, for the action $\alpha \mid H$ of H, the assumptions in the above proposition are also satisfied. First we note:

4.2. Lemma. $\Gamma(\hat{\alpha}) = G$.

Proof. This follows from a characterization of $\Gamma(\hat{\alpha})$ (in terms of ideals of A) (see [16]).

Now we shall show the proposition in the two special cases; first when G is discrete, and second when G is compact.

Discrete case. When G is (countable) discrete, $\hat{G}$ is compact. Since $\Gamma_2(\alpha) = \hat{G}$ (or equivalently all α_t with $t \neq 1$ are properly outer in this case), there is a faithful family of 'anti-covariant' irreducible representations of A (i.e., those π with $\Delta(\pi, \alpha) = \hat{G}$), and hence by the duality there is a faithful family of $\hat{\alpha}$-covariant irreducible (induced) representations of $A \times_\alpha G$.

We claim that for any non-zero $a \in (A \times_\alpha G)^{\hat{\alpha}} = A$, there is an ($\hat{\alpha}$-covariant) irreducible induced representation ρ of $A \times_\alpha G$ such that $\mathrm{Sp}(V \mid [\rho(a)H_\rho]) = G$ where V is the canonical unitary group on H_ρ that implements $\hat{\alpha}$.

To show this let J be the ideal of A generated by a. Then it follows that for any finite subset F of G,

$$\cap \{\alpha_s(J) : s \in F\}$$

is essential in J. By a construction method similar to that given in [7], there is an irreducible representation π of A such that $\pi \mid \cap\{\alpha_s(J) : s \in F\} \neq 0$ for any finite subset F of G and $\pi \circ \alpha_s$ is disjoint from π for any $s \neq 1$. The representation ρ of $A \times_\alpha G$ on $L^2(G, H_\pi)$ induced by π is irreducible, and the unitary representation V of $\hat{G}$ defined by

$$(V_p\xi)(t) = \langle t, p\rangle \xi(t), \qquad \xi \in L^2(G, H_\pi)$$

implements $\hat{\alpha}$ and satisfies

$$\mathrm{Sp}(V \mid [\rho(a)H_\rho]) = G$$

(since $\pi \circ \alpha_t(a) \neq 0$ for any $t \in G$).

We can prove Glimm's type theorem or Theorem 2.1 in [5] in this circumstance, by a method which is more like the one given in 6.7 of [16] (using Kadison's transitivity theorem in an appropriate $\hat{\alpha}$-covariant irreducible representation whose existence is claimed above). Thus, 'embedding' an infinite tensor product type action into $(A \times_\alpha G, \hat{G}, \hat{\alpha})$ it is easy to construct an irreducible representation ρ of $A \times_\alpha G$ with $\Delta(\rho, \hat{\alpha}) = G$ (see e.g. [6]). From the method of embedding, we can arrange that the kernel of ρ does not contain a given arbitrary non-zero ideal. Hence there is a faithful family of irreducible representations ρ of $A \times_\alpha G$ with $\Delta(\rho, \hat{\alpha}) = G$. This concludes the proof of $\Gamma_2(\hat{\alpha}) = G$, by 3.2.

Compact case. When G is compact, $\hat{G}$ is (countable) discrete. It suffices to prove that for each $p \in \hat{G} \setminus \{1\}$ $\hat{\alpha}_p$ is properly outer on $A \times_\alpha G$.

4.3. Lemma. *Let $\{I_n\}$ be a decreasing sequence of ideals of $A \times_\alpha G$ such that I_n is essential in I_1 for any n. Then there is an $\hat{\alpha}$-covariant irreducible representation ρ of $A \times_\alpha G$ such that $\rho \mid I_n \neq 0$ for any n.*

Proof. Let J_n be the $\hat{\alpha}$-invariant ideal generated by I_n, and let K_n be the (α-invariant) ideal of A that generates J_n as an ideal of $A \times_\alpha G$. Then K_n is essential in K_1 for any n. There is an 'anti-covariant' irreducible representation π of A such that $\pi \mid K_n \neq 0$ for any n (see [11]). Let ρ be the representation of $A \times_\alpha G$ induced by π, which is $\hat{\alpha}$-covariant. Then by the 'anti-covariantness' of π, ρ is irreducible. Since $\rho \mid J_n \neq 0$ and ker ρ is $\hat{\alpha}$-invariant, it follows that $\rho \mid I_n \neq 0$.

Suppose that $\hat{\alpha}_{p_0}$ is not properly outer. Then there is an $\hat{\alpha}_{p_0}$-invariant non-zero ideal I of $A \times_\alpha G$ such that $\hat{\alpha}_{p_0} \mid I$ is universally weakly inner. Let $\tilde{I}$ be the $\hat{\alpha}$-invariant ideal generated by I. Then by a routine argument it follows that $\hat{\alpha}_{p_0} \mid \tilde{I}$ is still universally weakly inner. Thus, by 2.1, there is a decreasing sequence $\{I_n\}$ of ideals such that I_n is essential in $\tilde{I}$ for any n and there is a unitary u in $\overline{\cup M(I_n)}$ with $\tilde{\alpha}_{p_0} \mid \tilde{I} = \operatorname{Ad} u$.

Let F be the family of all finite intersections of $\hat{\alpha}_p(I_n)$, $p \in G$, $n = 1, 2, \ldots$. Then any ideal in F is essential in $\tilde{I}$. Note that F forms a downward directed set and that there is a decreasing sequence $\{J_n\}$ in F such that for any $J \in F$ there is an n with $J \supset J_n$. Thus one has that

$$B \equiv \overline{\bigcup_{J \in F} M(J)} = \overline{\cup M(J_n)}.$$

In this setting $\hat{\alpha} \mid \tilde{I}$ uniquely extends to an action β on B. We claim:

4.4. Lemma. *$\beta_p(u) = u$ for any $p \in \hat{G}$.*

Proof. Let J be a non-zero $\hat{\alpha}$-invariant ideal of $\tilde{I}$. Let π be an $\hat{\alpha}$-covariant irreducible representation of $A \times_\alpha G$ as given in 4.3 with the sequence $\{J \cap I_n\}$ (instead of $\{I_n\}$). Since $\pi(I_n) \neq (0)$ (or the identity of $\pi(I_n)''$ is 1), π uniquely extends to a representation of $M(I_n)$ for any n. Thus $\pi \mid \tilde{I}$ extends to a representation $\tilde{\pi}$ of B. Since π is $\hat{\alpha}$-covariant, it follows that $\tilde{\pi}$ is β-covariant. Since $\tilde{\pi}$ is irreducible it easily follows that $\tilde{\pi}(\beta_p(u)) = \tilde{\pi}(u)$ for any $p \in \hat{G}$. Denoting by $\tilde{\pi}_J$ this $\tilde{\pi}$ as J was given in the first place, let K be the intersection of ker $\tilde{\pi}_J$ with all those J. Then K must be zero: Because ker $\tilde{\pi}_J \not\supset J$, and if K is non-zero, $K \cap M(J_n)$ must be non-zero for large n, which, in turn, implies that $K \cap J_n \neq (0)$, a contradiction. This concludes the proof.

Recall that $\{J_n\}$ is the sequence defining B. Let $\tilde{J}_n$ be the $\hat{\alpha}$-invariant ideal of $A \times_\alpha G$ generated by J_n, and let K_n (resp. K) be the (α-invariant) ideal of A corresponding to J_n (resp. I). Then K_n is essential in K as well as $\tilde{J}_n$ is in $\tilde{I}$.

Let π be an irreducible representation of A such that $\pi(K_n) \neq (0)$ for any n. Let ρ be the representation of $A \times_\alpha G$ induced by π (on the space $L^2(G, H_\pi)$). Then it follows that $\rho(J_n) \neq (0)$ because $\rho(\tilde{J}_n) \neq (0)$ and ker ρ is $\hat{\alpha}$-invariant.

Since the identity of $\rho(\tilde{I})''$ is 1 and $\hat{\alpha}_{p_0} \mid \tilde{I}$ is universally weakly inner, there is a unitary $U \in \rho(\tilde{I})''$ such that $\rho \circ \hat{\alpha}_{p_0} = \operatorname{Ad} U$. By using the above lemma, we can assume that $V_q U V_q^* = U$ where V is the canonical unitary representation of G implementing $\hat{\alpha}$ (see 3.4 in [2]). (Or: If P is an ideal of $A \times_\alpha G$ such that $J_n \not\subset P$ for any n, then $\hat{\alpha}_{p_0}$ on the quotient $\tilde{I}/P$, which is universally weakly inner, is induced by a unitary $v \in M^\infty(\tilde{I}/P)$ which is left invariant under the action of $\hat{G}$ on $M^\infty(\tilde{I}/P)$ induced by $\hat{\alpha}$, as v can be constructed by using u in Lemma 4.4. When $P = \ker \rho$, ρ uniquely extends to a representation of $M^\infty(\tilde{I}/P)$.) Thus $U \in \rho(A)''$. As $\rho(\lambda(s)) U \rho(\lambda(s)^*) = \langle \overline{s, p_0} \rangle U$, $s \in G$, it follows that

$$U = \bar{p}_0 \otimes W$$

on $L^2(G) \otimes H_\pi$ where W is a unitary. As $U\rho(x)U^* = \rho(x)$, $x \in A$, it follows that W must be a scalar, i.e. U is in the center of $\rho(A)''$.

This shows, in particular, that for any $s \in G$ with $\langle s, p_0 \rangle \neq 1$, and for any irreducible representation π of A such that $\pi(K_n) \neq (0)$ for all n, $\pi \circ \alpha_s$ is disjoint from π.

Suppose that $p_0 \neq 1$ and let $s \in G$ be such that $\langle s, p_0 \rangle \neq 1$. If s is of finite order let n be its order; otherwise let $n = \infty$. Let $\mathbf{Z}_n = \mathbf{Z}/n\mathbf{Z}$ for $n < \infty$ and let $\mathbf{Z}_\infty = \mathbf{Z}$. Define an action β of $\mathbf{Z}_n$ by $\beta_k = \alpha_{s^k}$. Then $\Gamma_2(\beta) = \hat{\mathbf{Z}}_n$ since β_k is properly outer for $k \neq 1$. Hence by the result in the discrete case one obtains that $\Gamma_2(\hat{\beta}) = \mathbf{Z}_n$. Thus, there is a β-covariant irreducible representation π of A such that $\pi(K_n) \neq 0$ for all n. This contradiction shows that $p_0 = 1$, completing the proof.

To proceed to the general case, we first note:

4.5. Lemma. *Let A be a separable C^*-algebra, G a separable locally compact abelian group, and α an action of G on A. Let H be a closed subgroup of G and $\beta = \alpha \mid H$. Then the following conditions are equivalent:*

(i) $\Gamma_2(\alpha) \supset H^\perp$ *and* $\Gamma_2(\hat{\beta}) = H$.

(ii) *There exists a faithful family of β-covariant irreducible representations π of A with $\Delta(\pi, \alpha) = H^\perp$ (or if A is prime, there is such a faithful representation).*

Proof. See [11,12].

General case. Let H be a closed subgroup of G. If H is compact or discrete, it follows that $\Gamma_2(\hat{\beta}) = H$ for the dual action $\hat{\beta}$ of $\hat{H}$ on $A \times_\beta H$ with $\beta = \alpha \mid H$. From the assumption, $\Gamma_2(\alpha) = \hat{G}$. Thus by the above lemma one has that $\Gamma_2(\hat{\alpha}) \supset H$. From this one can conclude that $\Gamma_2(\hat{\alpha}) = G$.

5. Proof of Theorem 1.2

Let H be a closed subgroup of G with $H \supset \Gamma(\alpha)^\perp$ and let us prove that (ii) implies (iii).

Let $\beta = \hat{\alpha} \mid H^\perp$. Since $H^\perp \subset \Gamma(\alpha)$, one has that $I \cap \beta_t(I) \neq (0)$ for any non-zero ideal I of $A \times_\alpha G$ and any $t \in H^\perp$. Since $\Gamma_2(\hat{\alpha}) = G$, it follows that $\Gamma_2(\beta) = (H^\perp)^\wedge = G/H$. Thus, by Proposition 4.1 one obtains that $\Gamma_2(\hat{\beta}) = H^\perp$. Since $\Gamma_2(\hat{\alpha}) = G$, Lemma 4.5 implies that there is a faithful family of $\hat{\alpha} \mid H^\perp$-covariant irreducible representations π of $A \times_\alpha G$ such that $\Delta(\pi, \hat{\alpha}) = H$. By duality, this implies (iii).

Suppose (iii). Then it follows that $\Gamma_2(\hat{\alpha}) \supset H$. Since $\Gamma_2(\hat{\alpha} \mid H^\perp) = (H^\perp)^\wedge$, Lemma 3.4 implies that $\Gamma_2(\hat{\alpha}) = G$, i.e., (ii).

Now we assume (iv) and let $H = \Gamma(\alpha)^\perp$ and let $\beta = \alpha \mid H$.

Then it follows that $\Gamma(\beta) = \{1\}$. Hence, for any $p \in \hat{H}$, there is a non-zero ideal I of $A \times_\beta H$ such that $\hat{\beta}_p(I) \cap I = (0)$. Since $A \times_\beta H$ is $\hat{\beta}$-prime, it follows that $\Gamma_2(\hat{\beta}) = \hat{H}$. Since $\Gamma_2(\alpha) = H^\perp = \Gamma(\alpha)$, one obtains (iii) with $H = \Gamma(\alpha)^\perp$ by Lemma 4.5.

Last we shall discuss condition (v).

If G is discrete, or $\hat{G}/\Gamma(\alpha)$ is compact, it might be possible to perturb α by universally weakly inner automorphisms to get an action γ with ker $\gamma = \Gamma(\alpha)^\perp$. When $G = \mathbf{Z}^n$ or $\mathbb{R}$ (and $\hat{G}/\Gamma(\alpha)$ is compact) this is actually possible by replacing the C^*-algebra A by a non-zero invariant hereditary C^*-subalgebra of A. Then one can apply Proposition 4.1 to this situation, showing that (v) implies (ii).

If G is compact we can elaborate the arguments given in the proof of 4.1. Instead of giving the details, let us conclude this talk with the following conjecture:

Conjecture. Let (A, G, α) be as in 4.1. Suppose that $\Gamma_2(\alpha \mid I) = \Gamma(\alpha \mid I)$ for any non-zero α-invariant ideal I of A, and that $I \cap \alpha_t(I) \neq (0)$ for any non-zero ideal I of A and $t \in G$. Then $\Gamma_2(\hat{\alpha}) = G$.

REFERENCES

[1] Akemann, C.A., Pedersen, G.K., *Central sequences and inner derivations of separable C^*-algebras,* Amer. J. Math. **101** (1979), 1047–1061.

[2] Bratteli, O., Elliott, G.A., Evans, D.E., Kishimoto, A., *Quasi-product actions of a compact abelian group on a C^*-algebra,* Tohoku Math. J. **41** (1989), 133–161.

[3] Bratteli, O., Elliott, G.A., Kishimoto, A., in preparation.

[4] Bratteli, O., Kishimoto, A., *Derivations and free group actions on C^*-algebras,* J. Operator Theory **15** (1986), 377–410.

[5] Bratteli, O., Kishimoto, A., Robinson, D.W., *Embedding product type actions into C^*-dynamical systems,* J. Functional Analysis **75** (1987), 188–210.

[6] Evans, D.E., Kishimoto, A., *Duality for automorphisms on a compact C^*-dynamical system,* Ergod. Th. and Dynam. Sys., **8** (1988), 173–189.

[7] Kishimoto, A., *Outer automorphisms and reduced crossed products of simple C^*-algebras,* Commun. Math. Phys. **81** (1981), 429–435.

[8] Kishimoto, A., *C^*-crossed products by R,* Yokohama Math. J. **30** (1982), 151–164.

[9] Kishimoto, A., *Automorphisms and covariant irreducible representations,* Yokohama Math. J. **31** (1983), 159–168.

[10] Kishimoto, A., *One-parameter automorphism groups of C^*-algebras,* Pitman Research Notes in Math. Series **123** (1986), 312–325.

[11] Kishimoto, A., *Type I orbits in the pure states of a C^*-dynamical system,* Publ. RIMS, Kyoto Univ. **23** (1987), 321–336.

[12] Kishimoto, A., *Type I orbits in the pure states of a C^*-dynamical system,* II. Publ. RIMS, Kyoto Univ. **23** (1987), 517–526.

[13] Kishimoto, A., *Outer automorphism subgroups of a compact abelian ergodic action,* J. Operator Theory, **20** (1988), 59–67.

[14] Kishimoto, A., *Compact group actions on C^*-algebras,* unpublished (1988).

[15] Olesen, D., Pedersen, G.K., *Applications of the Connes spectrum to C^*-dynamical systems,* III, J. Functional Analysis **45** (1982), 357–390.

[16] Pedersen, G.K., *C^*-algebras and their automorphism groups,* Academic Press, 1979.

College of General Education
Tohoku University
Sendai, Japan

Present address:
Department of Mathematics
Hokkaido University
Sapporo 060, Japan

Proper Actions of Groups on C*-Algebras

MARC A. RIEFFEL*

Recently I have been attempting to formulate a suitable C^*-algebraic framework for the subject of deformation quantization of Poisson manifolds [1,13]. Some of the main examples which I have constructed within this framework [27] involve "proper" actions of groups on C^*-algebras, where "proper" actions are to be defined as a generalization of proper actions of groups on locally compact spaces. Much of the material on proper actions which I have developed for this purpose is of a general nature which may be useful in other situations, as it has seemed appropriate to give a separate exposition of it, in the present article.

The notion of "proper" action which we introduce in this article is closely related to various notions of "integrable" actions which are discussed in the literature [5,7,15]. The main difference is that our notion of "proper" action emphasizes a natural inner-product having values in the crossed product algebra for the action. It turns out that because of this, our notion of "proper" actions is more closely related to reduced crossed products than full crossed products.

Section 1 of this article is devoted to the definition and basic properties of "proper" actions, especially a strong Morita equivalence between a certain ideal of the crossed product algebra and the generalized fixed-point algebra which we associate to a proper action. Section 2 contains several general examples. The first of these, which provides some clarifying

---

*Supported in part by NSF grant DMS-8601900.

counter-examples, consists of the action of a group by conjugation by its left regular representation on the algebra of compact operators. The second, for Abelian groups, consists of the dual action of the dual group on a crossed product algebra. The Morita equivalence alluded to above provides in this case what can be considered to be another manifestation of Takesaki duality as it was extended to C^*-algebras by Takai (as in 7.9 of [15]). In Section 3 we use the Morita equivalence alluded to above to study when the field of generalized fixed-point algebras corresponding to a continuous field of proper actions will be continuous. The results so obtained play a key role in the construction of the examples of deformation quantization discussed in [27]. In particular, [27] contains further interesting examples of proper actions.

§1. Proper Actions on C^*-Algebras

We recall that an action, α, of a locally compact group G on a locally compact space M is said to be proper if the map from $G \times M$ to $M \times M$ defined by $(x, m) \mapsto (x, \alpha_x(m))$ is proper, in the sense that preimages of compact sets are compact. (A recent paper concerning proper actions, containing references to earlier papers, is [17].) A basic fact about proper actions is that the space of orbits, X/α, with the quotient topology, is again locally compact and Hausdorff (see propositions 3 and 9 in 3.4 of [2]).

Let $A = C_\infty(M)$, the algebra of continuous complex-valued functions on M vanishing at infinity, and let α also denote the action of G on A defined by

$$(\alpha_x f)(m) = f(\alpha_x^{-1}(m)).$$

Let $A^\alpha = C_\infty(X/\alpha)$. How are A and A^α related via α? Well, the elements of A^α can be viewed as continuous bounded functions on X which are constant on α-orbits. Thus A^α is a subalgebra of the multiplier algebra, $M(A)$, of A. The action α on A defines a corresponding action on $M(A)$ (which need not be strong-operator continuous). Let $M(A)^\alpha$ denote the subalgebra of fixed points in $M(A)$ for this action. Then it is easily seen that

$$A^\alpha \subseteq M(A)^\alpha.$$

But how do we characterize A^α as a subalgebra of $M(A)$? Intuitively, one obtains elements of A^α by "averaging," that is by integrating, elements of A over G. But if G is not compact, only elements in the dense subalgebra $A_0 = C_c(M)$ of A consisting of functions of compact support can be so integrated, and even then, the integration is not with respect to the norm topology but rather with respect to the strict topology. That is, if $f, g \in A_0$, then the function $x \mapsto \alpha_x(f)$ is not norm-integrable, but rather the function $x \mapsto g\alpha_x(f)$ is, since it has compact support.

Generalizations of the above situation to the case in which A is non-commutative have been considered by various authors, generally under some variation of the name "integrable" actions. (See 7.84 of [15] with the note at the end of [14], definition II.2.1 of [5], and [7].) However, there is another important aspect of the commutative case $A = C_\infty(M)$ which does not seem to have been considered in the generalizations of properness to non-commutative A, but which we need to stress here, and which, when it is present, will lead us to use the term "proper" action instead of "integrable" action. This aspect is that for α a proper action of G on M, there is an inner-product on $C_c(M)$ with values in the transformation group C^*-algebra

$$E = C^*(G, M).$$

If for simplicity we assume for the moment that G is unimodular, this inner-product is defined by

$$\langle f, g\rangle_E(x, m) = f(m)\bar{g}(\alpha_x^{-1}(m)),$$

where the properness of α ensures that $\langle f, g\rangle_E$ has compact support, and so is in $L^1(G, C_\infty(M))$. This kind of inner-product already plays a key role in various commutative situations, such as those in [24]; and variants of it have appeared in related contexts, such as equation 2.3 of [20] and theorem 6.3 of [17]. In the case where G is compact, so that every action of G should be proper, this kind of inner-product has also been used for actions on non-commutative C^*-algebras; see section 7.1 of [16] and the references given there.

Suppose now that G is not necessarily compact, and that α is an action of G on some C^*-algebra A. We will need to consider a dense subalgebra of A, but we do not want to insist on an analogue of compact support (e.g.

using the Pedersen ideal [15]), because of interesting examples that do not permit this, such as that of theorem 2.18 of [25], and the second example of the next section. But then we must be careful about the treatment of modular functions in case G is not unimodular. Guidance for this can be obtained from §2 of [20], which treats a special case of the situation we will consider here. Anyway, our first crucial assumption is that there is a dense α-invariant subalgebra, A_0, of A such that for any $a, b \in A_0$ both the functions $x \mapsto a\alpha_x(b^*)$ and

$$\langle a, b\rangle_E(x) = \Delta(x)^{-1/2} a\alpha_x(b^*)$$

are norm integrable on G as A-valued functions, where Δ denotes the modular function of G.

As our notation suggests, we wish to view $\langle\ ,\ \rangle_E$ as an inner-product on A_0 with values in a C^*-algebra completion of $L^1(G, A)$. For this inner-product to be useful, we need to know that $\langle a, a\rangle_E$ is appropriately positive for any $a \in A_0$, Now if (π, U) is any covariant representation of (G, A) on a Hilbert space Ξ, then for $\xi \in \Xi$ we have, using module notation for the integrated form of (π, U),

$$\begin{aligned}\langle\langle a, a\rangle_E\xi, \xi\rangle &= \int \langle\pi(a\alpha_x(a^*))U_x\xi, \xi\rangle\, \Delta(x)^{-1/2} dx \\ &= \int \langle U_x\pi(a^*)\xi, \pi(a^*)\xi\rangle\, \Delta(x)^{-1/2} dx.\end{aligned}$$

If G is unimodular, the integrand is clearly a function of positive type on G which is integrable. But when G is not amenable, it is easy to construct functions of positive type and of compact support whose integral over G is strictly negative. To see this, suppose for simplicity that G is discrete, and recall [15] that if G is not amenable then the trivial representation, τ, is not contained in the left regular representation, λ. This means that we can find $a \in C^*(G)$ such that $\|a\| = 1$, $a = a^*$ and $\tau(a) > 0$ while $\lambda(a) = 0$. Thus we can find $f \in C_c(G)$ such that $\|f\| = 1$, $f = f^*$ and $\|a - f\| < \tau(a)/4$. This means that

$$\|\lambda(f)\| < \tau(a)/4$$

while

$$\tau(f) \geq 3\tau(a)/4.$$

Let

$$g = (\tau(a)/2)\delta_e - f$$

where δ_e is the delta-function at the identity element. Then $g \in C_c(G)$, and $\lambda(g)$ is positive so that g is of positive type. But

$$\int_G g = \tau(g) \leq -\tau(a)/4.$$

Thus the integral we are examining above does not appear to be automatically non-negative. Suppose, however, that the covariant representation (π, U) is induced from a representation ρ of A on a Hilbert space H, as described in 7.7.1 of [15], so that $\Xi = L^2(G, H)$. Consider any $\xi \in \Xi$ which is actually in $C_c(G, H)$. Then

$$\begin{aligned}
(1.1)\int &\langle U_x\pi(a^*)\xi, \pi(a^*)\xi\rangle\, \Delta(x)^{-1/2}dx \\
&= \int\int \langle (U_x\pi(a^*)\xi(y), (\pi(a^*)\xi)(y)\rangle dy\, \Delta(x)^{-1/2}dx \\
&= \int\langle\int \rho(\alpha_{y^{-1}}(\alpha_x(a^*)))\xi(x^{-1}y)\Delta(x)^{-1/2}dx, \rho(\alpha_{y^{-1}}(a^*))\xi(y)\rangle dy
\end{aligned}$$

$$= \langle \int \rho(\alpha_x(a^*))\xi(x^{-1})\Delta(x)^{-1/2}dx, \int \rho(\alpha_y(a^*))\xi(y^{-1})\Delta(y)^{-1/2}dy\rangle$$
$$\geq 0.$$

For this particular calculation we did not need to assume that $a, b \in A_0$. But if we do assume this, then

$$x \mapsto \Delta(x)^{-1/2} a\alpha_x(a^*)$$

is integrable, so $\langle a, a\rangle_E$ defines a bounded operator on Ξ. Then from the above calculation we conclude that this bounded operator is positive on Ξ. Thus we see that the appropriate place to view the values of $\langle\ ,\ \rangle_E$ is in the reduced C^*-algebra $C_r^*(G, A)$, since this algebra is defined in terms of these induced representations [15]. (Our notation will not explicitly indicate the action α involved, as there will be no ambiguity about this in what follows.) A simple calculation shows that

$$\langle a, b\rangle_E^* = \langle b, a\rangle_E.$$

We want the linear space, E_0, of finite linear combinations of elements of the form $\langle a, b\rangle_E$ to be a subalgebra of $C_r^*(G, A)$, and we want this subalgebra to act on A_0 on the left. But none of this is evident without further hypotheses.

To see what these hypotheses should be, let us remark first that A-valued functions f on G act on the left on A, the appropriate formula being

$$fa = \int f(x)\alpha_x(a)\Delta^{-1/2}(x)dx,$$

when this makes sense. In particular, for $a, b, c \in A_0$ we have

$$\langle a, b \rangle_E c = \int a\alpha_x(b^* c)dx,$$

which does make sense by our hypotheses on A_0. We need to assume that such integrals are again in A_0, so that A_0 has a chance of being a left E_0-module. Since on A_0 we already have defined an E_0-valued inner product, A_0 will then be a left E_0-rigged space in the terminology of 2.8 of [21]. We can then look for its imprimitivity algebra, D_0, in the sense of definition 6.4 of [21] except using left modules. This will be generated by the operators $\langle b, c \rangle_D$ acting on the right on A_0 and defined by the formula

$$a\langle b, c \rangle_D = \langle a, b \rangle_E c.$$

We want these operators to be nicely related to the situation. By slight abuse of notation, let $M(A_0)$ denote the subalgebra of $M(A)$ consisting of the multipliers which carry A_0 into itself, and let $M(A_0)^\alpha$ denote its subalgebra of α-invariant elements. We will require that $\langle b, c \rangle_D$ come from an element of $M(A_0)^\alpha$. Thus:

1.2 *Definition.* Let α be an action of a locally compact group G on a C^*-algebra A. We say that α is a *proper* action if there is a dense α-invariant $*$-subalgebra A_0 of A such that

1) for any $a, b \in A_0$ the function

$$\langle a, b \rangle_E(x) = \Delta(x)^{-1/2} a\alpha_x(b^*)$$

is in $L^1(G, A)$, as is the function $x \mapsto a\alpha_x(b^*)$.

2) For any $a, b \in A_0$ there is a (uniquely determined) element $\langle a, b \rangle_D$ of $M(A_0)^\alpha$ such that for every $c \in A_0$ we have

$$\int c\alpha_x(a^*b)dx = c\langle a,b\rangle_D.$$

Under these hypotheses we can now show that E_0, as defined earlier, is an algebra. Indeed, for $a, b, c, d \in A_0$, we have

$$\begin{aligned}
\langle a,b\rangle_E\langle c,d\rangle_E(x) &= \int \langle a,b\rangle_E(y)\alpha_y(\langle c,d\rangle_E(y^{-1}x))dy \\
&= \int \Delta(y)^{-1/2}a\alpha_y(b^*)\alpha_y(\Delta(y^{-1}x)^{-1/2}c\alpha_{y^{-1}x}(d^*))dy \\
&= \int \Delta(x)^{-1/2}\int a\alpha_y(b^*c)dy\,\alpha_x(d^*) \\
&= \langle a\langle b,c\rangle_D, d\rangle_E(x).
\end{aligned}$$

Since $a\langle b,c\rangle_D$ is by hypotheses again in A_0, it follows that E is an algebra. But, from a slightly earlier calculation, the above calculation can be restated as giving

$$\langle\langle a,b\rangle_E c, d\rangle_E = \langle a,b\rangle_E\langle c,d\rangle_E.$$

From this it follows that for any $e \in E_0$ we have

$$\langle ea,b\rangle_E = e\langle a,b\rangle_E.$$

If we requip E_0 with the norm from $C_r^*(G,A)$, so that E_0 is a pre-C^*-algebra, then the above observations show that A_0 is a left E_0 rigged space in the terminology of definition 2.8 of [21].

As discussed in §2 of [21], we can define a norm on A_0 by

$$||a||_E = ||\langle a,a\rangle_E||^{1/2},$$

where the norm on the right-hand side is that of E_0, and so of $C^*_{\mathrm{r}}(G, A)$. We let $\bar{A}_0$ denote the completion of A_0 with respect to this norm. Let E denote the closure of E_0 in $C^*_{\mathrm{r}}(G, A)$. Then the action of E_0 on A_0 defined above extends by continuity to an action of E on $\bar{A}_0$. A simple argument, given in lemma 6.13 of [21], shows that this action is non-degenerate in the sense that $E\bar{A}_0$ is dense in $\bar{A}_0$.

We will show now that E is an ideal in $C^*_{\mathrm{r}}(G, A)$. For any $a \in A$ let m_a denote a viewed as a multiplier of $C^*_{\mathrm{r}}(G, A)$. Then for $a, b, c \in A_0$ a simple calculation shows that

$$m_a\langle b, c\rangle_E = \langle ab, c\rangle_E.$$

It follows by continuity that $m_a E \subseteq E$ for every $a \in A$. For any $y \in G$ let δ_y denote y viewed as a multiplier of $C^*_{\mathrm{r}}(G, A)$. Another simple calculation shows that

$$\delta_y\langle a, b\rangle_E = \langle \alpha_y(a), b\rangle_E$$

for any $a, b \in A_0$. It follows by continuity that $\delta_y E \subseteq E$ for every $y \in G$. But for any $f \in C_{\mathrm{c}}(G, A)$ and any $\eta \in C^*_{\mathrm{r}}(G, A)$ their product in $C^*_{\mathrm{r}}(G, A)$ can be written as

$$f\eta = \int m_{f(y)}\delta_y \eta \, dy,$$

so that if $\eta \in E$ then $f\eta \in E$. It follows by continuity that E is a left ideal, and so a two-sided ideal since it is a $*$-subalgebra. Since $\bar{A}_0$ is a non-degenerate left E-module, the action of E on $\bar{A}_0$ extends uniquely to an action of $C^*_{\mathrm{r}}(G, A)$ on $\bar{A}_0$.

We now consider further the imprimitivity algebra for the situation. Let $d \in M(A_0)^\alpha$. Then from calculation (1.1), with the notation used there, we find that for any $a \in A_0$,

$\langle\langle ad, ad\rangle_E\xi, \xi\rangle =$

$$\langle\rho(d^*)\int\rho(\alpha_x(a^*))\xi(x^{-1})\Delta(x)^{-1/2}dx, \rho(d^*)\int\rho(\alpha_y(a^*))\xi(y^{-1})\Delta(y)^{-1/2}dy\rangle$$

$\leq \|d\|^2 \langle\langle a, a\rangle_E\xi, \xi\rangle.$

Consequently, as elements of E,

$$\langle ad, ad\rangle_E \leq \|d\|^2\langle a, a\rangle_E,$$

so that d is a bounded operator on A_0 in the sense of definition 2.3 of [21], because d^* is easily seen to serve as an adjoint for d with respect to $\langle\ ,\ \rangle_E$. Consequently, d extends to a bounded operator on $\bar{A}_0$. We show next that the norm of this operator is the same as the norm of d in $M(A)$. To this end, choose ρ acting on H and a unit vector $v \in H$ such that $\|\rho(d)v\|$ is close to $\|d\|$. Then choose an $a \in A_0$ close to a suitable element of an approximate identity for A. Finally, let ξ be a unit vector in $\Xi = L^2(G, H)$ supported in a small neighborhood of the identity element of G and constant there with value a multiple of v. Then the calculation made above shows that $\langle\langle ad, ad\rangle_E\xi, \xi\rangle$ is close to

$$\|d\|^2\langle\langle a, a\rangle_E\xi, \xi\rangle.$$

Consequently the norm of d as a bounded operator on $\bar{A}_0$ is $\|d\|$. We have thus established:

1.3 **Lemma.** *With notation as above, each element $d \in M(A_0)^\alpha$ determines an element of the algebra, $L(\bar{A}_0)$, of bounded operators on A_0, defined by $a \mapsto ad$ for $a \in A_0$. The corresponding anti-homomorphism of $M(A_0)^\alpha$ into $L(\bar{A}_0)$ is isometric.*

The imprimitivity algebra, D, of $\bar{A}_0$ is by definition the closure in $L(\bar{A}_0)$ of the linear span, D_0, of the operators $\langle a, b\rangle_D$ defined by

$$c\langle a, b\rangle_D = \langle c, a\rangle_E b$$

for $a, b, c \in A_0$. The norm on D_0 is that from $L(\bar{A}_0)$. Notice that D_0 is already a *-subalgebra of $L(\bar{A}_0)$, by simple calculations using the relation of $\langle\ ,\ \rangle_D$ to $\langle\ ,\ \rangle_E$, as indicated in proposition 6.3 of [21]. But by hypothesis, the operators $\langle a, b\rangle_D$ for $a, b \in A_0$ are all in $M(A_0)^\alpha$, which we have just seen is isometrically embedded in $L(A_0)$. Thus we can now simply view D as the closure in $M(A)$ of the linear span of the $\langle a, b\rangle_D$'s for $a, b \in A_0$.

In view of the fact that for $a, b, c \in A_0$ we have

$$c\langle a, b\rangle_D = \int c\alpha_x(a^*b)dx,$$

it is natural to write symbolically that

$$\langle a, b\rangle_D = \int \alpha_x(a^*b)dx,$$

even though the integral on the right will not converge unless G is compact. But this suggests that the elements of D should be viewed as the generalized fixed-points for α, much as happens for proper actions on locally compact spaces, as discussed at the beginning of this section. Thus we make:

1.4 *Definition.* Let α be a proper action of G on A. Then by the *generalized fixed-point algebra* of α we will mean the closure, D, in $M(A)$ of the linear span, D_0, of the elements $\langle a, b\rangle_D$ for $a, b \in A_0$. When convenient we will, by slight abuse of notation, denote D by A^α.

The above discussion together with proposition 6.6 of [21] shows that

A_0 is an $E_0 - D_0$-imprimitivity bimodule, as defined in 6.10 of [21]. Taking completions, we obtain a strong Morita equivalence as defined in [23]:

1.5 **Theorem.** *Let α be a proper action of a locally compact group G on a C^*-algebra A. Then, with notation as above, A^α is strongly Morita equivalent to the ideal E of $C_r^*(G, A)$ defined above, with $\bar{A}_0$ serving as an equivalence (i.e. imprimitivity) bimodule.*

We remark that E can easily fail to be all of $C_r^*(G, A)$. For instance, if G is compact and α is the trivial action, then E will consist of exactly the A-valued functions on G which are constant. If G is a finite group acting on a compact space M, and thus on

$$A = C(M),$$

then it is easily checked that

$$E = C^*(G, A)$$

exactly if the action is free. Since there are other possible ways to try to generalize the notion of freeness, as discussed in great detail in [16] (see also §10.8 of [17]), we will not use the term "free" here, but will rather use the following terminology, which is consistent with that used for compact groups as discussed in §7.1 of [16]:

1.6 *Definition.* Let α be a proper action of a locally compact group G on a C^*-algebra A. We will say that α is *saturated* if $E = C_r^*(G, A)$, in the notation used above.

1.7 **Corollary.** *Let α be a saturated proper action of a locally compact* *group G on a C^*-algebra A. Then, with notation as above, A^α is strongly Morita equivalent to $C_r^*(G, A)$.*

In §2 we will give some interesting examples of saturated proper actions.

In concluding this section, let us remark that it is not very clear how the results of this section depend on the choice of the dense subalgebra A_0. It would be desirable to have a more intrinsic definition of proper actions, which produces the subalgebra A_0 by some canonical construction. It is also not clear how often the integral actions implicit in 7.8.4 of [15] will be proper.

§2. Examples

We now give several general examples of proper actions. The first of these will clarify the following issue. We have been careful in §1 to respect the distinction between $C^*(G, A)$ and $C_r^*(G, A)$. Of course, if G is amenable then this distinction disappears (theorem 7.7.7 of [15]). But actually, Phillips has shown (theorem 6.1 of [17]) that if α is any proper action of an arbitrary G on any locally compact space M, then

$$C^*(G, A) = C_r^*(G, A)$$

for

$$A = C_\infty(M).$$

This suggests that this might also happen for A non-commutative. But we now give a class of examples which show that this is not always the case, and that our earlier emphasis on the reduced algebras was appropriate. (Let me record here my thanks to Chris Phillips for helpful discussions about this matter.)

2.1 *Example.* Let G be any locally compact group, and let λ denote the left regular representation of G on $L^2(G)$. Let

$$A = K(L^2(G)),$$

the algebra of compact operators on $L^2(G)$, and let α be the action of G on A consisting of conjugation by λ. We show now that α is a proper action. Let A_0 be the subalgebra of A consisting of compact operators defined by kernels $F \in C_c(G \times G)$, where

$$(F\xi)(x) = \int F(x,y)\xi(y)dy$$

for $\xi \in L^2(G)$. It is easily calculated that

$$(\alpha_z(F))(x,y) = F(z^{-1}x, z^{-1}y),$$

so α carries A_0 into itself. For any F, $F' \in A_0$ we have

$$\begin{aligned}\langle F, F'\rangle_E(z)(x,y) &= \Delta(z)^{-1/2}(F\alpha_z(F'^*))(x,y) \\ &= \Delta(z)^{-1/2}\int F(x,w)(\alpha_z(F'^*))(w,y)dw \\ &= \Delta(z)^{-1/2}\int F(x,w)\bar{F}'(z^{-1}y, z^{-1}w)dw,\end{aligned}$$

which is easily seen to have compact support in all variables, as will be true also when the modular function is omitted. Thus condition 1 of Definition 1.2 holds. Next, we can prove a property slightly stronger than condition 2, namely, that for any $F \in A_0$ there is a $\Phi \in M(A_0)^\alpha$ such that

$$\int F'\alpha_z(F)dz = F'\Phi$$

for all $F' \in A_0$ (where the juxtaposition means product of operators, not pointwise product). As above, the integrand has compact support, so the integral is well-defined since the integrand is easily seen to be norm continuous. The integral will be a compact operator, and it is reasonable to hope that it is given by a kernel function. If we calculate pointwise at the level of functions, we find that

$$\begin{aligned}\left(\int F'\alpha_z(F)dz\right)(x,y) &= \int\int F'(x,w)F(z^{-1}w, z^{-1}y)dw\,dz \\ &= \int F'(x,w)\int F(z^{-1}, z^{-1}w^{-1}y)dz\,dw.\end{aligned}$$

Set

$$f(u) = \int F(z^{-1}, z^{-1}u)dz.$$

Then it is easily seen that $f \in C_c(G)$, while, of course,

$$\Big(\int F'\alpha_z(F)dz\Big)(x,y) = \int F'(x,w)f(w^{-1}y)dw.$$

Applying this to a $\xi \in L^2(G)$, still at the pointwise level,

$$\begin{aligned}\Big(\Big(\int F'\alpha_z(F)dz\Big)\xi\Big)(x) &= \int\Big(\int F'(x,w)f(w^{-1}y)dw\Big)\xi(y)dy \\ &= \int F'(x,w)\Big(\int f(y)\xi(wy)dy\Big)dw \\ &= (F'(\rho_f\xi))(x),\end{aligned}$$

where ρ denotes the *right* regular representation of G on $L^2(G)$. This makes sense since

$$M(A) = B(L^2(G)),$$

the algebra of all bounded operators on $L^2(G)$, and we expect to obtain something in $M(A)^\alpha$, that is, in this case, an operator commuting with λ. But the operators commuting with λ are exactly those in the von Neumann algebra generated by the *right* regular representation, ρ, of G. Anyway, we have found above, symbolically, that

$$\int \alpha_x(F) = \rho_f.$$

Now that we see what the answer might be, it is straightforward to justify the above calculations at the level of operator norms, rather than just pointwise. (One technique for this is given in the next example.) In this way condition 2 is verified. Furthermore, it is not difficult to see that every f can be approximated arbitrarily closely in the L^1-norm by functions of the form

$$u \mapsto \int F(z^{-1}, z^{-1}u)dz$$

for $F \in C_c(G \times G)$. It follows that A^α, as defined earlier, will be all of $C_r^*(G)$, acting by the right regular representation on $L^2(G)$.

It is easily checked that E_0 contains an approximate identity for $C_r^*(G, A)$, so that

$$E = C_r^*(G, A),$$

and α is saturated. Since α is an inner action, that is, comes from a representation of G into the group of unitary elements of $M(A)$, and since inner actions are not usually viewed as being analogues of free actions, this indicates some of the limitations of viewing saturation as an extension of freeness (even when G is compact). However, the above α certainly acts in some respects more like a free action than does, say, the trivial action, even though these actions are closely related, as we next discuss.

It is well known that the crossed product algebra for an inner action is isomorphic to the corresponding crossed product for the trivial action, the isomorphism being given at the level of functions by the map

$$f \mapsto (x \mapsto f(x)\lambda_x)$$

for $f \in C_c(G, A)$, where λ is the unitary representation defining the action. This also works for reduced crossed products, as is seen by applying the mapping

$$\xi \mapsto (x \mapsto \lambda_x \xi(x))$$

to vectors $\xi \in L^2(G, H)$ where H is the Hilbert space of a faithful representation of A, and $L^2(G, H)$ is the Hilbert space of the corresponding induced representation. Thus in our specific situation where $A = K$, we find that

$$C^*(G, A) \cong C^*(G) \otimes K$$
$$C_r^*(G, A) \cong C_r^*(G) \otimes K.$$

From this it is clear that if G is not amenable, then the full and reduced crossed product algebras do not coincide for the action α of conjugation by the left regular representation. For example, if G is the free group on two generators, Powers has shown [18] that $C_r^*(G)$ is simple, whereas $C^*(G)$ certainly is not simple since G has finite dimensional unitary representations. Consequently, theorem 6.1 of [17] does not generalize to non-commutative A. Since we saw above that the generalized fixed point algebra is $C_r^*(G)$, this example also supports the emphasis which we put on reduced crossed products in the definition of proper actions, since if matters had worked out using full crossed products, we would have found that $C^*(G)$ is strongly Morita equivalent to $C_r^*(G)$, which fails here since strong Morita equivalence preserves simplicity (by theorem 3.1 of [22]). These considerations show that, in part, it is our insistence that the generalized fixed-point algebra be a closed subalgebra of $M(A)$ which is forcing us to use reduced crossed products.

Our next example involves G which are Abelian, and can be considered to be another manifestation of Takesaki–Takai duality [15]. For G Abelian,

$$C_r^*(G, A) = C^*(G, A)$$

for any action α of G on a C^*-algebra A, and so for simplicity of notation we will here denote this crossed product algebra by $A \times_\alpha G$. Let $\hat{G}$ denote the dual group of G, and let $\hat{\alpha}$ denote the dual action [15] of $\hat{G}$ on $A \times_\alpha G$. For $f \in L^2(G, A)$ this is defined by

$$\hat{\alpha}_t(f)(x) = \langle x, t \rangle f(x)$$

for $x \in G$ and $t \in \hat{G}$, where $\langle\ ,\ \rangle$ denotes here the duality between G and $\hat{G}$. For simplicity of exposition we will actually assume that G is elementary in the sense of no. 11 of [28], that is, that G is a compactly generated Abelian Lie group, so of the form $R^p \times Z^q \times T^m \times F$ where R, Z, T and F denote respectively the reals, integers, circle group, and a finite Abelian group. There seems to be little doubt that the results obtained here can be extended to all locally compact Abelian groups by using their Schwartz space as defined by Bruhat [3] and employed in [28] and [25], but I have not checked this carefully. We state this example as:

2.2 **Theorem.** *Let α be an action of an elementary locally compact Abelian group G on a C^*-algebra A. Then the dual action, $\hat{\alpha}$, is proper. Furthermore, $\hat{\alpha}$ is saturated, and the generalized fixed-point algebra of $\hat{\alpha}$ is naturally identified with A. Thus $(A \times_\alpha G) \times_{\hat{\alpha}} \hat{G}$ is strongly Morita equivalent to A.*

Proof. Notation is now a bit confusing because here the roles of G, A and α in Definition 1.2 are played by $\hat{G}$, $A \times_\alpha G$ and $\hat{\alpha}$. To verify the conditions of Definition 1.2 we must first pick a suitable dense subalgebra of $A \times_\alpha G$. In preparation for this and for later needs, we consider the following slightly more general situation. Let H be another elementary group (which may also be G), and let $C_\infty(H, A)$ denote the Banach space of continuous A-valued functions on H vanishing at infinity, with the supremum norm. Define a strongly continuous action β on $G \times H$ on $C_\infty(H, A)$ by

$$\beta_{(x,y)}(f)(z) = \alpha_x(f(z - y)).$$

We let $S_\alpha(H, A)$ denote the space of elements of $C_\infty(H, A)$ which are infinitely differentiable for the action β, and which, with all of their higher partial derivatives for either G or H, vanish more rapidly at infinity than any polynomial on H grows. Here derivatives are taken in the R and T directions of G and H, whereas polynomials are taken with respect to the R and Z directions of H. Of course,

$$S_\alpha(H, A) \subseteq L^1(H, A),$$

and the values of functions in $S_\alpha(H, A)$ are in A^∞, the space of C^∞-vectors for the action α. Furthermore, there are plenty of elements in $S_\alpha(H, A)$, since any function of form $a\phi$ for $a \in A^\infty$ and $\phi \in S(H)$ will be in $S_\alpha(H, A)$.

The algebra which will play the role of the A_0 of Definition 1.2 is $S_\alpha(G, A)$. Some straightforward calculations show that $S_\alpha(G, A)$ is, in fact, a $*$-subalgebra of $A \times_\alpha G$, and that $S_\alpha(G, A)$ is carried into itself by the dual action $\hat{\alpha}$.

We must now examine the functions on $\hat{G}$ of the form

$$\langle f, g \rangle_E(t) = f\hat{\alpha}_t(g)$$

for $f, g \in S_\alpha(G, A)$. As functions of both t and x we have

$$\begin{aligned}(f\hat{\alpha}_t(g))(x) &= \int f(y)\alpha_y(\hat{\alpha}_t(g(x-y)))dy \\ &= \int f(x-y)\alpha_{x-y}(g(y))\langle y, t\rangle dy,\end{aligned}$$

which we recognize as just the partial Fourier transform (for appropriate conventions) with respect to the second variable, of the function

$$\Phi(x, y) = f(x-y)\alpha_{x-y}(g(y)).$$

A simple linear change of coordinates converts this function to the function

$$(x, y) \mapsto f(x)\alpha_x(g(y)),$$

which is easily seen to be in $S_\alpha(G\times G, A)$, and from this it follows easily that Φ itself is in $S_\alpha(G \times G, A)$. Standard arguments show, much as indicated at the top of page 159 of [28], that the partial Fourier transform of Φ in its second variable will then be in $S_\alpha(G \times \hat{G}, A)$, that is, the function

$$(x, t) \mapsto (f\hat{\alpha}_t(g))(x)$$

is in $S_\alpha(G \times \hat{G}, A)$. In particular, it is in $L^1(G \times \hat{G}, A)$, and so $\langle f, g\rangle_E(t)$ is in

$$S_\alpha(G, A) \subseteq L^1(G, A)$$

for each fixed t. It follows from the Fubini theorem that

$$\langle f, g\rangle_E \in L^1(\hat{G}, A \times_\alpha G).$$

We have thus verified condition 1 of Definition 1.2.

To verify condition 2 we wish to evaluate

$$\int_{\hat{G}} f\hat{\alpha}_t(g)dt$$

for any $f, g \in S_\alpha(G, A)$. For this purpose, let $LC(G, A)$ be the Banach space

$$L^1(G, A) \cap C_\infty(G, A)$$

with the norm $\|\cdot\|_1 + \|\cdot\|_\infty$. Note that

$$S_\alpha(G, A) \subseteq LC(G, A),$$

and that the injection of $LC(G, A)$ into $L^1(G, A)$, and so into $A \times_\alpha G$, is continuous. The utility of the above norm is that evaluations at points of G are clearly continuous for it. To take advantage of this we must also notice that if $f \in LC(G, A)$, then $t \mapsto \hat{\alpha}_t(f)$ is continuous not only for $\|\cdot\|_1$, but also for $\|\cdot\|_\infty$. But this is true essentially by the definition of the topology on $\hat{G}$. Note also that convolution is clearly at least separately continuous for the above norm, so that

$$t \mapsto f\hat{\alpha}_t(g)$$

is continuous for this norm. Thus, to calculate the integral of this function with values in $A \times_\alpha G$, under the assumption that

$$f, g \in S_\alpha(G, A),$$

it suffices to view its values as being in $LC(G, A)$, and calculate pointwise. But, calculating as we did somewhat earlier, we see that for $x \in G$ we have

$$\begin{aligned}\int\left(\int f(y)\alpha_y(\hat{\alpha}_t(g)(x-y))dy\right)dt \\ = \int\left(\int f(x-y)\alpha_{x-y}(g(y))\langle y, t\rangle dy\right)dt.\end{aligned}$$

For our fixed x, define a function h by

$$h(y) = f(x-y)\alpha_{x-y}(g(y)).$$

It is easily seen that $h \in S_\alpha(G,A)$. Furthermore, the above calculation shows that

$$\left(\int_{\hat{G}} f\hat{\alpha}_t(g)dt\right)(x) = (\hat{h})^\vee(0),$$

where $^\vee$ denotes the inverse Fourier transform. Thus we need to know that the Fourier inversion formula can be applied to h. But by composing h with continuous linear functionals on A, we obtain complex-valued functions in $S(G)$, and it is well known [28] that the Fourier inversion theorem applies to these. It follows that it applies to h, and so we find that

$$\left(\int_{\hat{G}} f\hat{\alpha}_t(g)dt\right)(x) = h(0) = f(x)\alpha_x(g(0)).$$

Now for $a \in A$ let m_a denote a viewed as an element of $M(A \times_\alpha G)$. Then we recognize the right-hand side above to be just $(fm_{g(0)})(x)$. Thus we have found that

$$\int_{\hat{G}} f\hat{\alpha}_t(g)dt = fm_{g(0)},$$

or, symbolically,

$$\int_{\hat{G}} \hat{\alpha}_t(g)dt = m_{g(0)},$$

(this latter integral not converging in norm unless $\hat{G}$ is compact). From this it is easily seen that condition 2 of Definition 1.2 is satisfied, and that for $f, g \in S_\alpha(G,A)$ we have

$$\langle f, g\rangle_D = m_{(f*g)(0)}.$$

By taking f and g of the form $a\phi$ for $a \in A^\infty$ and $\phi \in S(G)$, and letting the corresponding ϕ's range over an approximate identity for $L^1(G)$, one sees easily that the elements of A of the form $\langle f, g\rangle_D$ are dense in A. Here we identify $a \in A$ with m_a, which is appropriate, as it is easily seen that the norms coming from A and $M(A \times_\alpha G)$ coincide. It follows that the generalized fixed-point algebra of $\hat{\alpha}$ is A.

The one fact which remains to be verified is the density of E_0, that is, the fact that $\hat{\alpha}$ is saturated. To this end we show first that E_0 is invariant under the action $\hat{\hat{\alpha}}$ dual to $\hat{\alpha}$. It suffices to verify this for elements of E_0 of the form $\langle f, g\rangle_E$ for

$$f, g \in S_\alpha(G, A).$$

But for $z \in G$,

$$\begin{aligned}(\hat{\hat{\alpha}}_z(\langle f, g\rangle_E))(t) &= \langle z, t\rangle\langle f, g\rangle_E(t)\\ &= \langle z, t\rangle f\hat{\alpha}_t(g^*)\\ &= \langle z, t\rangle(f\delta_z^*)\delta_z\hat{\alpha}_t(g^*).\end{aligned}$$

But $\hat{\alpha}_t(\delta_z) = \langle z, t\rangle\delta_z$, so the above

$$= (f\delta_z^*)\hat{\alpha}_t((g\delta_z^*)^*) = \langle f\delta_z^*, g\delta_z^*\rangle_E(t).$$

Since $f\delta_z^*$ and $g\delta_z^*$ are both again in $S_\alpha(G, A)$, it follows that E_0 is $\hat{\hat{\alpha}}$-invariant. Of course, then also E is $\hat{\hat{\alpha}}$-invariant.

The proof of Theorem 2.2 will thus be complete once we have given a proof of the following proposition, which returns to the notation of Definition 1.2 rather than that used just above.

2.3 **Proposition.** *Let α be a proper action of a locally compact Abelian group G on a C^*-algebra A. Let E be the ideal of $A \times_\alpha G$ defined as above.*

Then the following conditions are equivalent:

1) α *is saturated.*

2) E *contains an approximate identity for* $A \times_\alpha G$.

3) E *is carried into itself by the dual action* $\hat{\alpha}$.

Proof. Since saturation means just that $E = A \times_\alpha G$, and since E is an ideal, it is clear that conditions 1 and 2 are equivalent. We have included condition 2 because it has been a frequently used method for showing this kind of thing — not only earlier in this paper, but also in, for example, the proof of situation 10 of [24], and lemma 2.4 of [20]. It is also clear that conditions 1 and 2 imply condition 3.

To show that condition 3 implies condition 1, we use proposition 6.3.9 of [16] (which for the separable case also appears as corollary 2.2 of [10]). This proposition 6.3.9 tells us that because E is $\hat{\alpha}$-invariant, it must be of the form $I \times_\alpha G$ for some α-invariant ideal I in A. Then the image of E in $(A/I) \times_\alpha G$ must be (0). But if $I \neq A$, we can find $a, b \in A_0 \setminus I$, and then the image of $\langle a, b\rangle_E$ in $(A/I) \times_\alpha G$ will clearly not be 0. Thus $I = A$, so $E = A \times_\alpha G$. $\square$

For the case of actions for which G is not Abelian, one has instead of a dual action of the dual group, a dual coaction of G itself, as discussed in [11] and the references therein. It seems to be an interesting challenge to formulate an appropriate definition of what it means for a coaction to be proper, and then to show that the dual coaction to an action will be proper.* Also, the dual action to a coaction should be proper. All this should also be possible for duality of twisted crossed products, as discussed in [19].

Let us also remark that for G Abelian the strong Morita equivalence of Theorem 2.2 fits into the framework of [4] and [6]. To be specific, with

---

* This has now been verified by Kevin Mansfield in *Induced representations of crossed products by coactions,* preprint.

notation as earlier, define an action $\bar{\alpha}$ of G on $A_0 = S_\alpha(G, A)$ by

$$\bar{\alpha}_z f = f\delta_z^*.$$

Then the calculation used above to show that E_0 is $\hat{\hat{\alpha}}$ invariant shows that

$$\langle \bar{\alpha}_z f, \bar{\alpha}_z g\rangle_E = \hat{\hat{\alpha}}_z(\langle f, g\rangle_E).$$

Furthermore, we have (identifying D with A)

$$\begin{aligned}\langle \bar{\alpha}_z f, \bar{\alpha}_z g\rangle_D &= (\bar{\alpha}_z f)^*(\bar{\alpha}_z g)(0) = (f\delta_z^*)^*(g\delta_z^*)(0) \\ &= \delta_z(f^*g)\delta_z^*(0) = \alpha_z((f^*g)(0)) \\ &= \alpha_z(\langle f, g\rangle_D).\end{aligned}$$

Thus the actions $\hat{\hat{\alpha}}$, $\bar{\alpha}$ and α on $(A \times_\alpha G) \times_{\hat{\alpha}} \hat{G}$, $\bar{A}_0$, and A respectively satisfy exactly the relations of theorem 1 of [6], that is, $\bar{\alpha}$ provides a Morita equivalence between $\hat{\hat{\alpha}}$ and α as defined in §3 of [4]. Thus we can conclude from these articles the unsurprising fact that

$$((A \times_\alpha G) \times_{\hat{\alpha}} G) \times_{\hat{\hat{\alpha}}} G$$

is strongly Morita equivalent to $A \times_\alpha G$.

Because the ideas involved will be useful in [27], let us see how condition 3) of Proposition 2.3 can be used to obtain the following well-known result (see [24], where it is seen to be true for any locally compact G).

2.4 **Corollary.** *Let G be a discrete countable Abelian group, and let α be a proper action of G on a locally compact space M. Let α also denote the corresponding proper action of G on the C^*-algebra $C_\infty(M)$. If the action α on M is free, then the proper action α on $C_\infty(M)$ is saturated.*

Proof. We will show that the ideal E is carried into itself by $\hat{\alpha}$. We assume the dense subalgebra is $C_c(M)$, but the same proof would work for various other choices, such as $C_c^\infty(M)$ if M is a manifold. Anyway, let

$$\phi, \psi \in C_c(M),$$

and let $t \in \hat{G}$. We need to show that $\hat{\alpha}_t(\langle\phi, \psi\rangle_E)$ is again in E. Since α is free and proper, and G is discrete, we can assume that all the translates of the closure of the support of ψ are disjoint, for if this is not the case, then we can express ψ as a finite sum of elements of $C_c(M)$ which do have this property. Likewise we can then assume that the support of ϕ meets the support of at most one translate of ψ. If it meets no translate, then

$$\langle\phi, \psi\rangle_E = 0$$

and so $\hat{\alpha}$ applied to it is in E. Otherwise, there is a unique $k_0 \in G$ such that

$$\langle\phi, \psi\rangle_G(k_0) \neq 0.$$

Let $\theta = \langle k_0, t\rangle\bar{\psi}$. Then for $x \in M$ and $k \in G$,

$$\begin{aligned}(\hat{\alpha}_t(\langle\phi, \psi\rangle_E))(x, k) &= \langle k, t\rangle\langle\phi, \psi\rangle_E(x, k) \\ &= \langle k, t\rangle\phi(x)\bar{\psi}(\alpha_k^{-1}(x)),\end{aligned}$$

which is 0 if $k \neq k_0$, and so can be rewritten as

$$= \phi(x)\langle k_0, t\rangle\bar{\psi}(\alpha_k^{-1}(x)) = \phi(x)\bar{\theta}(\alpha_k^{-1}(x)) = \langle\phi, \theta\rangle_E(x, k)$$

as desired. □

Our next two examples are somewhat less interesting since they explicitly involve proper actions on spaces. Hence we present them in somewhat sketchy fashion.

2.5 *Example.* This one comes from theorem 2 of [6]. The set-up used there, and in situation 10 of [24], consists of a locally compact space P and locally compact groups G and H having commuting proper and free actions on P, which we now denote by juxtaposition. One can then define an action α of G on the (non-commutative) transformation group C^*-algebra $A = C^*(H, P)$ by

$$\alpha_x(f)(t,p) = f(x, x^{-1}p)$$

for $f \in C_c(H,P) = A_0$. Because of the properness of the action of G on P, it is easily seen that $x \mapsto f\alpha_x(g)$ has compact support for $f, g \in A_0$, so that condition 1 of Definition 1.2 is satisfied. Furthermore, it is not difficult to check that for $f, g \in A_0$,

$$\int_G f\alpha_x(g)dx = f\tilde{g}$$

where $\tilde{g} \in C_c(H, P/G)$ is defined by

$$\tilde{g}(t,p) = \int_G g(t, x^{-1}p)dx,$$

and elements of $C_c(H, P/G)$ act as multipliers on $C^*(H,P)$ in the evident way. Thus α is a proper action. But what norm does $C_c(H, P/G)$ obtain when it acts as multipliers on $C^*(H, P)$? Now

$$C^*(H,P) = C_r^*(H,P)$$

according to theorem 6.1 of [17]. Thus we can choose a faithful representation of $C_\infty(P)$ on some Hilbert space Ξ, and then the corresponding induced representation on $L^2(H, \Xi)$ will give a faithful representation of $C^*(H, P)$, and so of $M(C^*(H, P))$. So $C_c(H, P/G)$ obtains the norm from its corresponding action on $L^2(H, \Xi)$. But the representation of $C_\infty(P)$ on Ξ extends to a faithful representation of $M(C_\infty(P))$, and so of

$$C_\infty(P/G) \subseteq M(C_\infty(P)).$$

And it is easily seen that if this faithful representation of $C_\infty(P/G)$ is induced up to H, one obtains just the representation of $C_c(H, P/G)$ considered above. Since it is reduced crossed products which are defined in terms of such induced representations [15], it is now clear that the norm on $C_c(H, P/G)$ is that of $C_r^*(H, P/G)$. It is not difficult to then show that the generalized fixed-point algebra for α is all of $C_r^*(H, P/G)$. But it takes some serious work, along the lines given in [24], to show that α is saturated. Once this is done, one obtains the fact that $C_r^*(G \times H, P)$ is strongly Morita equivalent to $C_r^*(H, P/G)$, since, using theorem 6.1 of [17],

$$\begin{aligned} C_r^*(G, C^*(H, P)) &= C_r^*(G, C_r^*(H, P)) \\ &= C_r^*(G \times H, P). \end{aligned}$$

This should be contrasted with theorem 2 of [6], which gives the corresponding conclusion for the full crossed products.

We remark that if we consider the special case in which $P = G = H$, so that G acts on the left and right on itself, then we obtain essentially the first example of this section. And one sees, as done there, that it must be the reduced crossed products which are involved here.

2.6 *Example.* This one comes from theorem 2.2 of [20]. The set-up used there consists of a locally compact space P with a free and proper action of a locally compact group G, as well as of an action β of G on a C^*-algebra B. Let

$$A = C_\infty(P, B),$$

and let α be the diagonal action of G on A defined by

$$\alpha_x(f)(p) = \beta_x(f(x^{-1}p))$$

for $f \in C_c(P, A) = A_0$. Then it is not difficult to verify that α is proper, with generalized fixed-point algebra $GC(P, A)^\alpha$ as defined in [20]. Again, it requires some serious work, along the lines found in [20], to show that the action is saturated, so that one obtains a strong Morita equivalence of $C_r^*(G, C_\infty(P, A))$ with $GC(P, A)^\alpha$. But in theorem 2.2 of [20] the assertion is that it is the full algebra of $C^*(G, C_\infty(P, A))$ which is strongly Morita equivalent to $GC(P, A)^\alpha$. Thus one can expect that

$$C_r^*(G, C_\infty(P, A)) = C^*(G, C_\infty(P, A)),$$

in mild generalizations of theorem 6.1 of [17].

§3. Continuous Fields of Proper Actions

In this section we consider continuous fields of proper actions, and how they lead to continuous fields of the corresponding generalized fixed-point algebras. We work in the setting of [26], especially theorems 3.1 and 3.4. Thus $\{A_\omega\}$ will be an upper semi-continuous field [8] of C^*-algebras over a locally compact Hausdorff space Ω, and A will be the corresponding maximal C^*-algebra of sections. For each $\omega \in \Omega$ the evaluation map from A to A_ω will be denoted by π_ω, and its kernel by K_ω. By an upper semi-continuous field of actions on $\{A_\omega\}$ we mean an action α of a locally compact group G on A which carries each K_ω into itself, and so defines an action α_ω on A_ω for each $\omega \in \Omega$.

3.1. *Definition.* Let α be an upper semi-continuous field of actions of G on the field $\{A_\omega\}$. We will say that α is an *upper semi-continuous field of proper actions* if there is given a dense *-subalgebra A_0 of A with respect to which α is a proper action, and if, in addition, the evident action of $C_\infty(\Omega)$ on A carries A_0 into itself.

A glance at Definition 1.2 shows that the action α_ω on each A_ω is then proper, where as dense subalgebra we take the image A_ω^0 of A_0 under π. The only point which needs a few moment's thought is that each π_ω extends to a homomorphism, $\tilde{\pi}_\omega$, from $M(A)$ into $M(A_\omega)$, which carries $M(A_0)^\alpha$ into $M(A_\omega^0)^\alpha$, so that for any $a, b \in A_0$ the element $\langle a, b\rangle_D$ in $M(A_0)^\alpha$ determines corresponding elements in each $M(A_\omega^0)^\alpha$, which depend only on the images of a and b in A_ω^0. Accordingly, for each ω we can define the corresponding generalized fixed-point algebra, D_ω, which from the above comments will just be the image of D under $\tilde{\pi}_\omega$. Then $\{D_\omega\}$ is a field of C^*-algebras over Ω, with D identified as an algebra of sections of this field, but without any obvious continuity properties.

We remark that since upper semi-continuity is associated with full crossed products, as seen in theorem 3.1 of [26], while proper actions involve reduced C^*-algebras as seen in §1, it can be expected that in order to obtain upper semi-continuity for a field of generalized fixed-point algebras, we will need to assume that the full and reduced crossed products agree. The next theorem will play a key role in our discussion of deformation quantization in [27], and so is the main result of the present paper. The term "Hilbert-continuous" used here is defined in definition 3.2 of [26].

3.2 **Theorem.** *Let α be an upper semi-continuous field of proper actions on a field $\{A_\omega\}$, and let D be the generalized fixed-point algebra for the action α on the maximal algebra A, of sections. Assume that*

$$C_r^*(G, A_\omega) = C^*(G, A_\omega)$$

for all ω. Then the field $\{D_\omega\}$ of generalized fixed-point algebras is upper semi-continuous, when D is used to define its continuity structure. Furthermore, D can be identified with the maximal algebra of sections of $\{D_\omega\}$. If, in addition, the field $\{A_\omega\}$ is actually Hilbert-continuous, then the field $\{D_\omega\}$ is continuous.

Proof. According to Theorem 3.1 of [26], $\{C^*(G, A_\omega)\}$ is an upper semi-continuous field, with $C^*(G, A)$ as its maximal C^*-algebra of sections. By exactly the same argument as in the second paragraph of the proof of Theorem 3.4 of [26], our assumption that

$$C_r^*(G, A_\omega) = C^*(G, A_\omega)$$

implies that

$$C_r^*(G, A) = C^*(G, A).$$

Consequently, in those places where $C_r^*(G, A)$ is needed in the treatment of proper actions, we will here write $C^*(G, A)$ instead. Accordingly, let E

denote the ideal in $C^*(G, A)$ spanned by the elements $\langle a, b\rangle_E$ for $a, b \in A_0$, just as in §1. For each ω let E_ω be the image of E in $C^*(G, A_\omega)$. It is clear that, equivalently, E_ω is the closed ideal of $C^*(G, A_\omega)$ spanned by the functions $\langle a, b\rangle_{E_\omega}$ for $a, b \in A_\omega^0$, in the way described in §1.

We need the following simple result, in part because it is crucial for us to know that

$$E_\omega \cong E/EI_\omega,$$

where, as in [26], I_ω is the ideal in $C_\infty(\Omega)$ of functions which vanish at ω.

3.3 **Proposition.** *Let $\{B_\omega\}$ be an upper semi-continuous field of C^*-algebras over Ω, and let B be its maximal C^*-algebra of sections. Let E be an ideal of B with the property that if $c \in C_\infty(\Omega)$ and if $cE = \{0\}$, then $c = 0$. For each ω let E_ω be the image of E in B_ω. Then $\{E_\omega\}$ is an upper semi-continuous field of C^*-algebras over Ω, which is continuous if $\{B_\omega\}$ is. In particular,*

$$E_\omega = E/EI_\omega$$

for each ω, and E is the maximal C^*-algebra of sections of $\{E_\omega\}$.

Proof. Multipliers of B act as multipliers of E. Because of the special hypotheses made on E, the corresponding homomorphism of $C_\infty(\Omega)$ into $M(E)$ is isometric, so that $C_\infty(\Omega)$ can be viewed as a subalgebra of $M(E)$, and the theory of §1 of [26] applies. Now the kernel of the evaluation map π_ω restricted to E is, of course, $E \cap BI_\omega$. But the latter is clearly an essential I_ω-module, and so

$$E \cap BI_\omega = EI_\omega.$$

Thus

$$E_\omega = E/EI_\omega.$$

The rest of the assertions now follow immediately from the results of §1 of [26]. □

We now return to the proof of Theorem 3.2. It is clear that, in this setting, $\{E_\omega\} \neq \{0\}$, because we tacitly assume that $A_\omega \neq \{0\}$. Thus Proposition 3.3 applies, and we conclude that $\{E_\omega\}$ is an upper semi-continuous field, with E as its maximal C^*-algebra of sections. Now E is strongly Morita equivalent to D according to Theorem 2.4, with $\bar{A}_0$ serving as equivalence bimodule. Using this strong Morita equivalence, we should be able to transfer the field structure of E to D. Recall from theorem 3.1 of [22] that if A and B are C^*-algebras, and if X is an $A-B$-equivalence bimodule (i.e. imprimitivity bimodule), then X establishes a canonical inclusion-preserving bijection between the (two-sided) ideals of A and those of B, which is given, for example, by sending the ideal J of A to $\langle X, JX\rangle_B$.

3.4 **Theorem.** *Let $\{A_\omega\}$ be an upper semi-continuous field of C^*-algebras over a locally compact space Ω, with A its maximal C^*-algebra of sections. Let B be a C^*-algebra which is strongly Morita equivalent to A via an $A-B$-equivalence bimodule X, and let h denote the corresponding bijection from ideals of A to ideals of B. For each ω let K_ω be the kernel of the evaluation map from A to A_ω, and let*

$$B_\omega = B/h(K_\omega).$$

Then $\{B_\omega\}$ is an upper semi-continuous field of C^*-algebras over Ω, with B as its maximal C^*-algebra of sections. If $\{A_\omega\}$ is actually continuous, then so is $\{B_\omega\}$.

Proof. From corollary 3.3 of [22], the restriction of h to primitive ideals gives a homeomorphism, also denoted by h, from Prim(A) to Prim(B). Now $C_\infty(\Omega)$ can be viewed as a subalgebra of bounded continuous functions on Prim(A), and so, by this homeomorphism, as a subalgebra of bounded continuous functions on Prim(B). The points of Ω then correspond to closed subsets of Prim(A) and Prim(B) which correspond under h. Then the commonly called Dauns–Hofmann theorem [9] (which, as discussed in [8], is just a special result on the way to the main theorem of Dauns and Hofmann) says that $C_\infty(\Omega)$ can be viewed as a central subalgebra of $M(B)$. Let I_ω denote, as before, the ideal in $C_\infty(\Omega)$ of functions vanishing at ω. Then $\{B/BI_\omega\}$ will be an upper semi-continuous field with maximal algebra B, by Proposition 1.2 of [26]. But it is easily seen from the way h is involved here that

$$BI_\omega = h(AI_\omega) = h(K_\omega),$$

thus demonstrating the first assertion. Suppose now that $\{A_\omega\}$ is lower semi-continuous, hence continuous. According to theorem 4 of [12] this is equivalent to the assertion that the evident continuous map p from Prim(A) to Ω coming from $C_\infty(\Omega) \subseteq A$ is open. But the corresponding map from Prim(B) to Ω is easily seen to be ph^{-1}, and h is a homeomorphism, so this map also is open. Then, again by theorem 4 of [12], $\{B_\omega\}$ will be continuous. □

We return now to the proof of Theorem 3.2. From the strong Morita equivalence of D with E, we obtain, by Theorem 3.4, a structure for D as an upper semi-continuous field over Ω. For any ω the ideal in D corresponding to EI_ω will be

$$\langle \bar{A}_0, EI_\omega \bar{A}_0\rangle_D = \langle \bar{A}_0, \bar{A}_0\rangle_D I_\omega = DI_\omega,$$

so the fiber algebra is D/DI_ω.

It might seem that we have now almost completed the proof, but in fact we have only just begun, for we must show that this field structure coincides with that given by the field of generalized fixed point algebras as in the statement of Theorem 3.2. The proof of this is somewhat slippery, because of the many identifications which one is tempted to assume, but which need to be verified. Thus we proceed with some care. Also, perhaps we should comment here that the reason for the approach we are taking is that it appears difficult to prove directly the upper semi-continuity of the field of generalized fixed-point algebras, whereas, as we have just seen, the upper semi-continuity of the field coming from the strong Morita equivalence follows from quite straightforward considerations.

The main identification which we must carefully check is the following

3.5 **Lemma.** *Let $\bar{A}_\omega^0$ and $\bar{A}_0$ denote the completions of A_ω^0 and A_0 for $\langle\ ,\ \rangle_{E_\omega}$ and $\langle\ ,\ \rangle_E$ respectively. Let $\bar{\pi}_\omega$ denote the restriction of π_ω to A_0, mapping onto A_ω^0. Then $\bar{\pi}_\omega$ extends to an E_ω-module homomorphism*

$$\bar{\pi}_\omega : \bar{A}_0/\bar{A}_0 I_\omega \longrightarrow \bar{A}_\omega^0,$$

which is an isometric isomorphism preserving the E_ω-valued inner products.

Let us remark that if we were to make precise the idea of a field of modules with C^*-algebra valued inner products, along lines similar to those used in §3 of [8], then this lemma would signify that the field $\{\bar{A}_\omega^0\}$ is upper semi-continuous.

Proof of Lemma 3.5. For $a, b \in A_0$ we have

$$\begin{aligned}\langle \bar{\pi}_\omega(a), \bar{\pi}_\omega(b)\rangle_{E_\omega}(x) &= \Delta(x)^{-1/2}\pi_\omega(a)\alpha_x(\pi_\omega(b^*))\\ &= \Delta(x)^{-1/2}\pi_\omega(a\alpha_x(b^*)).\end{aligned}$$

If we let $\tilde{\pi}_\omega$ denote the quotient map from E to E_ω coming from π_ω, then the above says that

$$\langle \bar{\pi}_\omega(a), \bar{\pi}_\omega(b) \rangle_{E_\omega} = \tilde{\pi}_\omega(\langle a, b \rangle_E),$$

that is, $\bar{\pi}_\omega$ respects inner products via $\tilde{\pi}_\omega$. In particular,

$$\|\bar{\pi}_\omega(a)\|_{\bar{A}^0_\omega} \leq \|a\|_{\bar{A}_0},$$

so that $\bar{\pi}_\omega$ extends to a map, again $\bar{\pi}_\omega$, between the completions, which will again respect the inner-products. Now

$$\pi_\omega(A_0 I_\omega) = 0,$$

and so by continuity

$$\bar{\pi}_\omega(\bar{A}_0 I_\omega) = 0.$$

Thus $\bar{\pi}_\omega$ drops to a map from $\bar{A}_0/\bar{A}_0 I_\omega$. Now $E_\omega = E/EI_\omega$ by Proposition 3.3, so that E_ω acts on $\bar{A}_0/\bar{A}_0 I_\omega$, and the inner-product there can be viewed as having values in E_ω. From this and the earlier calculations, it is easily checked that $\bar{\pi}_\omega$ is an E_ω-module homomorphism which preserves the E_ω-valued inner-products, and hence is isometric. Since $\bar{\pi}_\omega$ clearly has dense range, it follows that $\bar{\pi}_\omega$ is a module isomorphism. □

We are now prepared to deal with D. As earlier, let $\hat{\pi}_\omega$ denote the homomorphism from D onto D_ω obtained by extending π_ω to a homomorphism $\hat{\pi}_\omega$ from $M(A)$ to $M(A_\omega)$. Since $\hat{\pi}_\omega$ clearly has I_ω in its kernel, $\hat{\pi}_\omega$ drops to a homomorphism, $\check{\pi}_\omega$, of D/DI_ω onto D_ω. Our objective is to show that $\check{\pi}_\omega$ is an isomorphism. Now D/DI_ω clearly acts on the right on $\bar{A}_0/\bar{A}_0 I_\omega$, as does D_ω on $\bar{A}^0_\omega$, while $\check{\pi}_\omega$ and $\bar{\pi}_\omega$ are easily seen to be compatible for these actions in the sense that

$$\bar{\pi}_\omega((x + \bar{A}_0 I_\omega)(d + DI_\omega)) = \bar{\pi}_\omega(x + \bar{A}_0 I_\omega)\, \check{\pi}_\omega\, (d + DI_\omega)$$

for $x \in \bar{A}_0$ and $d \in D$. But $\bar{\pi}_\omega$ is an isomorphism by Lemma 3.5. Thus the kernel of $\check{\pi}_\omega$ must act as zero operators on $\bar{A}_0/\bar{A}_0 I_\omega$. But D/DI_ω is exactly the imprimitivity algebra for $\bar{A}_0/\bar{A}_0 I_\omega$ as left E_ω-rigged space by corollary 3.2 of [22], and so D/DI_ω is faithfully represented on $\bar{A}_0/\bar{A}_0 I_\omega$. Thus $\check{\pi}_\omega$ must be an isomorphism, as desired. Consequently, $\{D_\omega\}$ is upper semi-continuous.

Suppose now that $\{A_\omega\}$ is Hilbert-continuous, with faithful representations σ_ω of A_ω on the fixed Hilbert space H. Then $\{C^*(G, A_\omega)\}$ is Hilbert-continuous by theorem 3.4 of [26] with faithful representations ρ_ω on $L^2(G, H)$ as in the proof of that theorem. From Proposition 3.3 it follows that $\{E_\omega\}$ is continuous. But then from Theorem 3.4 it follows that $\{D_\omega\}$ is continuous, because of the fact that we have shown earlier in this proof that $D_\omega = D/DI_\omega$. (It is not clear to me whether $\{D_\omega\}$ will always, in fact, be Hilbert-continuous.) □

No examples are given here of the application of Theorem 3.2, because such examples will be given in [27].

REFERENCES

[1] F. Bayen, M. Flato, C. Fronsdal, A. Lichnerowicz, and D. Sternheimer, *Deformation theory and quantization*, I, II, Ann. Physics **110** (1978), 61–110, 111–151.

[2] N. Bourbaki, *Topologie Générale*, ch. III and IV Hermann, Paris, 1963.

[3] F. Bruhat, *Distributions sur un group localement compact et applications à l'étude des représentations des groupes p-adiques*, Bull. Soc. Math. France **89** (1961), 43–75.

[4] F. Combes, *Crossed products and Morita equivalence*, Proc. London Math. Soc. **49** (1984), 289–306.

[5] A. Connes and M. Takesaki, *The flow of weights on factors of type III*, Tohoku Math. J. **29** (1977), 473–575.

[6] R.E. Curto, P.S. Muhly and D.P. Williams, *Cross products of strongly Morita equivalent C^*-algebras*, Proc. AMS **90** (1984), 528–530.

[7] D. DeSchreye, *Integrable-ergodic C^*-dynamical systems on Abelian groups*, Math. Scand. **57** (1985), 189–205.

[8] M.J. Dupré and R.M. Gillette, *Banach bundles, Banach modules and automorphisms of C^*-algebras*, Research Notes in Math. **92** Pitman Publishing, London, 1983.

[9] G.A. Elliott and D. Olesen, *A simple proof of the Dauns–Hofmann theorem*, Math. Scand. **34** (1974), 231–234.

[10] E.C. Gootman and A.J. Lazar, *Crossed products of type I AF algebras by Abelian groups*, Israel J. Math. **56** (1986), 267–279.

[11] ———, *Applications of non-commutative duality to crossed product algebras determined by an action or coaction*, Proc. London Math. Soc., to appear.

[12] R.-Y. Lee, *On the C^*-algebras of operator fields*, Indiana U. Math. J. **25** (1976), 303–314.

[13] A. Lichnerowicz, *Deformations and quantization*, in *Geometry and Physics* (ed. M. Modugno), pp. 103–116, Pitagora Editrice, Bologna, 1983.

[14] D. Olesen and G.K. Pedersen, *Application of the Connes spectrum to C^*-dynamical systems,* II, J. Funct. Anal. **36** (1980), 18–32.

[15] G.K. Pedersen, *C^*-algebras and their automorphism groups,* London Math Soc. Monographs **14** Academic Press, London, 1979.

[16] N.C. Phillips, *Equivariant K-theory and freeness of group actions on C^*-algebras,* Lecture Notes in Math. **1274** Springer-Verlag, 1987.

[17] ———, *Equivariant K-theory for proper actions,* Pitman Research Notes Math. **178** (1988).

[18] R.T. Powers, *Simplicity of the C^*-algebra associated with the free group on two generators,* Duke Math. J. **42** (1975), 151–156.

[19] J.C. Quigg, *Duality for reduced twisted crossed products,* Indiana U. Math. J. **35** (1986), 549–571.

[20] I. Raeburn and D.P. Williams, *Pull-backs of C^*-algebras and crossed products by certain diagonal actions,* Trans. AMS **287** (1985), 755–777.

[21] M.A. Rieffel, *Induced representations of C^*-algebras,* Adv. Math. **13** (1974), 176–257.

[22] ———, *Unitary representations of group extensions; an algebraic approach to the theory of Mackey and Blattner,* Studies in Analysis, Adv. Math. Suppl. Series 4 (1979), 43–82.

[23] ———, *Morita equivalence for operator algebras,* in *Operator Algebras and Applications* (R.V. Kadison ed.), pp. 285–298; Proc. Symp. Pure Math. **38** Amer. Math. Society, Providence, 1982.

[24] ———, *Applications of strong Morita equivalence to transformation group C^*-algebras,* in *Operator Algebras and Applications* (R.V. Kadison, ed.), pp. 299–310; Proc. Symp. Pure Math. **38** American Math. Society, Providence, 1982.

[25] ———, *Projective modules over higher dimensional non-commutative tori,* Canadian J. Math. **40** (1988), 257–338.

[26] ———, *Continuous fields of C^*-algebras coming from group cocycles and actions,* Math. Ann. **283** (1989), 631–643.

[27] ———, *Deformation quantization for Heisenberg manifolds,* Comm. Math. Phys. **122** (1989), 531–562.

[28] A. Weil, *Sur certains groupes d'opérateurs unitaire,* Acta Math. **111** (1964), 143–211.

Department of Mathematics
University of California
Berkeley, California 94720

On The Baum–Connes Conjecture

HIROSHI TAKAI

1. Introduction

Recently, Connes [3] initiated a new index theory for both dynamical systems and foliated manifolds, which is quite useful to cases with pathological ambient spaces for which the Atiyah–Singer index theory is no longer applicable. The main ideas of his theory are based on both K-theory and cyclic cohomology of algebras, whose validity is illustrated in many papers.

Among others, Baum–Connes [1] has conjectured the existence of a K-theoretic index formula between geometric and analytic K-theory of differentiably dynamical systems and foliated manifolds. It may be viewed as an ultimate version of a generalization of Atiyah–Singer index theory. Precisely speaking, their conjecture says that given a foliated manifold or a smooth dynamical system, its geometric K-theory is isomorphic to its analytic one under the K-index mapping. If it is affirmative, as corollaries are deduced the conjectures of Novikov, Gromov–Lawson–Rosenberg and (generalized) Kadison in topology, differential geometry and C^*-algebras respectively. Moreover the conjecture is also related to the vanishing problems of various characteristic classes for ambient systems.

In this note, we present a survey of the Baum–Connes conjecture for foliated manifolds and smooth dynamical systems. Especially, we explain it in more detail for the generalized Anosov foliations of infra-homogeneous manifolds.

2. Construction

Let (M, F) be a foliated manifold and G its holonomy groupoid. Let $\Omega^{1/2}$ be the half density bundle over G tangential to $s^*(F) \otimes r^*(F)$ where s, r are the source, range map from G to M respectively. We denote by $C_c(\Omega^{1/2})$ the *-algebra consisting of all continuous sections with compact support equipped with the following algebraic operations:

$$(fg)(\gamma) = \int_{\gamma=\gamma_1\gamma_2} f(\gamma_1)g(\gamma_2)$$
$$f^*(\gamma) = \overline{f}(\gamma^{-1})$$

where $\overline{f}(\gamma) = \overline{f(\gamma)}$. Given any $x \in M$, Let H_x be the Hilbert space of all L^2-sections of $\Omega^{1/2}$ over $s^{-1}(x)$. Let π_x be a *-representation of $C_c(\Omega^{1/2})$ on H_x defined by

$$(\pi_x(f)\xi)(\gamma) = \int_{\gamma=\gamma_1\gamma_2} f(\gamma_1)\xi(\gamma_2)$$

for every $x \in M$. We denote by $C_r^*(M,F)$ the completion of $C_c(\Omega^{1/2})$ with respect to a C^*-norm $\|\cdot\|$ defined by

$$\|f\| = \sup_{x\in M} \|\pi_x(f)\|.$$

It is called a foliation C^*-algebra associated with (M,F). We then consider the K-theory $K(C_r^*(M,F))$ of $C_r^*(M,F)$ which is denoted by $K_a(M,F)$. We call it the *analytic* K-theory of (M,F).

On the other hand, we offer a purely geometric way of defining its K-theory according to Baum–Connes [1]. Let X be a proper G-manifold and $\tilde{\nu}$ the dual bundle of the normal bundle $\tilde{\nu}$ of the G-orbital foliation of X. We denote by ρ the canonical G-equivariant mapping from X to M and take $\rho^*(\nu^*)$ the pull back of the dual bundle ν^* of the normal bundle ν of F. We consider a pair (X,ξ) of X and a G-vector bundle ξ over $\tilde{\nu}\oplus\rho^*(\nu^*)$, which is called a *$K$-cocycle* of (M,F). We denote by $\Gamma(M,F)$ the set of all K-cocycles of (M,F). Let us introduce an equivalence relation $\sim$ on $\Gamma(M,F)$ by the following fashion: $(X_1,\xi_1) \sim (X_2,\xi_2)$ if and only if there exist a proper G-manifold X and G-mappings φ_j from X_j into X with the property that

$$(1) \quad \rho_j = \rho\cdot\varphi_j \ (j=1,2) \quad \text{and} \quad (2) \quad \varphi_1!(\xi_1) = \varphi_2!(\xi_2),$$

where ρ, ρ_j are the canonical G-mappings from X, X_j to M respectively, and $\varphi_j!$ mean the Thom–Gysin mappings from G-vector bundles over $\tilde{\nu}_j^* \oplus \rho_j^*(\nu^*)$ to those over $\tilde{\nu}^* \oplus \rho^*(\nu^*)$. We denote by $K_g(M,F)$ the set of all equivalence classes in $\Gamma(M,F)$ with respect to $\sim$. Then it is an abelian group equipped with the disjoint union of G-vector bundles. We call it the *geometric* K-theory of (M,F).

In what follows, we explain how to define the *K-index* mapping μ from $K_g(M,F)$ to $K_a(M,F)$. Given any $(X,\xi)\in\Gamma(M,F)$, let us take the

G-map j from X to $X \times M$ by $j(x) = (x, \rho(x))$ and π the projection from $X \times M$ onto M. Then $\rho = \pi \cdot j$. Let $\tilde{j}$ be the canonical G-mapping from $\tilde{\nu}_X^* \oplus \rho^*(\nu^*)$ to $\tilde{\nu}_{X \times M}^* \oplus \pi^*(\nu^*)$ associated with j. Then it is K-oriented with respect to the G-structure. Let $\tilde{j}!$ be the Gysin map of $\tilde{j}$ from G-vector bundles over $\tilde{\nu}_X^* \oplus \rho^*(\nu^*)$ to those over $\tilde{\nu}_{X \times M}^* \oplus \pi^*(\nu^*)$ and put $\tilde{\xi} = \tilde{j}!(\xi)$. Let τ_m be the cotangent bundle $T^*(\pi^{-1}(m))$ of $\pi^{-1}(m)$ for each $m \in M$ and put $\tau = \cup_{m \in M} \tau_m$. Since π is a submersion, the G-space $\tilde{\nu}_{X \times M}^* \oplus \pi^*(\nu^*)$ is the total space of the bundle over τ under the canonical projection $\tilde{\pi}$ whose fibres are $\nu^* \oplus \nu^*$. Therefore it follows from Thom–Gysin's theorem that $\tilde{\xi}$ can be viewed as a G-vector bundle over τ under the Gysin map $\tilde{\pi}!$ of $\tilde{\pi}$. Let $\tilde{\xi}_m = \tilde{\xi} \mid \tau_m$ be the restriction of $\tilde{\xi}$ to τ_m. By the definition of $\tilde{\xi}$, there are elliptic differential operators D_m on $\pi^{-1}(m)$ whose symbols $\sigma(D_m)$ are equal to $\tilde{\xi}_m$ themselves. Let D be the G-equivariant field of $\{D_m\}_{m \in M}$ which is considered as a G-invariant differential operator on $X \times M$ such that

(1) D_m are elliptic on $\pi^{-1}(m)$ $(m \in M)$,

(2) $\tilde{\xi}$ is the symbol $\sigma(D)$ of D.

Let us take the K-theoretic index ind D of D in $K_a(M, F)$ by

$$\operatorname{ind} D = [\operatorname{Ker} D] - [\operatorname{Ker} D^*]$$

where $[\cdot]$ means a $C_r^*(M, F)$-module generated by $\cdot$. We then define a homomorphism μ from $K_g(M, F)$ into $K_a(M, F)$ by

$$\mu(X, \xi) = \operatorname{ind} D.$$

Actually, it depends only on the equivalence class of (X, ξ). We now state the first Baum–Connes conjecture as follows:

Baum–Connes Conjecture I. *Given a foliated manifold (M, F), the K-index map μ is an isomorphism from $K_g(M, F)$ onto $K_a(M, F)$.*

On the other hand, let (M, G, φ) be a smooth dynamical system where φ is free. Then the family F of all G-orbits becomes a foliation of M, whose C^*-algebra $C_r^*(M, F)$ is nothing but the C^*-crossed product $C(M) \times_\varphi G$ of $C(M)$ of φ. It then implies $K_a(M, F) = K(C(M) \times_\varphi G)$. Moreover $K_g(M, F)$ is isomorphic to the abelian group $K_g(M, G)$ defined by the following manner: Let $\Gamma(M, G)$ be the set of all triplets (X, ξ, π) where X is a proper G-manifold, π is a G-map from X to M and ξ is a G-vector bundle over $T^*(X) \oplus \pi^*(T^*(M))$. Then it has a similar equivalence relation as before. Namely, $(X_1, \xi_1, \pi_1) \sim (X_2, \xi_2, \pi_2)$ if and only if there exist a proper G-manifold X and G-mappings π, ρ_j such that

$$(1) \quad \pi_j = \pi \cdot \rho_j \quad (j = 1, 2) \quad \text{and} \quad (2) \quad \rho_1!(\xi_1) = \rho_2!(\xi_2)$$

where $\rho_j!$ are the Thom–Gysin mappings from the groups generated by all G-vector bundles over $T^*(X_j) \oplus \pi_j^*(T^*(M))$ to the group by those over $T^*(X) \oplus \pi^*(T^*(M))$. We denote by $K_g(M, G)$ the set of all equivalence classes in $\Gamma(M, G)$ with respect to $\sim$. Then it is an abelian group with the canonical sum. Thanks to the conjecture I, we also state the second conjecture due to Baum–Connes [1]:

Baum–Connes Conjecture II. *Given any smooth dynamical system (M, G, φ), the K-index map μ is an isomorphism from $K_g(M, G)$ to $K_a(M, G) = K(C(M) \rtimes_\varphi G)$.*

Remark. Let BG be the classifying space of G and EG the total space of the universal principal G-bundle over BG. Let τ be the vector bundle over $M \times_G EG$ whose fibers are $T^*(M)$. We denote by $K^\tau(M \times_G EG)$ the K-group $K(B\tau/S\tau)$ of the Thom space $B\tau/S\tau$ of τ. Then there exists a homomorphism δ from $K^\tau(M \times_G EG)$ into $K_g(M, G)$ with the property that $\mu \cdot \delta$ is the Kasparov β-map when M is a point. Suppose G is discrete, then δ is $\mathbf{Q}$-injective. If G is torsion free, then δ is bijective.

3. Miscellaneous Results

Let (A, G, α) be a C^*-dynamical system where G is a simply connected solvable Lie group. By the result of Iwasawa, G is a multisemidirect product of $\mathbb{R}$. Using the duality for C^*-crossed products, Connes showed that $K_i(A \times_a \mathbb{R})$ is isomorphic to $K_{i+1}(A)$ under a Thom isomorphism. Since crossed products are compatible with semidirect products, the next theorem then follows.

Theorem 3.1 (Connes). *Let (A, G, α) be a C^*-dynamical system where G is a simply connected solvable Lie group. Then $K_i(A \rtimes_\alpha G)$ is isomorphic to $K_{i+\dim G}(A)$ $(i = 1, 2)$.*

Given any dynamical system (M, G, φ) where G is simply connected solvable, it follows from Theorem 3.1 that $K_a(M, G)$ is isomorphic to $K_{\dim G}(C(M))$ via the Thom map. On the other hand, since G is simply connected, it implies from Baum–Connes [1] that $K_g(M, G)$ is equal to the K-group $K(B\tau/S\tau)$ of the Thom space $B\tau/S\tau$ of $\tau = T^*(M) \times_G EG$. By the assumption of G, there exists a strong retraction from $M \times_G EG$ to $M \times \mathbb{R}^{\dim G}$ with respect to which $K_g(M, G)$ is isomorphic to $K^{\dim G}(M)$. We then have the following corollary:

Corollary 3.2. *Let (M, G, φ) be a differentiably dynamical system* *where G is a simply connected solvable Lie group. Then the conjecture* II *holds for the triplet.*

Suppose G is a compact Lie group, then the situation is quite simple. In other words, the conjecture II is nothing more than the Atiyah–Singer index theory:

Proposition 3.3 (Atiyah–Singer). *Let (M, G, φ) be a differentiably dynamical system where G is a compact Lie group. Then the conjecture* II *is affirmative.*

Thanks to the proposition cited above, we next study the case where G is a noncompact semisimple Lie group. Let H be its maximal compact subgroup. If G/H has a G-invariant spin^c structure, we can see due to Baum–Connes [1] that

$$K_g(M, G) = K_g^{\dim G/H}(M, H).$$

By Proposition 3.3, it follows that

$$K_g(M, H) = K_a(M, H)$$

up to the K-index map. Thus it suffices to show that

$$K_a(M, G) = K_a^{\dim G/H}(M, H).$$

The following result is one case supporting the above equality:

Proposition 3.4 (Kasparov). *Let G be a connected Lie group and H its maximal compact subgroup with the property that G/H has a G-invariant* spin^c *structure. If there is an amenable normal subgroup L of G such that G/L is locally isomorphic to the finite product of* $\mathrm{SO}_0(n, 1)$ *and compact groups, then*

$$K_a(M, G) = K_a^{\dim G/H}(M, H).$$

In particular, suppose M is one point, then the conjecture II may be verified affirmatively for wider classes of G.

Proposition 3.5 (A. Wasserman). *Let G be a connected reductive Lie group and H its maximal compact subgroup such that G/H has a G-invariant* spin^c *structure. Then we have that*

$$K_a(pt, G) = K_a^{\dim G/H}(pt, H).$$

When G is a discrete group, there is no theorem supporting the conjecture II affirmatively except finite groups at the moment. The following examples are important and of independent interest due to Kasparov [8], Natsume [11] and Pimsner-Voiculescu [13] respectively.

Proposition 3.6. *$K_g(pt, G) = K_a(pt, G)$ up to the K-index map for $G =$ the fundamental group $\pi_1(M_g)$ of the compact Riemann surface M_g with genus $g \geq 2$, $SL(2, \mathbf{Z})$ or the free group F_n with n-generators ($n \geq 2$).*

We now discuss the conjecture I which is verified affirmatively only for few cases. In what follows, we list up all the examples that we know:

Proposition 3.7 (Torpe). *The Reeb foliations on 2-torus or 3-sphere enjoy the conjecture* I *affirmatively.*

Proposition 3.8 ([16], [17]). *Conjecture* I *holds for the foliations from Anosov flows on infrahomogeneous manifolds.*

Suppose an ambient manifold has an Anosov flow, its rank is one automatically. The next case supports Conjecture I positively for foliated manifolds with an arbitrary rank:

Proposition 3.9 ([18] or §4). *Given any $n \geq 1$, there exists a foliated manifold (M_n, F_n) such that*

$$(1) \quad \operatorname{rank} M_n = n, \; GV(F_n) \neq 0 \quad \textit{and} \quad (2) \quad K_g(M_n, F_n) = K_a(M_n, F_n)$$

where $GV(F_n)$ is the Godvillon–Vey class of F_n.

Although the foliations cited above have nontrivial holonomy in general, the next two cases are examples without holonomy:

Proposition 3.10 (Natsume). *Conjecture* I *is true for all codimension one foliations without holonomy on smooth manifolds.*

Proposition 3.11 ([16]). *Conjecture* I *is affirmative for the foliations derived from the Anosov diffeomorphisms on infrahomogeneous manifolds.*

Proposition 3.12 (Connes). *Conjecture* I *holds for orbital foliations derived from smooth free actions of simply connected solvable Lie groups on smooth manifolds.*

Question 1. Given a K-oriented foliation whose leaves are contractible, does Conjecture I hold affirmatively?

Question 2. Can we show Conjecture I for all foliated bundles?

Question 3. Is it true that Conjecture I holds for all foliated manifolds whose fundamental groups are $\mathrm{SL}(n, \mathbf{Z})$?

4. Generalized Anosov Foliations

In this section we shall examine Conjecture I for generalized Anosov foliations on infrahomogeneous manifolds.

Let (M, G, φ) be a differentiably dynamical system. The action φ is called *Anosov* if there exist an element $g \in G$ and subbundles E^i ($i = s, u, c$) of $T(M)$ such that

(1) $T(M) = E^s \oplus E^c \oplus E^u$, $d\varphi_g(E^i) = E^i$,

(2) E^i are integrable, $E^c = T(\varphi(G))$ ($i = s, u$),

(3) $\|d\varphi_g^n(\xi)\| \leq \lambda^n \|\xi\|$ ($\xi \in E^s$), $\mu^n \|\xi\| \leq \|d\varphi_g^n(\xi)\|$ ($\xi \in E^u$), $\lambda^n \|\xi\| \leq \|d\varphi_g^n(\xi)\| \leq \mu^n \|\xi\|$ ($\xi \in E^c$) ($n \geq 0$) for some $0 < \lambda < 1 < \mu$.

Then there exist foliations F^i of M such that $E^i = T(f^i)$ for $i = s, u, c$. Each leaf W_x^i in F^i ($i = s, u, c$) is given by the following

$$\begin{aligned} W_x^s &= \{y \in M \mid d(\varphi_g^n(x), \varphi_g^n(y))\lambda^{-n} \to 0 \quad (n \to +\infty)\}, \\ W_x^u &= \{y \in M \mid d(\varphi_g^{-n}(x), \varphi_g^{-n}(y))\mu^n \to 0 \quad (n \to +\infty)\}, \\ W_x^c &= \varphi(G)x. \end{aligned}$$

Let us now take a connected semisimple Lie group G with finite center and H its maximal compact subgroup. We denote by G' the Lie algebra of G. Let $G' = H' + P'$ be a Cartan decomposition of G' and A' a maximal abelian subalgebra of P'. If Λ is the root system of A', then we have the root space decomposition of G' with respect to Λ as follows:

$$G' = M' \oplus A' \oplus \sum_{\lambda \in \Lambda} G'_\lambda,$$

where M' is the centralizer of A' in H' and G'_λ is the λ-eigenspace of $\mathrm{ad}(A')$ in G'. Given a regular element $a \in A = \exp A'$, we define two subsets $\Lambda_a^{+(-)}$ of Λ by

$$\Lambda_a^{+(-)} = \{\lambda \in \Lambda \mid \lambda(\log a) > 0\ (< 0)\}$$

respectively where $\log a$ is the element of A' such that $\exp(\log a) = a$. Let us define $N'_{+(-)}$ as the direct sum of G'_λ $(\lambda \in \Lambda_a^{+(-)})$ respectively. Let $M = \exp M'$ and consider the diffeomorphism φ_a of G/M defined by $\varphi_a(gM) = gaM$. Then we have the following lemma:

Lemma 4.1. *The action φ is Anosov of A on G/M whose E^s_M, E^u_M are N'_+, N'_- respectively.*

Remark. If $a \in A$ is singular, then the decomposition of G'/M' with respect to a is obtained as follows:

$$G'/M' = N'_- \oplus A' \oplus N'_+ \oplus \sum_{\lambda(\log a)=0} G'_\lambda.$$

Therefore $d\varphi_a$ is no longer Anosov in general.

Let Γ be a torsion free uniform lattice of G and define an action ψ of A on $\Gamma \setminus G/M$ by $\psi_a(\Gamma gM) = \Gamma\varphi_a(gM) = \Gamma gaM$. We then show the following lemma:

Lemma 4.2. *The action ψ is Anosov of A on $\Gamma \setminus G/M$.*

Except for the foliations F^j of $\Gamma\setminus G/M$ with respect to ψ $(j = s, u, c)$, there exist other foliations F^j $(j = sc, uc)$ such that

$$T(F^{jc}) = E^j \oplus E^c \quad (j = s, u).$$

Each leaf $W^j_x \in F^j$ $(j = sc, uc)$ has the following form:

$$W^{jc}_x = \bigcup_{y\in\varphi(G)x} W^j_y \quad (j = s, u).$$

Let π_Γ be the canonical projection from G/M onto $\Gamma \setminus G/M$ and put $P^{+(-)} = MAN^{+(-)}$. Identifying G/M with $(G/H) \times (G/P^{+(-)})$ by the map $gM \to (gH, gP^{+(-)})$, we obtain the following lemma:

Lemma 4.3. *The Anosov system $(\Gamma \setminus G/M, A, \psi)$ lifts five foliations F^j $(j = s, u, c, sc, uc)$ whose leaves $W^j_{\Gamma gM}$ are given by*

$$W^s_{\Gamma gM} = \pi_\Gamma((N^+H/H) \times \{gP^+\}) = \Gamma \setminus (N^+H/H) \times \{gP^+\}),$$
$$W^u_{\Gamma gM} = \Gamma \setminus ((N^-H/H) \times \{gP^-\}), \quad W^c_{\Gamma gM} = \psi(A)(\Gamma gM),$$
$$W^{sc}_{\Gamma gM} = \Gamma \setminus ((G/H) \times \{gP^+\}), \quad W^{uc}_{\Gamma gM} = \Gamma \setminus ((G/H) \times \{gP^-\}).$$

Remark. The next observation can be viewed as a geometric approach to the above lemma. According to Oshima [12], there exists a real analytic closed manifold $(G/H)^\sim$ containing G/H as an open submanifold and G/P^+ as the boundary of G/H. For the Iwasawa decomposition $G = N^-AH$, we know that $N^- \times \mathbb{R}^\ell$ can be embedded in $(G/H)^\sim$ and $N^- \times \mathbb{R}^\ell_+$ is diffeomorphic to G/H via the map defined by

$$(n^-, \exp -\lambda_1(\log a), \cdots, \exp -\lambda_\ell(\log a)) \to n^- aK$$

where $\ell = \operatorname{rank}_{\mathbb{R}} G$ and $\{\lambda_j\}_{j=1}^{\ell}$ is a restricted positive simple root system of Λ. Moreover G/P^+ is identified with $N^- \times \{0\}^\ell$. Using the fact that $\{g \exp(t \log a)H\}_{t\geq 0}$ and $\{\exp(t \log a)H\}_{t\geq 0}$ are asymptotically approaching each other ($t \to +\infty$) if and only if $gM \in W^{sc}_{hM}$. On the other hand, W^s_{gM} are interpreted as the horosphere whose boundary passes through gP^+. The leaves W^{uc}, W^u are similarly translated as W^{sc}, W^s.

We then study the structure of F^j ($j = s, u, sc, uc$) in more detail. Since $G/M = (G/H) \times (G/P^{+(-)})$ G-equivariantly, it follows that $\Gamma \backslash G/M$ is a $G/P^{+(-)}$-bundle over $\Gamma \backslash G/H$. Applying Lemma 4.3, we deduce the following lemma:

Lemma 4.4. *The foliated manifold* $(\Gamma\backslash G/M, F^{sc})$ *is a foliated* G/P^+*-bundle over* $\Gamma \backslash G/H$ *whose holonomy group is the image of the left translation action* λ *of* Γ *on* G/P^+*. The same is true for* $(\Gamma \backslash G/M, F^{uc})$ *and* P^-*.*

Let us now consider the principal M-bundle $\Gamma \backslash G$ over $\Gamma \backslash G/M$ and π_M its canonical projection. Applying Lemma 4.3, we can show the next lemma:

Lemma 4.5. *The pull back foliations* $\pi_M^*(F^s)$, $\pi_M^*(F^u)$ *of* F^s, F^u *by* π_M *are* MN^+, MN^-*-orbital by the right translation action* ρ *of* G *on* $\Gamma \backslash G$ *respectively.*

Since Hausdorff are the holonomy groupoids of F^j ($j = s, u, sc, uc$), we have the next result using Lemma 4.4 and [16]:

Lemma 4.6. *Concerning* $(\Gamma \backslash G/M, F^j)$ $(j = sc, uc)$,

$$C^*_r(\Gamma \backslash G/M, F^{sc}) = (C(G/P^+) \times_\lambda \Gamma)_r \otimes BC(L^2(\Gamma \backslash G/H)),$$
$$C^*_r(\Gamma \backslash G/M, F^{uc}) = (C(G/P^-) \times_\lambda \Gamma)_r \otimes BC(L^2(\Gamma \backslash G/H))$$

up to isomorphisms where $(\cdot)_r$ *means reduced crossed products and* $BC(H)$ *is the* C^**-algebra of all compact operators on* H.

By Rieffel's work on Morita equivalence, $(C(G/P^+) \times_\lambda \Gamma)_r$ is stably isomorphic to $(C(\Gamma \backslash G) \times_\rho P^+)_r$. Since P^+ is amenable, the latter is nothing but $C(\Gamma \setminus G) \times_\rho P^+$. As $N^- = \theta(N^+)$ for the Cartan involution θ of G, it follows that $C_r^*(\Gamma \setminus G/M, F^{sc})$ is stably isomorphic to $C_r^*(\Gamma \setminus M, F^{uc})$. By Hilsum–Skandalis [6], we have the following lemma:

Lemma 4.7. *If* F^j *are nontrivial* ($j = sc, uc$), $C_r^*(\Gamma \setminus G/M, F^{sc})$ *is isomorphic to* $C_r^*(\Gamma \setminus G/M, F^{uc})$.

By Lemma 4.5, we also see the next lemma:

Lemma 4.8. *Concerning* $(\Gamma \setminus G, \pi_M^*(F^j))$ $(j = s, u)$, *we deduce that* $C_r^*(\Gamma \setminus G, \pi_M^*(F^{s(u)}))$ *is isomorphic to* $C(\Gamma \setminus G) \times_\rho MN^{+(-)}$ *resp.*

Let (X, F) be a foliated manifold and ξ a bundle over X whose fibers are a compact manifold C. Let us take the pull back $\pi^*(F)$ of F by the projection π from ξ to X. We then check the following lemma:

Lemma 4.9. $C_r^*(\xi, \pi^*(F))$ *is isomorphic to* $C_r^*(M, F) \otimes BC(L^2(X))$.

Combining Lemma 4.8 and 4.9, the next observation is naturally deduced:

Lemma 4.10. *Concerning* $(\Gamma \setminus G/M, F^j)$ $(j = s, u)$,

$$C_r^*(\Gamma \setminus G/M, F^{s(u)}) \otimes BC(L^2(M)) = C(\Gamma \setminus G) \times_\rho MN^{+(-)}$$

up to isomorphisms respectively.

Applying Hilsum–Skandalis [6] again, it follows from Lemma 4.10 that

Corollary 4.11. *If* F^j *are nontrivial* ($j = s, u$), $C_r^*(\Gamma \setminus G/M, F^s)$ *is isomorphic to* $C_r^*(\Gamma \setminus G/M, F^u)$.

In what follows, we shall compute $K_a(\Gamma \setminus G/M, F^j)$ $(j = s, u, sc, uc)$ using Lemma 4.9∼4.10. It certainly implies that

$$K_a(\Gamma \setminus G/M, F^{s(u)}) = K_a(\Gamma \setminus G, MN^{+(-)})$$

$$K_a(\Gamma \setminus G/M, F^{s(u)c}) = K_a(G/P^{+(-)}, \Gamma) = K_a(\Gamma \setminus G, P^{+(-)})$$

respectively.

To analyze the right-hand sides of the above equalities, we prepare a generalized Thom isomorphism essentially due to Kasparov [7].

Lemma 4.12. *Let (A, G, α) be a C^*-dynamical system where G is the semidirect product $\mathbb{R}^n \rtimes_s C$ of $\mathbb{R}^n$ by a compact group C. Then there exists a Thom isomorphism between $K_a(A, G)$ and $K^n_{a,C}(A)$ where $K_{a,C}$ means the analytic C-equivariant K-theory.*

Remark. If C is trivial, the above lemma is due to Connes and if $n = 0$, it is due to Julg.

Since $P^{+(-)} = N^{+(-)} \rtimes_s MA$ respectively, Lemma 4.12 tells us that

$$\begin{aligned} K_a(\Gamma \setminus G, P^{+(-)}) &= K_a(C(\Gamma \setminus G) \times_\rho MN^{+(-)}, A) \\ &= K_a^{\dim A}(\Gamma \setminus G, MN^{+(-)}) \\ &= K_a^{\dim AN^{+(-)}}(\Gamma \setminus G) \end{aligned}$$

respectively. Since Γ is torsion free, $M \cap \Gamma = \{e\}$. Hence ρ is a free action of M on $\Gamma \setminus G$. By Segal [19], we have that

$$K_{a,M}^{\dim AN^{+(-)}}(\Gamma \setminus G) = K^{\dim AN^{+(-)}}(\Gamma \setminus G/M)$$

respectively. This implies the next lemma:

Lemma 4.13. *$K_a(\Gamma \setminus G/M, F^{s(u)c}) = K^{\dim AN^{+(-)}}(\Gamma \setminus G/M)$ up to Thom isomorphisms respectively.*

Similarly it follows from Lemma 4.10 and 4.11 that

Lemma 4.14. *$K_a(\Gamma \setminus G/M, F^{s(u)}) = K^{\dim N^{+(-)}}(\Gamma \setminus G/M)$ up to Thom isomorphisms respectively.*

We shall next compute the geometric K-theory $K_g(\Gamma \setminus G/M, F^{s(u)c})$ of $(\Gamma \setminus G/M, F^{s(u)c})$. Let us look at the leave structure of $F^{s(u)c}$ in what follows. Since we know that G/H is contractible and

$$W^{s(u)c}_{\Gamma gM} = \Gamma \setminus ((G/H) \times \{gP^{+(-)}\}) \quad (g \in G)$$

respectively by Lemma 4.6, all of them are $K(\Gamma, 1)$-spaces. Since Γ is torsion free, so are $\tilde{G}_j = \mathrm{Hol}(F^j)$ $(j = sc, uc)$ respectively. Let τ^j be the vector bundle over $B\tilde{G}_j$ whose fibers are the dual normal bundle ν^* of F^j $(j = sc, uc)$ respectively. We obtain from Baum–Connes [1] that

Lemma 4.15. *$K_g(\Gamma \backslash G/M, F^j) = K^{\tau^j}(B\tilde{G}_j)$ $(j = sc, uc)$ resp.*

By definition, $\tilde{G}_{s(u)c}$ is isomorphic to $(G/P^{+(-)}) \times_\lambda \Gamma) \times (B\Gamma \times B\Gamma)$ as a Borel groupoid by Natsume–Takai respectively. However they are no longer isomorphic to each other as a topological groupoid in general. Let us study this correspondence more carefully. Consider the map Φ from $\tilde{G}_{s(u)c}$ to $B\Gamma \times B\Gamma$ by taking $\Phi(\gamma) = (\pi_{B\Gamma} \cdot s(\gamma), \pi_{B\Gamma} \cdot r(\gamma))$. Then one checks that the groupoids $\Phi^{-1}(x, y)$ $(x, y \in B\Gamma)$ are isomorphic to the principal one $(G/P^{+(-)}) \times_\lambda \Gamma$ respectively. Namely we have that

$$(G/P^{+(-)}) \times_\lambda \Gamma \xrightarrow{\iota} \tilde{G}_{s(u)c} \xrightarrow{\Phi} B\Gamma \times B\Gamma$$

respectively. Taking the classifying spaces of the above spaces, we have that

$$B((G/P^{+(-)}) \times_\lambda \Gamma) \xrightarrow{B\iota} B\tilde{G}_{s(u)c} \xrightarrow{B\Phi} B(B\Gamma \times B\Gamma)$$

respectively. Since $B(B\Gamma \times B\Gamma)$ is homotopic to a point, we know that $B\tilde{G}_{s(u)c}$ are homotopic to $B((G/P^{+(-)}) \times_\lambda \Gamma)$ under $B\iota$ respectively. Let σ be the pull back $(B\iota)^*(\tau)$ of τ by $B\iota$. Since $B\iota$ is a homotopic isomorphism, we obtain the following lemma:

Lemma 4.16. *$K^\sigma(B((G/P^{+(-)}) \times_\lambda \Gamma)) = K^{\tau^{s(u)c}}(B\tilde{G}_{s(u)c})$ respectively.*

By definition, ν^* is equal to $T^*(G/P^{+(-)})$ resp. Let $E\Gamma$ be the total space of the universal principal Γ-bundle over $E\Gamma$. Since we see that $(G/P^i) \times_\Gamma E\Gamma$ are the base spaces of principal $(G/P^i) \times_\lambda \Gamma$-bundles, there exist the classifying maps f^i from $(G/P^i) \times_\Gamma E\Gamma$ to $B((G/P^i) \times_\lambda \Gamma)$ which realize the above bundles $(i = +, -)$ respectively. Let us take the pull backs $f^{i*}(\sigma)$ of σ by f^i $(i = +, -)$. By definition, it can be shown that $E((G/P^i) \times_\lambda \Gamma)$ is nothing but $(G/P^i) \times E\Gamma$ up to Γ-equivariant homotopy $(i = +, -)$ (cf. Segal [21]). Therefore f^i are homotopic to id. We then have the following lemma:

Lemma 4.17. *$K^\sigma(B((G/P^i) \times_\lambda \Gamma)) = K^{f^{i*}(\sigma)}((G/P^i) \times_\Gamma E\Gamma)$ $(i = +, -)$ respectively.*

Since $\nu^* = T^*(G/P^i)$ $(i = +, -)$, it follows that

Lemma 4.18. $K^{f^i_{ast}(\sigma)}((G/P^i) \times_\Gamma E\Gamma) = K^{\delta^i}((G/P^i) \times_\Gamma E\Gamma)$ ($i = +,-$) *respectively where* $\delta^i = T^*(G/P^i) \times_\Gamma E\Gamma$.

Combining Lemmas 4.15∼4.18, we obtain the next lemma:

Lemma 4.19. $K_g(\Gamma \setminus G/M, F^{s(u)c}) = K_g(G/P^{+(-)}, \Gamma)$ *respectively.*

Let H_i be two closed subgroups of G ($i = 1,2$). We compare the geometric K-groups $K_g(G/H_1, H_2)$ and $K_g(H_2 \setminus G, H_1)$ of $(G/H_1, H_2, \lambda)$ and $(H_2 \setminus G, H_1, \rho)$ respectively. We expect the same property as in the case of K_a. Actually, we have the following lemma:

Lemma 4.20. *If H_1 is amenable and H_2 is torsion free with $H_1 \cap H_2 = \varphi$, then*

$$K_g(G/H_1, H_2) = K_g(H_2 \setminus G, H_1)$$

up to isomorphisms.

Applying the above lemma to $H_1 = P^i$ ($i = +,-$) and $H_2 = \Gamma$, then we obtain by Lemma 4.19 that

Lemma 4.21. $K_g(\Gamma \setminus G/M, F^{s(u)c}) = K_g(\Gamma \setminus G, P^{+(-)})$ *respectively.*

Since $P^i = N^i \times_s MA$ ($i = +,-$), it follows from the next lemma which can be viewed as equivariant Thom isomorphisms:

Lemma 4.22. $K_g(\Gamma \setminus G, P^i) K_M^{\dim AN^i}(\Gamma \setminus G)$ ($i = +,-$) *respectively.*

Summing up all the arguments discussed above, we conclude the main theorem:

Theorem 4.23. *The conjecture* I *is affirmative for the foliated manifolds* $(\Gamma \setminus G/M, F^j)$ $(j = s, u, c, sc, uc)$.

In fact, the other cases are also computed by the same method.

Remark. The above theorem can be applicable to the cases where Kamber–Tondeur [22] discussed the Godvillon–Vey classes of foliated manifolds.

REFERENCES

[1] P. Baum and A. Connes, *Geometric K-theory for Lie groups and foliations,* preprint (1982).
[2] P. Baum and A. Connes, *Chern character for discrete groups,* A Fete de Topology, (1988).
[3] A. Connes, *A survey of foliations and operator algebras,* Proc. Symp. Pure Math. **38** (1982), Part 1, 521–628.
[4] A. Connes, *Noncommutative differential geometry,* Chapter I, II, Publ. Math. IHES **62** (1986), 257–360.
[5] A. Connes, *Cyclic cohomology and the transverse fundamental class of a foliation,* Pitman Res. Notes Math. Ser. **123** (1986), 52–144.
[6] M. Hilsum and G. Skandalis, *Stabilité des C^*-algebres de feuilletage,* Ann. Inst. Fourier, Grenoble 33, **3** (1983), 201–208.
[7] G.G. Kasparov, *Group C^*-algebras and higher signatures,* Chernogolovka, preprint (1981).
[8] G.G. Kasparov, *The index of invariant elliptic operators, K-theory and Lie group representations,* preprint (1982).
[9] A.S. Miscenko and T. Fomenko, *The index of elliptic operators over C^*-algebras,* Izv. Akad. Nauk, Ser. Math. **43** (1979), 831–859.
[10] T. Natsume, *Topological K-theory for codimension* 1 *foliations without holonomy,* Adv. Stud. Pure Math. **5** (1985), 15–27.
[11] T. Natsume, *On* $K_*(C^*(\mathrm{SL}_2(\mathbf{Z})))$, J. Operator Theory **13** (1985), 103–118.
[12] T. Oshima, *A realization of Riemannian symmetric spaces,* J. Math. Soc. Japan **30** (1978), 117–132.
[13] J. Rosenberg, *C^*-algebras, positive scalar curvature and the Novikov conjecture,* Publ. Math. IHES. **58** (1983), 197–212.
[14] J. Rosenberg, *C^*-algebras, positive scalar curvature and the Novikov conjecture* II, Res. Notes Math. Pitman **123** (1986).
[15] J. Rosenberg, *C^*-algebras, positive scalar curvature and the Novikov conjecture* III, Topology **25** (1986), 319–336.
[16] H. Takai, *C^*-algebras of Anosov foliations,* Lec. Notes Math., Springer **1132** (1985), 509–516.
[17] H. Takai, *KK-theory for the C^*-algebras of Anosov foliations,* Res. Notes Math. Ser., Pitman (1986), 387–399.
[18] H. Takai, *Baum–Connes conjectures and their applications,* World Sci. Adv. Ser. Dynamical Systems **5** (1987), 89–116.
[19] A.M. Torpe, *K-theory for the leaf space of foliations by Reeb components,* Jour. Func. Anal. **61** (1985), 15–71.
[20] G. Segal, *Equivariant K-theory,* Publ. Math. IHES. **34** (1968), 129–151.

[21] G. Segal, *Classifying spaces and spectral sequences,* Publ. Math. IHES. **34** (1968), 105–112.

[22] F.W. Kamber and P. Tondeur, *Foliated bundles and characteristic classes,* Lec. Notes Math., Springer **493** (1975).

Department of Mathematics
Tokyo Metropolitan University
Japan

On Primitive Ideal Spaces of C*-Algebras over Certain Locally Compact Groupoids

SHIGERU YAMAGAMI

Abstract

Let Γ be a locally compact Hausdorff second countable groupoid with a left Haar system $\{\nu^x\}_{x\in X}$ in the sense of [9] ($X =$ the unit space of Γ). By analogy with Fell's algebraic bundles over groups, we define the notion of C^*-algebras over Γ and, given a C^*-algebra A over Γ, we can form a C^*-algebra $C^*(\Gamma, A)$ as the completion of the cross sectional algebra of A. In this note, under some stringent assumptions on Γ, we present a concrete realization of the primitive ideal space of $C^*(\Gamma, A)$. This is a C^*-version of [12].

1. Assumptions

We impose a series of conditions on Γ. First we assume that

(A1) Γ is minimal, i.e., for all $x \in X$, $[x] \equiv \{y \in X$; there exists $\gamma \in \Gamma$ such that $s(\gamma) = x$, $t(\gamma) = y\}$ is dense in X.

(A2) Stabilizers of Γ are discrete and uniformly abelian, i.e., there are a discrete abelian group G and a family of topological isomorphisms $\{\iota_x : G \to \Gamma_x^x\}_{x\in X}$ such that for $g \in G$, $X \ni x \mapsto \iota_x(g) \in \Gamma$ is continuous and the equality $\iota_{t(\gamma)}(g)\gamma = \gamma\iota_{s(\gamma)}(g)$ holds for $\gamma \in \Gamma$.

(A3) Γ is amenable, i.e., for any quasi-invariant measure μ in X, there is a net $\{f_i\}_{i\in I}$ of functions in $C_c(\Gamma)$ such that

(i) for all $x \in X$ and all $i \in I$, $\int \nu^x(d\gamma)|f_i(\gamma)|^2 = 1$.

(ii) the function $\gamma \mapsto \int \nu^{s(\gamma)}(d\gamma')f_i(\gamma')f_i(\gamma\gamma')$ converges to 1 as $i \to \infty$, relative to the weak* topology of $L^\infty(\Gamma, \mu \circ \nu)$.

Comment. The amenability in (A3) is due to [9]. (A2) means that Γ is a central extension of a principal groupoid by an abelian group. We write $\iota_{t(\gamma)}(g)\gamma$ and $\gamma\iota_{s(\gamma)}(g)$ as $g\gamma$ and γg in the following.

Definition 1. Let $A = \{A_\gamma\}_{\gamma\in\Gamma}$ be a separable field of Banach spaces over Γ. A is called a C^*-algebra over Γ if the following structures are specified:

(i) For $(\gamma_1, \gamma_2) \in \Gamma$, a multiplication $A_{\gamma_1} \times A_{\gamma_2} \ni (a_1, a_2) \mapsto a_1 a_2 \in A_{\gamma_1\gamma_2}$ is defined. It is associative and satisfies $\|a_1 a_2\| \le \|a_1\|\,\|a_2\|$.

(ii) For $\gamma \in \Gamma$, a conjugate linear map ${}^*\colon A_\gamma \to A_{\gamma^{-1}}$ is defined and satisfies $(a^*)^* = a$, $\|a^* a\| = \|a\|^2$, and $(a_1 a_2)^* = a_2^* a_1^*$.

(iii) For $a \in A_\gamma$, $a^* a$ is a positive element in $A_{s(\gamma)}$.

(iv) If φ_1, φ_2 are continuous sections of A, then

$$\Gamma^{(2)} \ni (\gamma_1, \gamma_2) \mapsto \varphi_1(\gamma_1)\varphi_2(\gamma_2) \in A_{\gamma_1\gamma_2}$$

is a continuous section of $m^*(A)$, where $m^*(A)$ is the pull back of A under the composition map $m\colon \Gamma^{(2)} \to \Gamma$.

(v) If φ is a continuous section of A, $\varphi^*(\gamma) = \varphi(\gamma^{-1})^*$ is also a continuous section of A.

Let $C_c(\Gamma, A)$ be the set of continuous sections of A with compact supports. We make $C_c(\Gamma, A)$ into a *-algebra in the following way:

$$(\varphi_1\varphi_2)(\gamma) = \int \nu^{t(\gamma)}(d\gamma')\varphi_1(\gamma')\varphi_2(\gamma'^{-1}\gamma),$$
$$\varphi^*(\gamma) = \varphi(\gamma^{-1})^*.$$

$C_c(\Gamma, A)$ is completed to the C^*-algebra $C^*(\Gamma, A)$ according to the maximal continuous C^*-norm (the topology of $C_c(\Gamma, A)$ is that of uniform convergence on compact subsets). Analogously as in [3], we define the multiplier bundle $\{M(A_\gamma)\}_{\gamma\in\Gamma}$. We assume

(A4) for all $\gamma \in \Gamma$, $U(A_\gamma) \equiv \{u \in M(A_\gamma); b^*b = 1_{s(\gamma)}, bb* = 1_{t(\gamma)}\}$ is not empty and there is a Borel section of $\{U(A_\gamma)\}_{\gamma\in\Gamma}$ in the sense that we can find a section $u(\gamma) \in U(A_\gamma)$ such that for any continuous section a of A, $\Gamma \ni \gamma \mapsto u(\gamma)^* a(\gamma) \in A_{s(\gamma)}$ is a Borel section.

Set $\Gamma(X) = \bigcup_{x\in X} \Gamma_x^x$. $\Gamma(X)$ is a closed subgroupoid of Γ and the restriction of A to $\Gamma(X)$ is a C^*-algebra over $\Gamma(X)$, so we can form the cross sectional C^*-algebra as in the above definition, which is denoted by $C^*(\Gamma(X), A)$. Similarly, from the restriction of A to Γ_x^x, we have a C^*-algebra $C^*(\Gamma_x^x, A)$. Let $G^*(x) = \operatorname{Prim} C^*(\Gamma_x^x, A)$ and set $G^* = \bigcup_{x\in X} G^*(x)$ (disjoint union).

Lemma 1 (cf. [4]).

$$\operatorname{Prim} C^*(\Gamma(X), A) = G^*.$$

The topology of G^* is defined by this identification. Let $u \in U(A_\gamma)$ and define an isomorphism from $C^*(\Gamma_x^x, A)$ onto $C^*(\Gamma_y^y, A)$ $(x = s(\gamma),\ y = t(\gamma))$ by $C_c(\Gamma_x^x, A) \ni \varphi \mapsto u\varphi \in C_c(\Gamma_y^y, A)$ where $(u\varphi)(\gamma') = u\varphi(\gamma^{-1}\gamma'\gamma)u^*$.

Now we define an equivalence relation $\simeq$ in G^* by $\omega_1 \simeq \omega_2 \Leftrightarrow$ there exists $\gamma \in \Gamma$ and $u \in U(A_\gamma)$ such that $\omega_1 \in G^*(s(\gamma))$, $\omega_2 \in G^*(t(\gamma))$, and $\omega_2 = u\omega_1$.

Definition 2 (cf. [1]). We define an equivalence relation $\sim$ in G^* by

$$\omega_1 \sim \omega_2 \Leftrightarrow \overline{[\omega_1]} = \overline{[\omega_2]}.$$

Here $[\omega]$ denotes the equivalence class of $\simeq$ containing ω and $\overline{[\omega]}$ is the closure of $[\omega]$ in G^*.

Given $\omega \in G^*$, we can construct the induced primitive ideal $\operatorname{ind}\omega$ of $C^*(\Gamma, A)$ (see, for example, [2]). Since $\operatorname{ind}\omega$ depends only on the equivalence class of $\sim$, we have obtained a map from $G^*/\sim$ into $\operatorname{Prim}(C^*(\Gamma, A))$, which is also denoted by ind.

Now the following is proved by the method of [5], [6], [10] (cf. [2], [4]).

Lemma 2.

$$\operatorname{ind} : G^*/\sim \longrightarrow \operatorname{Prim}(C^*(\Gamma, A))$$

is a homeomorphism.

Our next problem is the concrete description of $G^*/\sim$. For that purpose, we need one more assumption. Through the isomorphism $\iota_x \colon G \to \Gamma_x^x$, $A|_{\Gamma_x^x}$ will be regarded as a C^*-algebraic bundle over G and then, by Theorem 9.1 in [3], we have an action of G on $\operatorname{Prim}(A_x)$. Now suppose that

(A5) for each $x \in X$, all G-orbits in $\operatorname{Prim}(A_x)$ is dense in $\operatorname{Prim}(A_x)$.

For example, if A_x is simple this condition is trivially satisfied.

2. Topological Decomposition of $C^*(\Gamma, A)$

For $g \in G$, we denote by A^g the pull-back bundle of A under the map $X \ni x \mapsto \iota_x(g) \in \Gamma$. Let $M(A^g) = \{M(A_{\iota_x(g)})\}_{x \in X}$ be the multiplier bundle of A^g. A section φ of $M(A^g)$ is called strictly continuous if for each continuous section f of A^g, $X \ni x \mapsto \varphi(x)f(x) \in A_{\iota_x(g)}$ is a continuous section of A^g. Let C_g be the set of bounded strictly continuous sections, say φ, of $M(A^g)$ satisfying

$$\varphi(t(\gamma))a = a\varphi(s(\gamma))$$

for all $\gamma \in \Gamma$ and all $a \in A_\gamma$.

Lemma 3.

(i) $\dim C_g \leq 1$.

(ii) *If* $C_g \neq 0$, *we can find* $\varphi \in C_g$ *such that* $\varphi(x)$ *is a unitary element in* $M(A_{\iota_x(g)})$ *for all* $x \in X$.

Set $S = \{g \in G; C_g \neq 0\}$.

Lemma 4. S *is a subgroup of* G.

Let $\Omega = \{\omega: S \ni g \mapsto \omega_g \in C_g$; for all $x \in X$, $\omega_g(x) \in U(A_{\iota_x(g)})$ and $\omega_{g_1}(x)\omega_{g_2}(x) = \omega_{g_1 g_2}(x)$ $(g, g_1, g_2 \in G)\}$. Then $\hat{S}$ (= the dual group of S) acts on Ω as a transformation group:

$$(\sigma\omega)_g = \langle \sigma, g \rangle \omega_g,$$

for $\sigma \in \hat{S}$, $\omega \in \Omega$, and $g \in S$.

Lemma 5. Ω *is a principal homogeneous space under the action of* $\hat{S}$ (*in particular,* $\Omega \neq \emptyset$).

By this lemma, we can transform the topology of $\hat{S}$ into Ω and Ω becomes a compact Hausdorff space. Let $\omega \in \Omega$ and define an action of S on A by

$$ga = \omega_g(t(\gamma))a,$$

for $g \in S$ and $a \in A_\gamma$. Taking the quotient, we obtain a C^*-algebra A^ω over Γ/S. Let $\varphi \in C_c(\Gamma, A)$ and set

$$\varphi_\omega(\gamma) = \sum_{g \in S} \omega_g(t(\gamma))\varphi(g^{-1}\gamma),$$

for $\gamma \in \Gamma$. Then φ_ω defines an element in $C_c(\Gamma/S, A^\omega)$ and we have

Lemma 6. $\{\{\varphi_\omega\}_{\omega \in \Omega}; \varphi \in C_c(\Gamma, A)\}$ *gives a continuous field structure for the family of* C^**-algebras* $\{C^*(\Gamma/S, A^\omega)\}_{\omega \in \Omega}$.

Theorem. *Let* Γ *be a locally compact Hausdorff second countable groupoid satisfying* (A1) $\sim$ (A3) *and* A *be a* C^**-algebra over* Γ *satisfying* (A4) *and* (A5). *Then*

(i) $C^*(\Gamma/S, A^\omega)$ *is simple for all* $\omega \in \Omega$,

(ii) $C_c(\Gamma, A) \ni \varphi \mapsto \{\varphi_\omega\} \in \{C^*(\Gamma/S, A^\omega)\}_{\omega\in\Omega}$ *is extended to the isomorphism between* $C^*(\Gamma, A)$ *and the cross section* C^**-algebra of* $\{C^*(\Gamma/S, A^\omega)\}_{\omega\in\Omega}$.

Sketch of the Proof. We define an action of $\hat{G}$ on $C^*(\Gamma(X), A)$ by

$$(\sigma\varphi)(g,x) = \langle \sigma, g\rangle \varphi(g,x).$$

Then the associated action of $\hat{G}$ on G^* preserves the equivalence relation $\sim$ and we can show that this action is transitive. So $G^*/\sim$ is identified with $\hat{G}/S'^{\perp}$, where S' is a subgroup of G satisfying $S'^{\perp} = \{\sigma \in \hat{G};\ \sigma|_{S'} = 1\}$. Now imbed $C(G^*/\sim)$ into $C_b(G^*)$ and regard it as a subalgebra of the center Z of the multiplier algebra $M(C^*(\Gamma(X), A))$ (by Dauns–Hofmann theorem). Then we can explicitly write down the condition for an element in Z to belong to $C(G^*/\sim)$, and this shows that $S' = S$ and $\hat{S} = \hat{G}/S'^{\perp} = \Omega$. The topological decomposition in (ii) follows from [11].

Corollary 1. $\Omega = \mathrm{Prim}(C^*(\Gamma, A))$.

Corollary 2. $C^*(\Gamma, A)$ *is simple if and only if* $S = \{e\}$.

Let G be a locally compact group and N be an open normal subgroup of G with G/N abelian. Let A be a separable C^*-algebra and $\alpha: G \to \mathrm{Aut}(A)$, $\rho: N \to U(A)$ be homomorphisms, which form a twisted covariance system in the sense of [6]. Set $S = \{g \in G;$ there exists $u \in U(A)$ such that $\alpha_{h^{-1}}(u)a = \rho(h^{-1}g^{-1}hg)\alpha_{g^{-1}}(a)u$ for all $h \in G$ and all $a \in A\}$. Due to Section 9 of [3], we can construct a C^*-algebraic bundle over G/N. Since algebraic bundles are special cases of C^*-algebras over groupoids, we have

Corollary 3. *S is a subgroup of G containing N and, if A is G-simple, the primitive ideal space of the twisted covariance algebra* $A \rtimes_N G$ *is homeomorphic to* $\widehat{(S/N)}$.

This is a supplementary result to [7], [8].

REFERENCES

[1] E.G. Effros and F. Hahn, *Locally compact transformation groups and C^*-algebras,* Mem. Amer. Math. Soc. **75** (1967).

[2] T. Fack and G. Skandalis, *Sur les representations et ideaux de la C^*-algebre d'un feuilletage,* J. Operator Theory **8** (1982), 95–129.

[3] J.M.G. Fell, *An extension of Mackey's method to Banach *-algebraic bundles,* Mem. Amer. Math. Soc. **90** (1969).

[4] J. Glimm, *Families of induced representations,* Pacific J. Math. **12** (1962), 885–911.

[5] E.C. Gootman and J. Rosenberg, *The structure of crossed product C^*-algebras: A proof of the generalized Effros–Hahn conjecture,* Invent. Math. **102** (1979), 283–298.

[6] P. Green, *The local structure of twisted covariance algebras,* Acta Math. **140** (1978), 191–250.

[7] A. Kishimoto, *Simple crossed products of C^*-algebras by locally compact abelian groups,* Yokohama Math. J. **28** (1980), 69–85.

[8] D. Olesen and G.K. Pedersen, *Applications of the Connes spectrum to C^*-dynamical systems,* III, J. Funct. Anal. **45** (1982), 357–390.

[9] J.N. Renault, *A groupoid approach to C^*-algebras,* Lecture Notes in Math. **793** (1980), Springer.

[10] J.-L. Sauvageot, *Ideaux primitifs de certains produits croises,* Math. Ann. **231** (1977), 61–76.

[11] J. Tomiyama, *Topological representations of C^*-algebras,* Tohoku Math. J. **14** (1962), 187–204.

[12] S. Yamagami, *On factor decompositions of ergodic groupoids,* preprint.

Department of Mathematics
College of General Education
Tohoku University
Sendai 980
Japan

On Sequences of Jones' Projections

MARIE CHODA

1. Introduction

In the index theory for finite factors introduced by Jones [3], the following sequence $\{e_i; i = 1, 2, \ldots\}$ of projections plays an important role:

(a) $e_i e_{i\pm1} e_i = \lambda e_i$ for some $\lambda \leq 1$,

(b) $e_i e_j = e_j e_i$ for $|i - j| \geq 2$,

(c) the von Neumann algebra P generated by $\{e_i; i = 1, 2, \ldots\}$ is a hyperfinite II_1-factor,

(d) $\mathrm{tr}(we_i) = \lambda\,\mathrm{tr}(w)$ if w is a word on 1, e_1, $e_2, \ldots, e_{i-1}$, where tr is the canonical trace of P and 1 is the identity operator.

If Q is a subfactor of P generated by $\{e_i; i = 2, 3, \ldots\}$, then the index $[P : Q]$ of Q in P is $1/\lambda$. Hence, by his basic construction, we have the family $\{e_i;\ i = \ldots, -2, -1, 0, 1, 2, \ldots\}$ of projections with the properties (a), (b), (c′) and (d′);

(c′) $\{e_i; i = 0, \pm1, \pm2, \ldots\}$ generates a hyperfinite II_1 factor M,

(d′) $\mathrm{tr}(we_i) = \lambda\,\mathrm{tr}(w)$ for the trace tr of M if w is a word on 1 and $\{e_j; j < i\}$ (cf. [5]).

We shall call this family $\{e_i; i = 0, \pm1, \pm2, \ldots\}$ the *two-sided Jones projections for* λ. In the case of the sequences of Jones projections, the von Neumann algebra Q is isomorphic to the factor P, so that Q is a subfactor of P. In the case of two-sided Jones projections for λ, it is not obvious that the von Neumann algebra generated by $\{e_i;\ i \neq 0\}$ is isomorphic to the factor M. First, we have that

Proposition 1. *Let* $\{e_i; i \in Z\}$ *be the two-sided Jones projections for* $\lambda \in \{1/4\sec^2(\pi/m);\ m = 3, 4, \ldots\} \cup [4, \infty)$. *Then the von Neumann algebra* N *generated by* $\{e_i; i \neq 0\}$ *is a subfactor of* M.

The Jones Problem. *What value is the index of N in M?*

Here, we shall give a partial answer to Jones Problem by the following theorem.

Theorem. *Let $\{e_i; i = 0, \pm 1, \pm 2, \ldots\}$ be the two sided Jones projections for $\lambda = (1/4)\sec^2(\pi/m)$ for some m ($m = 3, 4, \ldots$). If M (resp. N) is the von Neumann algebra generated by $\{e_i; i = 0, \pm 1, \pm 2, \ldots\}$ (resp. $\{e_i; i = \pm 1, \pm 2, \ldots\}$), then the subfactor N has the index*

$$[M : N] = (m/4)\mathrm{cosec}^2(\pi/m),$$

and the relative commutant of N in M is trivial, that is, $N' \cap M = \mathbf{C}1$.

2. Notations and Preliminaries

Let B be a subfactor of a II_1 factor A. Then Jones defined in [3] the index $[A : B]$ of B in A using the coupling constants of A and B due to Murray and von Neumann ([4]) and he (also, Pimsner–Popa in [5]) derived some methods for finding the number $[A : B]$. In [6], Wenzl derives another method for computing $[A : B]$ in the case where A and B are σ-weak closures of the union of increasing sequences of finite dimensional algebras that satisfy some special conditions.

In this note, we shall use results in [6] and give a proof of the theorem stated in Section 1.

(2.1) Let A be a finite dimensional von Neumann algebra. Then A is decomposed into the direct sum $\sum_{i=1}^{m} \oplus A_i$, where A_i is the algebra of all $a(i) \times a(i)$ matrices over the complex numbers. The row vector $a = (a(i))$ is called the *dimension vector* of A, following Wenzl [6]. Each trace φ on the algebra A is determined by a column vector $w = (w(i))$ that satisfies $\varphi(x) = \sum_{i=1}^{m} w(i)\mathrm{Tr}(x_i)$ for $x \in A$, where $x = \Sigma \oplus x_i$ ($x_i \in A_i$) and Tr is the usual nonnormalized trace on the matrix algebra. The column vector w is called the *weight vector* of the trace φ. Let B be a von Neumann subalgebra of A with the direct summand $B = \sum_{i=1}^{n} \oplus B_i$ of the $b(i) \times b(i)$ matrix algebras B_i. The inclusion of B in A is specified up to conjugacy by an $n \times m$ matrix $[g_{i,j}]$, where $g_{i,j}$ is the number of simple components of a simple A_j module viewed as a B_i module. The matrix $[g_{i,j}]$ is called the *inclusion matrix* of B in A, which we denote by $[B \to A]$. Let $b = (b(i))$ be the dimension vector of B and v the weight vector of the restriction of φ to B, then

(e) $b[B \to A] = a$ and $[B \to A]w = v$.

(2.2) The factor M is the σ-weak closure of the union of the increasing sequence of the following von Neumann algebras $\{M_k\colon k = 1,2,\ldots\}$:

$$M_1 = \mathbb{C}1, \quad M_{2m} = \{e_j; |j| \leq m-1\}'', \quad M_{2m+1} = \{M_{2m}, e_m\}''.$$

The subfactor N of M is generated by the following increasing sequence of $\{N_k\colon k = 1,2,\ldots\}$:

$$N_1 = N_2 = \mathbb{C}1, \quad N_{2m} = \{e_j; 0 \neq |j| \leq m-1\}'', \quad N_{2m+1} = \{N_{2m}, e_m\}''.$$

The algebras M_k and N_k are all finite dimensional ([3]). We denote by a_k (resp. b_k) the dimension vector of M_k (resp. N_k). In the case where M_k is the direct sum of d_k matrix algebras, we say that d_k is the *dimension of the dimension vector* a_k.

(2.3) Every N_k is a subalgebra of M_k. Let $E(B)$ be the conditional expectation of M onto the von Neumann subalgebra B of M conditioned by $\mathrm{tr}(xE(B)(y)) = \mathrm{tr}(xy)$ for $x \in B$ and $y \in M$. Then we have;

Lemma 2. *$E(N_{k+1})E(M_k) = E(N_k)$ and $E(N)E(M_k) = E(N_k)$ for all k.*

(2.4) Let (A_k) and (B_k) be sequences of finite dimensional von Neumann algebras such that $B_k \subset A_k$ for all k. Following [6], we write $(A_k)_k \supset (B_k)_k$ if $(A_k)_k$ (resp. $(B_k)_k$) generates a II_1-factor A (resp. a subfactor B of A) and satisfies the property of Lemma 2. So, by (c′), Proposition 1 and Lemma 2, we have $(N_k) \subset (M_k)$. Such a sequence (M_k) is said to be *periodic* with period r if there is a number m such that $[M_{n+r} \to M_{n+r+i}] = [M_n \to M_{n+i}]$ for $n \geq m$ $(i = 1,2,\ldots)$ and the matrix $[M_n \to M_{n+k}]$ is primitive for $n \geq m$. The sequences $(M_k)_k \supset (N_k)_k$ are *periodic* if both (M_k) and (N_k) are periodic with the same period r and $[N_{n+r} \to M_{n+r}] = [N_n \to M_n]$ for a large enough n ([6]). In Section 6, we show the periodicity of $(N_k)_k \subset (M_k)_k$.

3. The Bratteli Diagram for (M_k) and Path Maps

As a convenient notation we use:

$$\text{(3.1)} \qquad \text{for a positive integer } k, p = \left[\frac{k}{2}\right] \text{ and } q = k - p.$$

In this section, we shall find, for the sequence $\{M_k\}$ in (2.3), the components of the inclusion matrix $[M_q \to M_k]$, which we need to obtain the inclusion matrix $[N_k \to M_k]$. Let $A_k = \{1, e_1, \ldots, e_k\}''$. Then M_k is *-isomorphic to A_{k-1} for $k \geq 2$. On the other hand there is a unitary u in

M_{2m} that satisfies $ue_iu^* = e_{-i}$ and $ue_{-i}u^* = e_i$ for all $i = 0, 1, \ldots, m-1$ ([2]). Hence $[M_k \to M_{k+1}] = [A_{k-1} \to A_k]$ for all $k \geq 2$. It is clear that $[M_1 \to M_2]$ is the 1×2 matrix $[1, 1]$. In [3], Jones gets the Bratteli diagram for (M_k). The dimension vector a_k of M_k, the dimension d_k of a_k and the weight vector w_k of the restriction of tr on M_k are as follows:

(3.2) If $\lambda \leq 1/4$, then

$$d_k = p+1, \quad a_k(i) = \begin{cases} \binom{k}{p+1-i} - \binom{k}{p-i} & \text{if } i = 1, 2, \ldots, d_k - 1 \\ 1 & \text{if } i = d_k \end{cases}$$

$$w_k(i) = \lambda^{p+1-i} P_{k-1-2p+2i}(\lambda),$$

where P_j is the polynomial defined in [2] by $P_1(x) = P_2(x) = 1$ and $P_{n+1}(x) = P_n(x) - xP_{n-1}(x)$.

$$[M_k \to M_{k+1}] = [\delta_{i,j} + \delta_{i+1,j}]_{i,j}, \text{ for Kronecker's } \delta_{i,j}.$$

where $i = 1, 2, \ldots, [\frac{k+1}{2}] + 1$ and

$$j = \begin{cases} 1, 2, \ldots, [\frac{k+1}{2}] + 1 & \text{if } k \text{ is even} \\ 1, 2, \ldots, \frac{k+3}{2} & \text{if } k \text{ is odd.} \end{cases}$$

(3.3) If $\lambda > 1/4$, then $\lambda = (1/4)\sec^2(\pi/n + 2)$ for some $n = 1, 2, \ldots$. The Bratteli diagram for $M_1 \subset M_2 \subset \cdots M_n$ has the same form as in the case of $\lambda \leq 1/4$ and the diagram for $M_{n+2i-1} \subset M_{n+2i}$ (resp. $M_{n+2i} \subset M_{n+2i-1}$) is the same as the one for $M_{n-1} \subset M_n$ (resp. the reverse form of the one for $M_{n-1} \subset M_n$), for all $i = 0, 1, 2, \ldots$. Hence $\{d_k, a_k, t_k\}$ follows after the movement of the diagram. For example,

$$d_k = \begin{cases} p+1 & \text{if } k < n-1, \\ [\frac{n}{2}] + 1 & \text{if } k \geq n-1 \text{ and } n \text{ is odd,} \\ \frac{n}{2} & \text{if } k \geq n-1, k \text{ is odd and } n \text{ is even,} \\ \frac{n}{2} + 1 & \text{if } k \geq n-1, k \text{ is even and } n \text{ is even.} \end{cases}$$

Now we consider the Bratteli diagram for (M_k) as a graph Λ, the set of vertices of which is the set of points where $a_k(i)$ ($k = 1, 2, \ldots$, $i = 1, 2, \ldots, d_k$) stand.

For example, the graph Λ of the sequence (M_k) for $\lambda < 1/4$ is as follows:

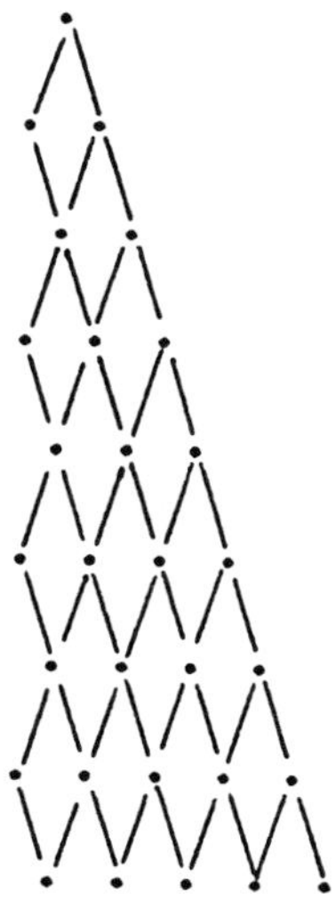

We denote the vertex in Λ corresponding to $a_k(i)$ by the same notation $a_k(i)$. We denote by $[a_k(i) \to a_{k+1}(j)]$ the edge from $a_k(i)$ to $a_{k+1}(j)$. A *path* on Λ is a sequence $\xi = (\xi_r)$ of edges such that $\xi_r = [a_{k(r)}(i_r) \to a_{k(r)+1}(j_r)]$ for some i_r, j_r and $k(r)$ such that $k(r+1) = k(r)+1$. The set of all paths in Λ with the starting point $a_k(i)$ and the ending point $a_r(j)$ is called a *polygon from the vertex* $a_k(i)$ *to the vertex* $a_r(j)$ and denoted by $[a_k(i) \to a_r(j)]$. Also the set of all paths in Λ with $a_k(i)$ as the starting point and for some j, $a_r(j)$ as the ending point is called a *path map from the vertex* $a_k(i)$ *to the floor* a_r and denoted by $[a_k(i) \to a_r]$. Let Ξ_m be the set of paths on Λ consisting of m edges. For a ξ in Ξ_1 and y in Ξ_m, let $\xi \circ y = \{\xi \circ \eta; \eta \in y\}$. Let $x \in \Xi_m$ be a polygon. If there are polygons y and z in Ξ_{m-1} such that, as sets of paths, x is either the union of $\xi \circ y$ and $\eta \circ z$ or the union of $y \circ \xi$ and $z \circ \eta$ for some ξ and η in Ξ_1, we say x is *the direct sum of* y *and* z and we write $x = y \oplus z$ or $y = x \ominus z$.

Remark 3. The i-th coordinate $a_k(i)$ of the dimension vector a_k represents a cardinal number of different paths in the polygon $[a_1(1) \to a_k(i)]$. In the following, we consider $a_k(i)$ as the polygon $[a_1(1) \to a_k(i)]$ and the dimension vector a_k as the path map $[a_1(1) \to a_k]$. also, for the path map $x = (x(1), \dots, x(m))$, we denote by the same x the path map $(x(1), \dots, x(m), 0, \dots, 0)$.

With this identification, we define the direct sum of path maps. Let $x = (x(1), \ldots, x(h))$, $y = (y(1), \ldots, y(m))$ and $z = (z(1), \ldots, z(n))$ be path maps. If $h = \max\{h, m, n\}$ and $x(i) = y(i) \oplus z(i)$ for all polygons $\{x(i), y(i), z(i)\}$, we say x is the *direct sum of y and z*, and we write $x = y \oplus z$.

Remark 4. If we use the method of path model in [4], a polygon corresponds a matrix algebra and a path map corresponds a multi-matrix algebra.

Example. (1) The polygon $a_6(1) = (a_1(1) \to a_6(1))$ and the path map $a_6 = (a_1(1) \to a_6)$ are as follows in the case of either $\lambda \leq 1/4$ or $n \geq 6$:

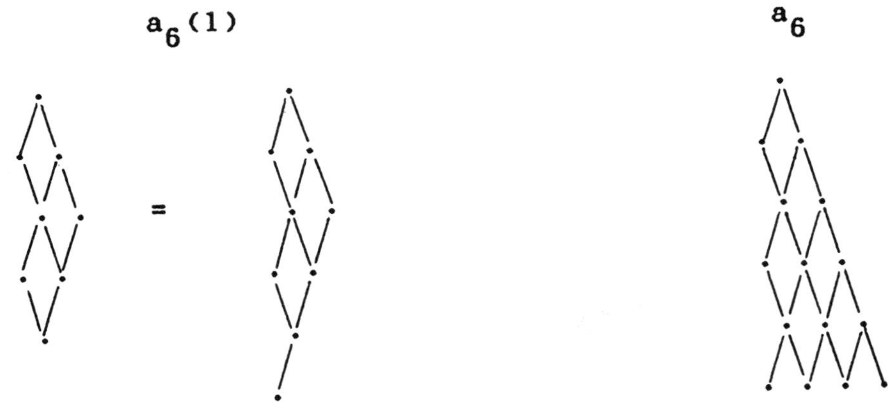

(2) Let $x \in \Xi_7$, $y \in \Xi_6$ and $z \in \Xi_6$ be polygons, then $x = y \oplus z$ are as follows:

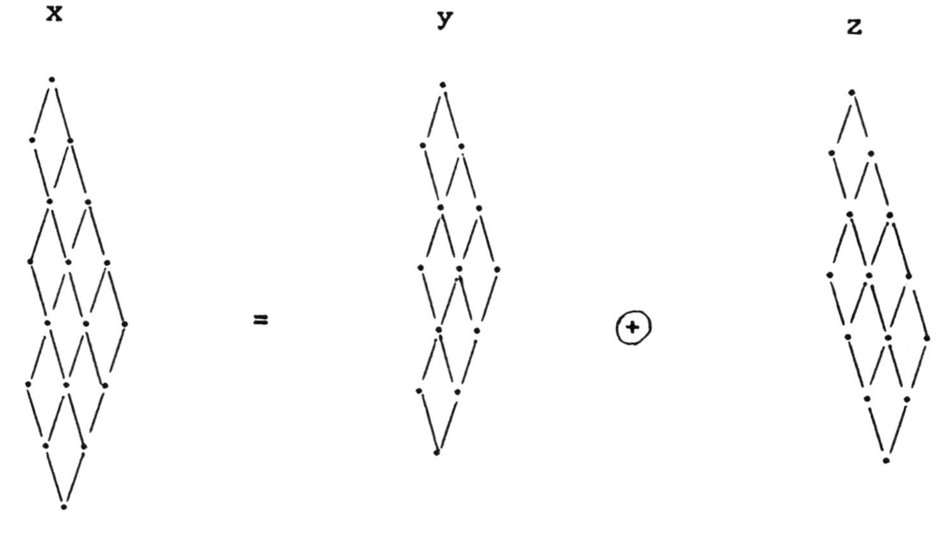

(3) Direct sum of path maps.

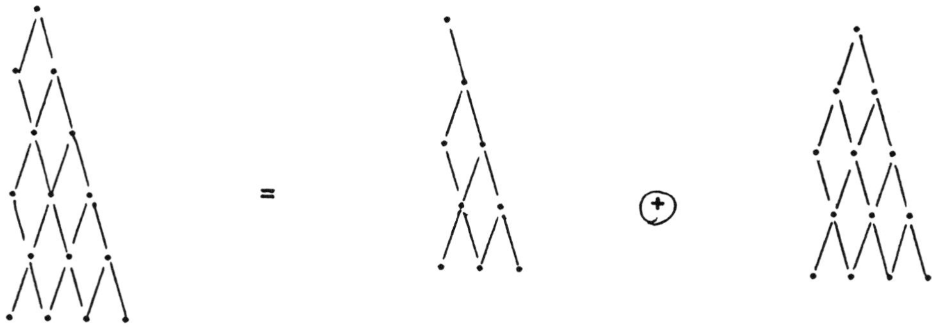

Now we discuss the inclusion matrix $[M_q \to M_k]$. It is obvious that the (i,j)-component of $[M_q \to M_k]$ means the cardinal number of $[a_q(i) \to a_k(j)]$. Hence the i-th row vector x_i of $[M_q \to M_k]$ is considered as the path map $[a_q(i) \to a_k]$.

Under the identification of vectors and path maps, we define the polynomials $f_i(m)$ of path maps on Λ by

$$f_i(0) = a_i, \;\; f_i(1) = a_{i+1} \quad \text{and} \quad f_i(m+1) = f_{i+1}(m) \ominus f_i(m-1).$$

Then for all positive integers i and m, $f_i(2m)$, (resp. $f_i(2m+1)$) is a polynomial on path maps $\{a_{i+2j};\ j = 0, 1, 2, \ldots, m\}$ (resp. $\{a_{i+2j+1};\ j = 0, 1, 2, \ldots, m\}$ with positive integers as coefficients. Then, by the identification of the vectors with the path maps, we have the following:

Lemma 5. *Let x_i be the i-th row vector of the inclusion matrix $[M_q \to M_k]$, for a triplet $\{k, p, q\}$ in (3.1). Then, the path map x_i is as follows for all i ($i = 1, 2, \ldots, d_q$);*

$$x_i = \begin{cases} f_p(2i-2) & \textit{if } q \textit{ is even} \\ f_p(2i-1) & \textit{if } q \textit{ is odd,} \end{cases}$$

under the identification for vectors that $(y(1), \ldots, y(m), 0, \ldots, 0) = (y(1), \ldots, y(m))$ or $y(j) \neq 0$ ($j = 1, \ldots, m$).

4. Bratteli diagram for (N_k)

Let (N_k) be the sequence in (2.3). Let $N_k(+) = \{e_i \in N_k;\ j \geq 1\}''$ and $N_k(-) = \{e_j \in N_k;\ j \leq -1\}''$. Then N_k is generated by the commuting pair $N_k(+)$ and $N_k(-)$. For a triplet $\{k, p, q\}$ in (3.1), $N_k(+)$ is isomorphic to M_q and $N_k(-)$ is isomorphic to M_p. Two dimension vectors and weight vectors of a finite dimensional von Neumann algebra are, respectively, conjugate by an inner automorphism. We may take a dimension vector b_k of N_k and the weight vector u_k for the restriction of the trace tr of M to N_k as

$$b_k = (a_p(1)a_q, a_p(2)a_q, \ldots, a_p(d_p)a_q) \tag{4.1}$$

and

$${}^t u_k = (w_p(1)\,{}^t w_q, t_p(2)\,{}^t w_q, \ldots, t_p(d_p)\,{}^t w_q), \tag{4.2}$$

where ${}^t y$ denotes the transposed vector of the vector y. Since we obtained the inclusion matrices for (M_k) in 3,

$$[N_k \to N_{k+1}] = \begin{cases} I_p \otimes [M_p \to M_{p+1}] & \text{if } k \text{ is odd} \\ [M_p \to M_{p+1}] \otimes I_q & \text{if } k \text{ is even,} \end{cases} \tag{4.3}$$

where I_k denotes the d_k by d_k identity matrix. It is easy to check that $[N_k \to N_{k+1}]$ satisfies the property (e) for b_k and u_k. The Bratteli diagram for (N_k) comes from the diagram for (M_k) using the above information.

In the case of $\lambda = (1/4)\sec^2(\pi/n+2)$ for some n ($n = 1, 2, \ldots$), the diagram for $N_1 = N_2 \subset N_3 \subset \cdots \subset N_{2n}$ has the same form as in the case of $\lambda \leq 1/4$, the diagram for $N_{2n+4i-2} \subset N_{2n+4i-1}$ (resp. $N_{2n+4i-1} \subset N_{2n+1}$) is similar to the one for $N_{2n-2} \subset N_{2n-1}$ (resp. $N_{2n-1} \subset N_{2n}$) and the diagram for $N_{2n+4i} \subset N_{2n+4i+1}$ (resp. $N_{2n+4i+1} \subset N_{2n+4i+2}$) has the reverse form of order of the one for $N_{2n-1} \subset N_{2n}$ (resp. $N_{2n-2} \subset N_{2n}$).

Example. In the case of $n = 4$, the diagram is as follows:

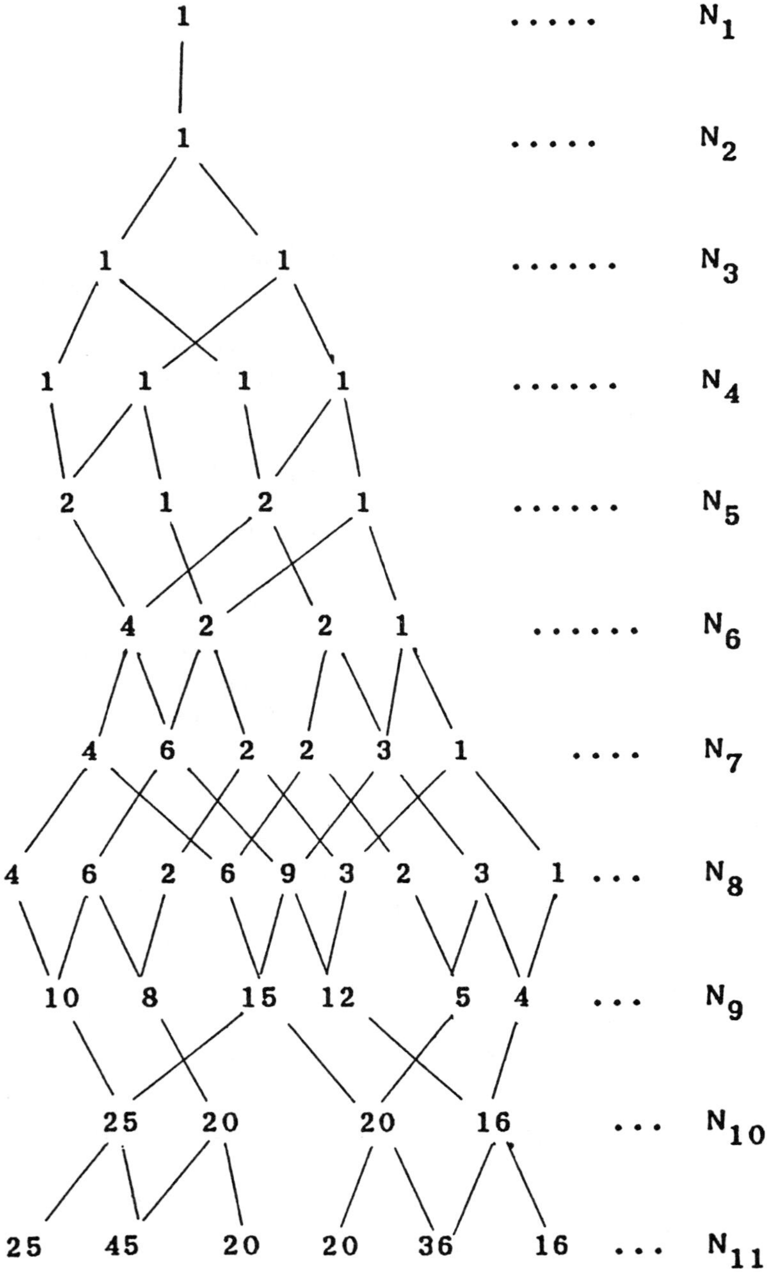

5. Inclusion Matrix of N_k in M_k

Let $\{k, p, q\}$ be a triplet in (3.1). Let the matrix $[M_q \to M_k]$ be

$$[M_q \to M_k] = \begin{bmatrix} x_1 \\ x_2 \\ \vdots \\ x_{d_q} \end{bmatrix} \quad \text{for vectors } x_i = (x_i(j)).$$

Here we consider $x_i(j)$ and x_i as a polygon and a path map in Ξ_p. By Lemma 5, the polygon $x_i(j)$ can be decomposed into the direct sum of polygons $\{a_{p+j}(i)\colon\ j = 0, 1, \ldots,\ i = 1, 2, \ldots, d_p\}$. Then we define the matrix $[a_p \to x_i] = [h(j,k)]$, where $h(j,k)$ is the number of times that $a_p(j)$ is contained in $x_i(k)$. We call the matrix $[a_p \to x_i]$ the *inclusion matrix of the path map a_p in the path map x_i.*

Remark 6. Let x, y, and z be path maps on Λ such that $[x \to y]$ and $[x \to z]$ are defined. Then, by the definition of the direct sum of path maps and the inclusion matrix for path maps, the matrix $[x \to (y \oplus z)]$ is defined and

$$[x \to (y \oplus z)] = [x \to y] \oplus [x \to z].$$

By this property and Lemma 5, the inclusion matrix $[a_p \to x_i]$ of the path map a_p in the path map x_i is defined from the inclusion matrices $[M_p \to M_r]$ $(r \geq p)$ by the natural method. The matrix $[a_p \to x_i]$ has the following forms in the specified cases:

$$\begin{array}{cccc}
[a_p \to x_1] & [a_p \to x_2] & [a_p \to x_3] & [a_p \to x_{d_q}] \\
\| & \| & \| & \| \\
\begin{bmatrix} 1 & & \\ & & \\ & & \\ & & \\ & & 1 \end{bmatrix} &
\begin{bmatrix} 01 & & \\ 1 & & \\ & & \\ & & 1 \\ & & 10 \end{bmatrix} &
\begin{bmatrix} 001 & & \\ 01 & & \\ 1 & & 1 \\ & & 10 \\ & & 100 \end{bmatrix} \quad \cdots &
\begin{bmatrix} & & 1 \\ & & \\ & & \\ & & \\ 1 & & \end{bmatrix}
\end{array}$$

In general, we have (by Lemma 5) the following lemma.

Lemma 7. *Let $\lambda = (1/4)\sec^2(\pi/n + 2)$ and x_i be the i-th row vector of $[M_q \to M_k]$. Assume $q \geq n$.*

(1) *If n is odd, then $[a_p \to x_i]$ is a $(q + [\frac{n}{2}])$-square matrix with the following form:*

(1.1) *If $p = q$ is an odd number, then*

$$[a_p \to x_i](j,r) = \begin{cases} 1, & 1-i \le 1-j \le i < j+1 \le n+2-i \\ 0, & \textit{otherwise.} \end{cases}$$

(1.2) *If $p+1 = q$ is even, then*

$$[a_p \to x_i](j,r) = \begin{cases} 1, & |1-j| \le i \le j+1 \le n+2-i \\ 0, & \textit{otherwise.} \end{cases}$$

(1.3) *If $p = q$ is even, then*

$$[a_p \to x_i](j,r) = \begin{cases} 1, & |1-j| < i < j+1 \le n+3-i \\ 0, & \textit{otherwise.} \end{cases}$$

(1.4) *If $p+1 = q$ is odd, then*

$$[a_p \to x_i](j,r) = \begin{cases} 1, & -i \le 1-j < i < j+1 \le n+2-i \\ 0, & \textit{otherwise.} \end{cases}$$

(2) *Let n be even.*

(2.1) *If $p = q$ is odd, then $[a_p \to x_i]$ is an $n/2 \times 1 + (n/2)$ matrix with*

$$[a_p \to x_i](j,r) = \begin{cases} 1, & 1-i \le 1-j \le i < j+1 \le n+2-i \\ 0, & \textit{otherwise.} \end{cases}$$

(2.2) *If $p+1 = q$ is even, then $[a_p \to x_i]$ is an $n/2$ square matrix with*

$$[a_p \to x_i](j,r) = \begin{cases} 1, & |1-j| < i \le j+1 \le n+2-i \\ 0, & \textit{otherwise.} \end{cases}$$

(2.3) *If $p = q$ is even, then $[a_p \to x_i]$ is a $1 + (n/2)$ square matrix with*

$$[a_p \to x_i](j,r) = \begin{cases} 1, & |1-j| < i < j+1 \le n+3-i \\ 0, & \textit{otherwise.} \end{cases}$$

(2.4) *If $p+1 = q$ is odd, then $[a_p \to x_i]$ is a $1 + (n/2) \times n/2$ matrix with*

$$[a_p \to x_i](j,r) = \begin{cases} 1, & -i \le 1-j < i < j+1 \le n+2-i \\ 0, & \textit{otherwise.} \end{cases}$$

Lemma 8. *For the weight vector w_k of the restriction of* tr *to M_k, we have*

$$[a_p \to x_i]w_k = w_q(i)w_p \qquad (i = 1, 2, \ldots, d_q).$$

Now we have that the inclusion matrix $[N_k \to M_k]$ is the $d_p d_q \times d_k$ matrix G_k, the $(d_q(j-1)+i)$-th row vector of which is the j-th row vector of the matrix $[a_p \to x_i]$, where $i = 1, 2, \ldots, d_q$, $j = 1, 2, \ldots, d_p$. That is, the transposed matrix ${}^tG_k$ of G_k is as follows:

$$ {}^tG_k = [G[1]_1, G[2]_1, \ldots, G[d_q]_1, G[1]_2, \ldots, G[d_q]_2, \\ \ldots, G[1]_{d_p}, \ldots, G[d_q]_{d_p}], $$

where $G[i]_j$ is the transposed vector of the j-th row vector of $[a_p \to x_i]$.

It is easy to check that the matrix $[N_k \to M_k]$ satisfies the following;

$$ b_k G_k = a_k,\ G_k w_k = u_k \quad \text{and} \quad G_k[M_k \to M_{k+1}] = [N_k \to N_{k+1}]G_{k+1}, $$

where a_k (resp. b_k) are dimension vectors of M_k (resp. N_k) and w_k (resp. u_k) are weight vectors of M_k (resp. N_k).

6. Periodicity of $(N_k) \subset (M_k)$ in the Case of $\lambda > 1/4$

In this section, we assume that $\lambda = (1/4)\sec^2 \pi/(n+2)$ for some n ($n = 1, 2, \ldots$). Then the sequence (M_k) is periodic with period 2 and the sequence (N_k) is periodic with period 4.

On the other hand, let x_i (resp. y_i) be the i-th column vector of $[M_q \to M_k]$ (resp. $[M_{q+2} \to M_{k+4}]$). If $q \geq n$, then

$$ [a_p \to x_i] = [a_{p+2} \to y_i] \qquad (i = 1, 2, \ldots, d_q). $$

Hence we have that the sequence $(N_k) \subset (M_k)$ is periodic.

7. Proof of the Theorem

Since the factors M and N are generated by the periodic sequences $(N_k) \subset (M_k)$ of finite dimensional algebras, we have that $[M : N] = \|u_k\|_2^2/\|w_k\|_2^2$ for a large enough k, by [6; Theorem 1.5], for the weight vectors w_k and u_k of the restriction tr to M_k and N_k. By (4.2),

$$ \|u_k\|_2^2 = \|w_p\|_2^2\,\|w_q\|_2^2 \quad \text{for a } \{k, p, q\} \text{ in (3.1)}. $$

Put $n = m - 2$. Then we have

$$ [M : N] = \|u_k\|_2^2/\|w_k\|_2^2 \quad \text{for all} \quad k \geq n-1. $$

Since $\|w_k\|_2^2/\|w_{k+1}\|_2^2 = 1/\lambda$ for all $k \geq n-1$,

$$ [M : N] = \|w_{n-1}\|_2^4/\|w_{2(n-1)}\|_2^2 = \|w_{n-1}\|_2^2/\lambda^{n-1}. $$

By (3.3),

$$||w_{n-1}||_2^2 = \Sigma_j \lambda^{2j} P_{n-2j}(\lambda)^2, \text{ where } j \text{ runs over } \{0, 1, \ldots, [\frac{n-1}{2}]\}.$$

On the other hand, by [3],

$$P_k((1/4)\sec^2\theta) = \sin k\theta / 2^{k-1} \cos^{k-1}\theta \sin\theta \text{ for all } k \text{ and } \theta.$$

Hence

$$\begin{aligned}[M : N] &= \Sigma_j \sin^2((n-2j)\pi/(n+2))/\sin^2(\pi/(n+2)) \\ &= (m/4)\text{cosec}^2(\pi/m).\end{aligned}$$

Let c be the dimension vector of $N' \cap M$. Since $(M_k) \supset (N_k)$ is periodic,

$$||c||_1 \leq \alpha = \min\{||G[i]_j||_1; k \geq 2n, i = 1, 2, \ldots, d_q, j = 1, 2, \ldots, d_p\},$$

by [6; Theorem 1.7], where $G[i]_j$ is the vector in Section 5. By Lemma 8, there are many $\{i, j\}$'s such that ${}^tG[i]_j = (1, 0, \ldots, 0)$. This implies that $\alpha = 1$. Hence $N' \cap M$ is 1-dimensional, so that $N' \cap M = \mathbb{C}1$.

Remark 9. (1) If $m = 3$ or 4, then $[M : N] = [P : Q]$ for the subfactor $Q = \{e_i;\ i = 2, 3, \ldots\}''$ of the factor $P = \{e_i\colon\ i = 1, 2, \ldots\}''$. That is, $[M : N] = 1$ if $m = 3$, and $[M : N] = 2$ if $m = 4$.

(2) If $m \geq 5$, then $[M : N] \neq [P : Q]$. If $m = 5$, then $[M : N] < 4$. Hence there is an integer k ($k \geq 3$) such that $[M : N] = 4\cos^2(\pi/k)$. H. Choda observed that $k = 10$.

On the other hand, by the proof of Lemma 14,

$$[M : N] = 4\cos^2(\pi/3) + 4\cos^2(\pi/5).$$

This implies the following equation;

$$\cos^2(\pi/3) + \cos^2(\pi/5) = \cos^2(\pi/10).$$

8. A Generalization

Choose a positive integer n. Let

$$L = \{\ldots, e_{-n-1}, e_{-n}, e_1, e_2, e_3, \ldots\}''.$$

In case $n = 1$, we have that $L = N$. By a proof like that of Lemma 1, L is a subfactor of M, for all n. Also, L is a subfactor of N and $[N : L] = 4\cos^2(\pi/m)$. Hence

$$[M : L] = (m/4)\mathrm{cosec}^2(\pi/m)\{4\cos^2(\pi/m)\}^{n-1}.$$

Let

$$L_1 = L_2 = \mathbb{C}1,\ L_{2i-1} = L_{2i} = \{e_i; i = 1, 2, \ldots, n-1\}'' \quad \text{if } i \leq n$$

and

$$L_{2i+1} = \{L_{2i}, e_i\}'',\ L_{2i+2} = \{e_{-i}, L_{2i+1}\}'' \quad \text{if } i \geq n.$$

The sequence (L_k) is periodic with period 4 and generates L. By a method similar to that used for $(N_k) \subset (M_k)$, we get the inclusion matrix $[L_k \to M_k]$. For a triplet $\{k, p, q\}$ in (3.1), we consider the matrix $[a_{p-(n-1)} \to x_i]$ for a large k, where x_i is the same as in Section 3, that is the i-th column vector of $[M_q \to M_k]$. Then $(N_k) \subset (M_k)$ is periodic. Let h be the dimension vector of $L' \cap M$.

If q is even, then $x_1 = a_p$, hence $[a_{p-(n-1)} \to x_1] = [a_{p-(n-1)} \to a_p]$.

If $n = 2$, we have $N' \cap M = \mathbb{C}1$, by the form of $[a_k \to a_{k+1}]$ for an odd k.

If $n \geq 3$, $\{e_{-n+2}, e_{-n+3}, \cdots, c_{-1}\}''$ is contained in $L' \cap M$ and is isomorphic to M_{n-1}. Hence we have

$$L' \cap M = \{e_{-n+2}, e_{-n+3}, \ldots, e_{-1}\}''.$$

REFERENCES

[1] O. Bratteli, *Inductive limits of finite dimensional C^*-Algebras,* Trans. A.M.S. **171**, (1972), 195–234.

[2] F. Goodman, P. de la Harpe, and V. Jones, *Coxter-Dynkin diagrams and towers of algebras,* preprint, I.H.E.S.

[3] V. Jones, *Index for subfactors,* Invent. Math. **72**, (1983), 1–25.

[4] F. Murray and J. von Neumann, *On rings of operators,* II, Trans. A.M.S. **41**, (1937), 208–248.

[5] M. Pimsner and S. Popa, *Entropy and index for subfactors,* Ann. Sci. El. Norm. Sup. **19**, (1986), 57–106.

[6] H. Wenzl, *Representations of Hecke algebras and subfactors,* Thesis, University of Pennsylvania.

Department of Mathematics
Osaka Kyoiku University
Tennoji, Osaka 543
Japan

The Powers' Binary Shifts on the Hyperfinite Factor of Type II_1

MASATOSHI ENOMOTO

Abstract

A unit preserving *-endomorphism σ on the hyperfinite II_1 factor R is called a shift if $\bigcap_{i=0}^{\infty} \sigma^k(R) = \{\lambda I;\ \lambda \in \mathbb{C}\}$. A shift σ is called a Powers binary shift if there is a self adjoint unitary u such that $R = \{\sigma^n(u);\ n \in \mathbb{N} \cup \{0\}\}''$ and $\sigma^k(u)u = \pm u\sigma^k(u)$ for $k \in \mathbb{N} \cup \{0\}$. Let $q(\sigma)$ be the number $\min\{k \in \mathbb{N};\ \sigma^k(R)' \cap R = \mathbb{C}I\}$. It is shown that the number $q(\sigma)$ is not the complete outer conjugacy invariant for a Powers' binary shift.

1. Introduction

Let R be the hyperfinite factor of type II_1. A unit preserving *-endomorphism σ on R is called a shift if $\bigcap_{i=0}^{\infty} \sigma^i(R) = \mathbb{C}I$. A shift σ is called a Powers' binary shift if there is a self adjoint unitary generator u_0 in R, such that R is generated by $\{u_n;\ n = 0, 1, 2, \ldots\}$, where u_n and u_m commute or anticommute and $u_n = \sigma^n(u_0)$. In [8], G. Price ingeniously found a non-binary shift with index two on R. Inspired by the construction of Price's non-binary shift with index two on R, in Section 2, we shall construct uncountably many non-binary shifts on R of index two.

Let $q(\sigma)$ be the number $\min\{k \in \mathbb{N};\ \sigma^k(R)' \cap R = \mathbb{C}I\}$. The numbers $q(\sigma)$ are outer conjugacy invariant for binary shifts σ. Powers [7] raised the following problem. Are the numbers $q(\sigma)$ the complete outer conjugacy invariant for binary shifts σ. In Section 3, we shall give a negative answer to this Powers' problem. In order to do this, we shall use the relative commutant algebras $\{\sigma^k(R)' \cap R;\ k = 0, 1, 2, \ldots\}$. Obviously, the set of the (isomorphism classes of) relative commutant algebras $\{\sigma^k(R)' \cap R;\ k = 0, 1, 2, \ldots\}$ is an outer conjugacy invariant for binary shifts. In Section 4, we shall describe the structure of relative commutant algebras of binary shifts.

This paper is a joint work with Y. Watatani and partly with M. Choda, M. Nagisa and H. Yoshida (cf. [3], [4], [5], [6]).

2. Uncountably Many Non-Binary Shifts

In this section, we shall show that there are uncountably many non-conjugate non-binary shifts on the hyperfinite II_1-factor.

The Powers' binary shifts are realized as follows [3].

Let $G = \coprod_{i=0}^{\infty} G_i$ be the restricted direct product of $G_i \cong Z_2 \cong \{0,1\}$. A function $a: Z \to \{0,1\}$ is called a signature sequence if $a(0) = 0$ and $a(n) = a(-n)$ for any integer $n \in \mathbf{Z}$ (cf. [7], [8], [9]). A signature sequence a is periodic if there exists an integer k such that $a(k+n) = a(n)$ for $n \in \mathbf{Z}$. In the following we shall always assume that the signature sequence a is aperiodic and identify the sequence $(a(i);\, i \in \mathbb{N} \cup \{0\})$ with $(a(i);\, i \in \mathbf{Z})$. Define the canonical shift σ on the group G by

$$(\sigma(x))(j) = x(j-1) \quad \text{for} \quad j \geq 1 \quad \text{and} \quad (\sigma(x))(0) = 0.$$

For $x = (x(i))$ and $y = (y(j))$ in G, let us define a multiplier

$$m_a(x,y) = (-1)^{\sum_{i>j} a(i-j)x(i)y(j)}.$$

Let $R_{m_a}(G)$ be the von Neumann algebra on $\ell^2(G)$ generated by $\{\lambda_{m_a}(x);\, x \in G\}$, where $(\lambda_{m_a}(x)\xi)(y) = m_a(x, x^{-1}y)\xi(x^{-1}y)$ for $\xi \in \ell^2(G)$. Price ([8], [9]) showed that a is not periodic if and only if $R_{m_a}(G)$ is the hyperfinite II_1-factor. Since $m_a(\sigma(x), \sigma(y)) = m_a(x,y)$, σ induces a shift σ on $R_{m_a}(G)$ such that $\sigma(\lambda_{m_a}(x)) = \lambda_{m_a}(\sigma(x))$ for $x \in G$. Then the shift σ on $R_{m_a}(G)$ is the Powers" binary shift with signature sequence a. Put $e_0 = (1,0,0,0,\ldots) \in G$ and $e_n = \sigma^n(e_0) \in G$. Similarly put $u_0 = \lambda_{m_a}(e_0)$ and $u_n = \sigma^n(u_0)$. Then $u_n u_m = (-1)^{a(n-m)} u_m u_n$ and the hyperfinite factor of type II_1, $R = R_{m_a}(G)$, is generated by $\{u_n;\, n = 0,1,2,\ldots\}$. Thus the shift $\sigma = \sigma_a$ on $R_{m_a}(G)$ is a Powers' binary shift with a signature sequence a.

Let $F = F_2$ be the finite field $\{0,1\}$. Let $F[t]$ be the polynomial ring over F. Take $p(t) = c_0 + c_1 t + c_2 t^2 + \cdots + c_k t^k$ such that $c_0 c_k \neq 0$. Define a mapping $\Psi: F[t] \to F[t]/p(t)$ by

$$\Psi(f(t)) = (p(t)f(t))/p(t).$$

Let $X = \coprod_{i=0}^{\infty} G_i$, $Y = \coprod_{i=0}^{\infty} H_i$, $G_i \cong H_i \cong Z_2$. Define a mapping $\theta: X \to F[t]$ by

$$\theta((x(i))) = \sum_{i \geq 0} x(i) t^i \in F[t].$$

Define $\gamma: Y \to F[t]/p(t)$ by

$$\gamma(y) = \left(\sum_{i \geq 0} y(t) t^i \right) /p(t) \quad \text{for} \quad y = (y(i)) \in Y.$$

Define $\phi_p: X \to Y$ by $\phi_p = \gamma^{-1}\Psi\theta$.

Then we have the following lemma.

Lemma 1. *Let $a: Z \to \{0,1\}$ be a non-periodic signature sequence and $p = c_0 + c_1 t + \cdots + c_k t^k \in F[t]$ such that $c_0 = c_k = 1$. Then there exists a non-periodic signature sequence $b: Z \to \{0,1\}$ such that*

$$m_b(\phi_p(x), \phi_p(y)) = m_a(x,y) \quad \textit{for any} \quad x, y \in X.$$

Take a non-periodic signature sequence a on Z and a sequence $p = (p_1, p_2, \dots)$ of polynomials $p_\ell(t) = c_{\ell,0} + c_{\ell,1}t + \cdots + c_{\ell,k(\ell)}t^{k(\ell)}$ over Z_2, with $c_{\ell,0} = c_{\ell,k(\ell)} = 1$. Put $X_\ell = \coprod_{i=0}^{\infty} G_i^\ell$, $G_i^\ell \cong Z_2$.

Using Lemma 1 inductively, we have injections $\phi_{p_\ell}: X_\ell \to X_{\ell+1}$ and a sequence $\{a_\ell;\ \ell = 1,2,\dots\}$ of non-periodic signature sequences a_ℓ that have compatible multipliers m_{a_ℓ} with respect to injections by the same considerations as above. The inductive limit of those X_ℓ is denoted by

$$X_{[p]} = \varinjlim(X_\ell, \phi_{p_\ell}).$$

Define the multiplier $m_{[a,p]}$ on $X_{[p]}$ by

$$m_{[a,p]}(x,y) = m_{a_\ell}(x,y) \quad \text{for} \quad x, y \in X_\ell.$$

Then $R_{m_{[a,p]}}(X_{[p]})$ is the hyperfinite II_1 factor. This fact comes from non-degeneracy of $m_{[a,p]}$. We denote those shifts constructed above by $\sigma_{[a,p]}$ and call them shifts of Price's type. Then we have the following.

Proposition 2. *Suppose two sequences of polynomials with non-zero constant terms, $p = (p_i)$ and $q = (q_i)$ for $i = 1,2,\dots$, and two non-periodic signature sequences a and b are given. If two shifts of Price's type $\sigma_{[a,p]}$ and $\sigma_{[b,q]}$ are conjugate, then shifts $(\sigma_{[p]}, X_{[p]})$ and $(\sigma_{[q]}, X_{[q]})$ are conjugate, where $\sigma_{[p]}$ is the induced shift on $X_{[p]}$ from $\sigma_{[a,p]}$.*

For an element $a \in \prod_{i=1}^{\infty} Z_2$ and a set $\{p_k;\ k = 1,2,\dots\}$ of irreducible polynomials over Z_2, let $X^a = \{g(t)/f(t);\ f(t)$ and $g(t)$ are polynomials over Z_2 such that $f(t)$ has no $p_k(t)$ factor if $a(k) = 0\}$. Consider the multiplication operator σ_t by t on X^a such that

$$\sigma_t(g(t)/f(t)) = (tg(t))/f(t) \quad \text{for} \quad g(t)/f(t) \in X^a.$$

We denote it by (σ^a, X^a). Then we have

Lemma 3. *Let a and b be non-periodic signature sequences. Then $a = b$ if and only if (σ^a, X^a) and (σ^b, X^b) are conjugate.*

Combining with Proposition 2 and Lemma 3, we have

Theorem 4. *There exist uncountably many non-conjugate non-binary shifts of index two on the hyperfinite II_1-factor.*

3. The Powers' Problem

In this section, we shall solve the problem by Powers [7] about outer conjugacy of binary shifts on the hyperfinite II_1-factor R. In the following we shall realize the relative commutant algebras $C_k(\sigma) = \sigma^k(R)' \cap R$ of binary shifts σ on R concretely.

Theorem 5. *Let a be a non-periodic signature sequence. Suppose that the set $\{i \in \mathbb{N};\ a(i) \neq 0\}$ is finite. Put $d = \max\{i \in \mathbb{N};\ a(i) \neq 0\}$. Let σ be the Powers' binary shift with a signature sequence a. Let $u_0 = \lambda_m(e_0)$ be the σ-generator. Put $u_n = \sigma^n(u_0)$. Then $\sigma^k(R)' \cap R = \mathbb{C}I$ if $0 \leq k \leq d$ and $\sigma^k(R)' \cap R = \{u_i;\ 0 \leq i \leq k-d-1\}''$ if $d+1 \leq k$.*

Remark 6. In [1], Bures and Yin considered independently the relative commutant algebras for group shifts abstractly and they proved the following:

Let G be a discrete abelian group and m a multiplier of G. Let $R_m(G)$ be the von Neumann algebra as in the above case where $m = m_a$. If H is a subgroup of G, then $R_m(H)' \cap R_m(G) = R_m(D_H)$, where D_H is the subgroup $\{g \in G;\ m(g,h) = m(h,g)$ for all $h \in H\}$ of G. Powers [6] defined the following outer conjugacy invariant $q(\sigma)$ for shifts σ. Put $q(\sigma) = \min\{k \in \mathbb{N};\ \sigma^k(R)' \cap R = \mathbb{C}I\}$. Then we have

Remark 7. Under the above situation of Theorem 5, $q(\sigma_a) = d+1$ where d is the number $\max\{i \in \mathbb{N};\ a(i) \neq 0\}$. This number is called the degree of a and is denoted by 'degree a.'

In [7], Powers raised the following problem (also [9]).

The Powers' Problem. If α and β are binary shifts and $q(\alpha) = q(\beta)$, then are α and β outer conjugate?

We give a negative answer to this problem using Theorem 5.

Corollary 8. *There exist binary shifts α and β such that $q(\alpha) = q(\beta)$*

but α and β are not outer conjugate.

Those shifts α and β are given by signature sequences a and b such that $a(2) = a(3) = 1$ and $a(i) = 0$ $(i \neq 2,3)$, $b(1) = b(3) = 1$ and $b(j) = 0$ $(j \neq 1,3)$.

Remark 9. Let a be a signature sequence such that the set $\{i \in \mathbb{N}; a(i) \neq 0\}$ is finite. Let order a be the number $\min\{n \in \mathbb{N}; a(n) \neq 0\}$. Then degree a and order a are outer conjugacy invariants for the Powers' binary shifts σ_a with degree $a < +\infty$. In fact, $d+1 = q(\sigma_a)$ and $d+r+1 = \min\{k \in \mathbb{N}; \sigma^k(R)' \cap R$ is not abelian$\}$. But orders and degrees are not complete outer conjugacy invariant. This is shown by the following example.

Example 10. Let a and b be signature sequences such that $a(1) = a(3) = 1$ and $a(i) = 0$ $(i \neq 1,3)$, $b(1) = b(2) = b(3) = 1$ and $b(j) = 0$ $(j \neq 1,2,3)$. Then obviously degree a = degree b and order a = order b. But σ_a and σ_b are not outer conjugate.

Remark 11. In [2], M. Choda also uses the numbers $\min\{k \in \mathbb{N}; \sigma^k(R)' \cap R = \mathbb{C}\}$ and $\min\{k \in \mathbb{N}; \sigma^k(R)' \cap R$ is not abelian$\}$ for projection shifts to show that there are at least a countable infinity of outer conjugacy classes among the projection shifts of R with the index $\lambda \in \{4\cos^2(\pi/n); n = 3,4,\ldots\} \cup [4,\infty)$.

Remark 12. Take signature sequences a and b such that $a(1) = 1$ and $a(i) = 0$ if $i \neq 1$, $b(j) = 1$ if $j \neq 0$. Then there exists a unitary $w \in R$ such that $\sigma_a = Adw \cdot \sigma_b$.

4. Relative Commutant Algebras of Binary Shifts

In this section, we shall consider structures of relative commutant algebras of binary shifts. Let $\sigma = \sigma_a$ be the shift on $R_{m_a}(G) = \{u_n; n = 0,1,2,\ldots\}''$ in Section 2.

At first we have the following.

Definition 13. Let a be a non-periodic signature sequence on $\mathbf{Z}$. The sequence a is called *essentially periodic* if there exist non-negative integers k and p such that, for any $n \geq k$, $a(n+p) = a(n)$.

Theorem 14. *Let a be a non-periodic signature sequence and σ_a be the associated shift of the hyperfinite II_1-factor R. The sequence a is essentially periodic if and only if there exists a non-negative integer r such* *that* $\sigma_a^r(R)' \cap R \neq \mathbb{C}I$.

If $u_0^{x(0)} u_1^{x(1)} \cdots u_n^{x(n)}$ is in the center of the C^*-algebras $C^*(u_0, u_1, \ldots, u_n)$ generated by $u_0, u_1, \ldots, u_n$, then we have

$$(*) \qquad \sum_{k=0}^{n} a(i-k)x(k) = 0 \quad \text{for} \quad 0 \leq i \leq n.$$

Therefore if we put

$$A(n) = \begin{pmatrix} a(0) & a(1) & \ldots & a(n) \\ a(1) & a(0) & \ldots & a(n-1) \\ \ldots & \ldots & \ldots & \ldots \\ a(n) & \ldots & \ldots & a(0) \end{pmatrix},$$

the equation (*) is equivalent to $A(n)x = 0$, where $x = {}^t(x(0), x(1), \ldots, x(n))$, $x(i) \in F_2$, $i = 0, 1, 2, \ldots$. Therefore the number of central unitary words consisting of $\{u_0, \ldots, u_n\}$ is equal to the number of solutions x of equation $A(n)x = 0$. Hence the dimension of the center of the algebra generated by $\{u_0, u_1, \ldots, u_n\}$ is the number of solutions of the equation $A(n)x = 0$ on F_2. We have the following result.

Proposition 15 (Nagisa and Yoshida).

$$\dim(\ker A(n+1)) = \dim(\ker A(n)) \pm 1.$$

Furthermore, the sequence $\{\dim(\ker A(n))\}$ *is periodic.*

By this Proposition 15 and Theorem 5, the C^*-algebra generated by $\{u_0, u_1, \ldots, u_n\}$ is isomorphic to $M_{2^k} \otimes C^{2^\ell}$ for some k and $\ell \geq 0$.

Proposition 16 (Nagisa and Yoshida). *Suppose the signature sequences* a *and* b *are such that* $a(2) = a(4) = 1$ *and* $a(i) = 0$ *if* $i \neq 2, 4$, *and* $b(2) = b(3) = b(4) = 1$ *and* $b(j) = 0$ *if* $j \neq 2, 3, 4$. *Then we have* $\sigma_a^k(R)' \cap R \cong \sigma_b^k(R)' \cap R$ *for any* $k \geq 0$.

We plan to discuss the structure results for relative commutant algebras in future publications.

Added in Proof. D. Bures and H.S. Yin recently showed our Remark 12 and Theorem 13 [4], independently, in their preprint (*Outer conjugacy of shifts on the hyperfinite* II_1*-factor*).

REFERENCES

[1] D. Bures and H.S. Yin, *Shifts on the hyperfinite factor of type* II_1, preprint, 1987.

[2] M. Choda, *Shifts on the hyperfinite* II_1*-factor,* J. Operator Theory, **17** (1987), 223–235.

[3] M. Enomoto and Y. Watatani, *Powers' binary shifts on the hyperfinite factor of type* II_1, Proc. Amer. Math. Soc., to appear.

[4] M. Enomoto and Y. Watatani, *A solution of Powers' problem on outer conjugacy of binary shifts,* preprint, 1987.

[5] M. Enomoto, M. Choda and Y. Watatani, *Generalized Powers' binary shifts on the hyperfinite* II_1*-factor,* Math. Japon, to appear.

[6] M. Enomoto, M. Choda and Y. Watatani, *Uncountably many non-binary shifts on the hyperfinite* II_1*-factor,* preprint, 1987.

[7] R.T. Powers, *An index theory for semigroups of* **-endomorphisms of* $B(H)$ *and type* II_1*-factors,* Can. J. Math., to appear.

[8] G. Price, *Shifts on type* II_1*-factors,* Can. J. Math. **39** (1987), 492–511.

[9] G. Price, *Shifts of integer index on the hyperfinite* II_1*-factor,* Pac. J. Math. **132** (1988), 379–390.

College of Business Administration and Information Science
Koshien University
Takarazuka, Hyogo, 665
Japan

Index Theory for Type III Factors

HIDEKI KOSAKI

1. Introduction

We describe the structure of (finite-index) inclusion of type III factors based on analysis of involved flows of weights. Roughly speaking, a type III index theory splits into a "purely type III" index theory and an (essentially) type II index theory. The factor flows constructed in [1] serve as the complete invariant for the former in the AFD case while the latter can be analyzed by paragroups or quantized groups (as announced in [7]). Therefore, classification of subfactors in an AFD type III factor reduces to classification of factor flows and an "equivariant" paragroup theory.

In his beautiful thesis [6], Loi has analyzed inclusion of type III_λ $(0 < \lambda < 1)$ factors. Some of the results here are closely related to his theorems.

Basic facts on index theory can be found in the Jones' fundamental article [4] (or in [5], [8]) while our standard reference on the theory of operator algebras is [9].

2. Common finite extension of flows of weights

We recall the construction in [1], which plays an important role in the present article. Throughout, let M be a type III factor with a type III subfactor N. Assume that there exists a normal conditional expectation E from M onto N satisfying Index $E < \infty$(see [5]).

Fixing a faithful normal state ϕ on N, we set $\psi = \phi \circ E$. Thanks to $\sigma^{\psi}\mid_N = \sigma^{\phi}$, we have the following inclusion of von Neumann algebras of type II_∞:

$$\widetilde{M} = M \rtimes \sigma^{\psi}\, \mathbf{R} \supseteqq \widetilde{N} = N \rtimes \sigma^{\phi}\, \mathbf{R}.$$

Their centers $Z(\widetilde{M})$ and $Z(\widetilde{N})$ are included in $Z(\widetilde{M} \cap \widetilde{N}')$. The dual action

$\{\theta_t\}_{t\in\mathbf{R}}$ restricted to $Z(\widetilde{M})$ (resp. $Z(\widetilde{N})$) is ergodic and gives rise to the flow of weights of M (resp. N). Note that the dual action is not necessarily ergodic on the larger algebra $Z(\widetilde{M}\cap\widetilde{N}')$. The Pimsner-Popa inequality ([8]) plays a crucial role in the proof of the following result:

Theorem 1 ([1]). *Let $(X_N, F_t^N), (X_M, F_t^M)$ be the flows of weights of N, M respectively. There exists a (not necessarily ergodic) common finite extension (X, F_t) in the sense that* (i) *X is isomorphic to $X_N \times \{1, 2, \ldots, n\}$ as a measure space for some $n \leq$ Index E, and the projection map $\pi_N : X \to X_N$ intertwines F_t and F_t^N.* (ii) *X is also isomorphic to $X_M \times \{1, 2, \ldots, m\}$ for some $m \leq$ Index E, and the projection map $\pi_M : X \to X_M$ intertwines F_t and F_t^M.*

A few remarks are in order: (i) The above construction does not depend upon a choice of ϕ. (ii) If the involved flows are periodic, for example, then we obtain $mn \leq$ Index E. (iii) It is possible to construct (purely type III)$E : M \to N$ from a given common finite extension of two ergodic (conservative) flows (see [2]).

3. Canonical decomposition

The Takesaki duality implies that study of $M \supseteqq N$ is the same as that of $\widetilde{M} \rtimes_\theta \mathbf{R} \supseteqq \widetilde{N} \rtimes_\theta \mathbf{R}$. The given conditional expectation E lifts up to $\hat{E} : \widetilde{M} \to \widetilde{N}$ and then to $(\hat{E})\hat{} : \widetilde{M} \rtimes_\theta \mathbf{R} \to \widetilde{N} \rtimes_\theta \mathbf{R}$, which is $E \otimes \mathrm{Id}_{B(H)}$ via the duality and conjugate to E. Since $\widetilde{M}$ might have a large center, a priori we do not know if $(\hat{E})^{-1}$ is a scalar. However, $(\hat{E})^{-1}(1) = (\text{Index } E) \times 1$ can be proved.

We now consider the following inclusion:

$$\widetilde{M} \supseteqq \widetilde{M} \cap Z(\widetilde{M} \cap \widetilde{N}')' \supseteqq \widetilde{N} \vee Z(\widetilde{M} \cap \widetilde{N}') \supseteqq \widetilde{N}.$$

The above $\hat{E}$ comes from the canonical trace on $\widetilde{M}$ (and the dual weight $\hat{\psi}$), and the trace (and $\hat{\psi}$) is semi-finite on each of the above algebras. Consequently, $\hat{E}$ splits to

$$\widetilde{M} \xrightarrow{\hat{F}} \widetilde{M} \cap Z(\widetilde{M} \cap \widetilde{N}')' \xrightarrow{\hat{G}} \widetilde{N} \vee Z(\widetilde{M} \cap \widetilde{N}') \xrightarrow{\hat{H}} \widetilde{N}.$$

These conditional expectations all intertwine the respective dual actions. By taking crossed products relative to the dual actions, we obtain

Theorem 2. *There exist von Neumann algebras $\mathfrak{A}$ and $\mathfrak{B}$ satisfying $M \supseteqq \mathfrak{A} \supseteqq \mathfrak{B} \supseteqq N$. Also the conditional expectation E is decomposed as*

follows:

$$M \xrightarrow{F} \mathfrak{A} \xrightarrow{G} \mathfrak{B} \xrightarrow{H} N.$$

Here, $\mathfrak{A}$ and $\mathfrak{B}$ have the same finite dimensional center. More precisely, the flows of weights of $\mathfrak{A}$ and $\mathfrak{B}$ are exactly the common finite extension (X, F_t) in Theorem 1.

The middle inclusion $\mathfrak{A} \supseteqq \mathfrak{B}$ is what we called the "type II part" in 1. Since $\tilde{\mathfrak{A}} \supseteqq \tilde{\mathfrak{B}}$ have the same center $Z(\widetilde{M} \cap \widetilde{N}')$, we can consider a family of type II_∞ indices by looking at the common central decomposition. Since $\hat{G}$ comes from the canonical trace, we are actually dealing with Jones' indices based on coupling constants. It can be shown that these index values are constant on each ergodic component of (X, F_t).

4. Purely type III index theory

Next we would like to analyze $M \supseteqq \mathfrak{A}$ and $\mathfrak{B} \supseteqq N$. They are completely "dual" to each other so that we will just state results on $\mathfrak{B} \supseteqq N$ in what follows.

We are mainly interested in inclusion of factors. However, our Index E does depend on a choice of E, which is slightly inconvenient. We thus use the following recent result due to F. Hiai:

Theorem 3 ([3]). *There exists a unique normal conditional expectation $E_0 : M \to N$ satisfying*

$$\textit{Index } E_0 = \textit{Min } \{\textit{Index } E;$$
$$E \textit{ is a normal conditional expectation from } M \textit{ onto } N\}.$$

This E_0 is characterized by the properties

(i) $E_0 \mid_{M \cap N'}$ *is a trace,*

(ii) $E_0^{-1} = (\text{Index } E_0) E_0$ *on* $M \cap N'$.

From now on we choose $E = E_0$ in our analysis. Then we can show $\{Z(\widetilde{M} \cap \widetilde{N}')\}_\theta = Z(M \cap N')$. Let $p_i, i = 1, 2, \ldots, k = \dim Z(M \cap N')$, be the central minimal projections in $M \cap N'$. For each i, the i-th ergodic component X_i of X (corresponding to p_i) itself gives rise to a common finite extension of the two flows of weights. We assume that on X_i the Jones' index described at the end of 3 is c_i and that π_N and π_M are n_i to one and m_i to one respectively. (Hence n_i's sum up to n and m_i's sum up to m).

Theorem 4. *With the above mentioned notations, we have*

$$\text{Index } E = \{\sum_{i=1}^{k}(m_i n_i c_i)^{1/2}\}^2$$

and

$$E(p_i) = (m_i n_i c_i)^{1/2} / \sum_{j=1}^{k}(m_j n_j c_j)^{1/2} \quad \text{for each } i.$$

Description of $\hat{F}$ is very simple while $\hat{H}$ can be expressed by using these $E(p_i)$'s. To avoid unnecessary complications, in what follows we will further assume $\dim Z(M \cap N') = 1$. (Otherwise, look at $p_i M p_i \supseteqq N p_i$ instead). This assumption makes $\mathfrak{a}$ and $\mathfrak{B}$ factors, and Index G is exactly the Jones' index c at the end of 3. We have $\mathfrak{B} \cap N' = \mathbb{C}1$ and a direct computation of $(\hat{H})^{-1}$ yields Index $H = n$.

In the case where the factor flow $(X, F_t) \overset{\pi_N}{\to} (X_N, F_t^N)$ is "not so complicated", we can find a finite group G of order n (and its outer action either on N or on $\mathfrak{B}$) such that $\mathfrak{B} = N \rtimes G$ (crossed product) or $N = \mathfrak{B}_N$ (fixed point subalgebra). This group G appears naturally as a certain Weyl group. Generally this is false (in the III_0 case), and we have the next handy criteria.

Theorem 5. *Assume that involved factors are AFD.*

(i) *Define the action $\{\theta_t\}_{t\in\mathbb{R}}$ on $M_n(\mathbb{C})\otimes Z(\tilde{N})$ (identified with the field of the $n \times n$ matrix algebras over X_N) as follows: a matrix unit e_{ij} on $\omega \in X_N$ is sent to $e_{i'j'}$ on $F_t^N(\omega) \in X_N$, where F_t sends $(\omega, i) \in X = X_N \times \{1, 2, \ldots, n\}$ to $(F_t^N(\omega), i') \in X$. Then $\mathfrak{B} = N \rtimes G$ (a finite group G of order n is acting outerly on N) if and only if the fixed point subalgebra $\{M_n(\mathbb{C}) \otimes Z(\tilde{N})\}_\theta$ is abelian and n dimensional.*

(ii) *Define the action $\{\theta_t\}_{t\in\mathbb{R}}$ on $\mathbb{C}^{n^2} \otimes Z(\tilde{N})$ as follows: f_{ij} on $\omega \in X_N$ is sent to $f_{i'j'}$ on $F_t^N(\omega) \in X_N$, where $\{f_{ij}\} i, j = 1, 2, \ldots, n$ is the canonical basis for $\mathbb{C}^{n^2}$ and i', j' are determined as in* (i). *Then $N = \mathfrak{B}_G$ if and only if $\{\mathbb{C}^{n^2} \otimes Z(\tilde{N})\}_\theta$ is n dimensional.*

The theorem probably remains valid for non-AFD factors.

The next result says that the common finite extension completely determines "purely type III" inclusion in the AFD case.

Theorem 6. *When involved factors are AFD, the factor flow $(X, F_t) \overset{\pi_N}{\to} (X_N, F_t^N)$ (up to isomorphism) is the complete invariant for the inclusion $\mathfrak{B} \supseteqq N$ (up to conjugacy).*

In other words, purely type III inclusion in the AFD case always looks like the "model inclusion" constructed in [2].

REFERENCES

[1] T. Hamachi and H. Kosaki, *Index and flow of weights of factors of type III,* Proc. Japan Academy, **64** (1988), 11–13.
[2] T. Hamachi and H. Kosaki, *Inclusion of type III factors constructed from ergodic flows,* Proc. Japan Academy, **64** (1988), 195–197.
[3] F. Hiai, *Minimizing indices of conditional expectations onto a subfactor,* Pub. RIMS, Kyoto Univ. , **24** (1988), 673–678.
[4] V. Jones, *Index for subfactors,* Inven. Math., **72** (1983), 1–25.
[5] H. Kosaki, *Extension of Jones' theory on index to arbitrary factors,* J. Funct. Anal., **66** (1986), 123–140.
[6] P.H. Loi, *On the theory of index and type III factors,* Thesis, Penn. State Univ., 1988.
[7] A. Ocneanu, *Quantized groups, string algebras and Galois theory for algebras,* Operator Algebras and Applications Vol. II, London Math. Soc. Lecture Note Series **136** , Cambridge Univ. Press, 1988.
[8] M. Pimsner and S. Popa, *Entropy and index for subfactors,* Ann. Sci. École Norm. Sup., **19** (1986), 57–106.
[9] S. Stratila, *Modular Theory of Operator Algebras,* Abacuss Press, Tunbgidge Wells (1981).

Hideki Kosaki
Department of Mathematics
College of General Education
Kyushu University
Fukuoka, 810, Japan

Relative Entropy of a Fixed Point Algebra

SATOSHI KAWAKAMI

Introduction

The relative entropy $H(M|N)$ for a pair $N \subset M$ of finite von Neumann algebras was introduced and studied by M. Pimsner and S. Popa in [7]. One of their important results was to clarify the relationship between $H(M|N)$ and the Jones index $[M : N]$ for a pair of finite factors ([2]). On the other hand, in [1], V. Jones succeeded in classifying actions of a finite group G on the hyperfinite type II_1 factor R, up to conjugacy, associated with normal subgroups of G, characteristic invariants and inner invariants.

The present article is devoted to a study of the relative entropy $H(M|M^\alpha)$ where M^α is the fixed point subalgebra of a von Neumann algebra M under an action α of a locally compact group G. Section 1 reports on the joint work of the author and H. Yoshida [3], [4]. Complete formulas for $H(M|M^\alpha)$ are given by applying Pimsner–Popa's deep results [7] and our complementary reduction theory [4]. Section 2 reports on the result of classification of actions α of G on R such that $H(R|R^\alpha) < +\infty$. Each conjugacy class of these actions is also decided with their invariants. This result is a generalization of Jones' [1]. An advantage of this work is the construction of a model action corresponding to each characteristic invariant and inner invariant. The technique of this construction stems from the viewpoint of representation theory developed by G.W. Mackey [5].

I would like to express my special thanks to Professor O. Takenouchi for his constant encouragements, to Professor Y. Katayama for his stimulating discussions, and to Professor C. Sutherland for his valuable suggestions.

1. Computations of $H(M|M^\alpha)$

Let M be a finite von Neumann algebra on a separable Hilbert space with a faithful normal normalized trace τ and N be a von Neumann subalgebra of M. For a positive element x of M, put $h(x) = \tau\eta E(x) - \tau\eta(x)$

where E is the unique τ-preserving conditional expectation of M to N and η is a continuous function for $t \geq 0$ such that $\eta(0) = 0$ and $\eta(t) = -t\log t$ if $t > 0$. Let $S(M)$ denote the family of all partitions of the unity in M and for each $\Delta = (x_i)_{i\in I}$ in $S(M)$; set $H_\Delta(M|N) = \sum_{i\in I} h(x_i)$. Pimsner–Popa's relative entropy $H(M|N)$ is now given by

$$H(M|N) = \sup\{H_\Delta(M|N); \Delta \in S(M)\}.$$

Fundamental properties and results on $H(M|N)$ are described in [3], [4], [7]; we omit the details.

Let α be an action of a second-countable locally compact group G on the von Neumann algebra M. M^α, or M^G if there is no need of mention of α, denotes the fixed point algebra of M under the action α. We shall give some formulas for the relative entropy $H(M|M^\alpha)$.

The action α induces an action of G on the center $Z(M)$ of M. Corresponding to $Z(M)^G$, (M, τ) is decomposed into a direct integral by general reduction theory. Namely, there exists a standard probability measure space (S, m) such that

$$(M, \tau) \cong \int_S^\oplus (M(s), \tau^s)dm(s) \qquad \text{and}$$

$$(Z(M)^G, \tau) \cong \{\text{diagonal operators}\} \cong L^\infty(S, m).$$

Moreover, for almost all $s \in S$, there exist actions α^s of G on the component algebra $M(s)$ such that the field $s \to \alpha^s$ of actions is measurable, and the relative entropy $H(M(s)|M(s)^G)$ are defined associated with the normalized traces τ^s of $M(s)$.

Proposition 1.1 *In the above situation, we get*

$$H(M|M^G) = \int_S (H(M(s)|M(s)^G)dm(s).$$

Here, we note that almost all actions α^s of G on $M(s)$ are centrally ergodic, namely, $Z(M(s))^G = \mathbb{C}$.

Proposition 1.2. *Suppose that an action α of G on M is centrally ergodic. Then, we get the following.*

(i) *If $H(M|M^G) < +\infty$, then $Z(M)$ is atomic.*

(ii) *When $Z(M)$ is atomic, $\{p_i\}_{i\in I}$ denotes the set of all atoms of $Z(M)$ and H denotes the stabilizer at p for a fixed projection p among p_i's, we have*

$$H(M|M^G) = \sum_{i\in I} \eta\tau(p_i) + H(M_p|M_p^H).$$

It remains to compute the relative entropy $H(M|M^G)$ in the case that M is a factor of finite type.

Let α be an action of a second countable locally compact group G on a factor M of type II_1 and let $K(\alpha)$ denote the subgroup $\{g \in G;\ \alpha_g$ is an inner automorphism of $M\}$ of G. $K(\alpha)$ is often abbreviated by K. Suppose that $H(M|M^\alpha) < +\infty$. Then, we get the following.

Lemma 1.3. *In the above situation,*

(i) $(M^\alpha)' \cap M$ *is atomic.*

(ii) K *is a closed normal subgroup of* G *such that* G/K *is a finite group.*

Hence, by the routine method of choosing a suitable Borel cross section, it is easily seen that there exists a Borel multiplier μ of K and a Borel μ-representation V of K such that $\alpha_k = \mathrm{Ad}\, V_k$ and $V_h V_k = \mu(h,k)V_{hk}$ ($V_e = 1$). Moreover, there exists a Borel $\mathbf{T}$-valued function λ of $G \times K$ satisfying $\alpha_g(V_k) = \lambda(g, gkg^{-1})V_{gkg^{-1}}(g \in G, k \in K)$. Here, we denote by $X(\mu)$, or simply by X, the set of all unitary equivalence classes of finite-dimensional, irreducible μ-representations of K and we define the action $\hat{\lambda}$ of G on $X(\mu)$ by $\hat{\lambda}_g(U_k) = \lambda(g, gkg^{-1})U_{gkg^{-1}}$ ($g \in G, k \in K$) for $[U] \in X(\mu)$. We denote by $\Omega(\lambda, \mu)$, or simply by Ω, the G-orbit space of $X(\mu)$ under this action $\hat{\lambda}$. For each orbit $\omega \in \Omega$, set $d_\omega = \dim x$ ($x \in \omega$) and $|\omega| =$ the number of $x \in \omega$. Denote by $\{f_x\}_{x\in X}$ the family of central minimal projections of $V(K)''$, which corresponds to the canonical central decomposition of V. Set $e_\omega = \sum_{x\in\omega} f_x$ for $\omega \in \Omega$.

Lemma 1.4. *In the above situations,* $\{f_x; x \in X, f_x \neq 0\}$ *is the family of all atoms of* $Z(M^K)$ *and* $\{e_\omega; \omega \in \Omega, e_\omega \neq 0\}$ *is the family of all atoms of* $Z(M^G)$. *Moreover,* $(M^G)' \cap M = (M^K)' \cap M = V(K)''$.

From these considerations we get the following.

Theorem 1.5. *Let* M *be a factor of type* II_1 *with the canonical trace* τ *and* α *be an action of a locally compact group* G *on* M. *If* $H(M|M^G) < +\infty$, *we have*

$$\begin{aligned} H(M|M^G) &= H(M|M^K) + H(M^K|M^G) \\ &= \log|G/K| + \sum_{\omega\in\Omega} \tau(e_\omega)\log(d_\omega^2|\omega|/\tau(e_\omega)). \end{aligned}$$

When G is a finite group and M is the hyperfinite type II_1 factor R, V. Jones constructed a model action $s_{G,K}^{(\lambda,\mu)}$ of G on $R \times_\mu K$ ($\cong R$) for each

$[\lambda, \mu] \in \Lambda(G, K)$ [1]. The next is an immediate consequence of Theorem 1.5.

Corollary 1.6. *Let α be an action of a finite group G on a factor M of type* II_1. *Then,* $0 \leq H(M|M^\alpha) \leq \log|G|$. *When $M = R$, $H(M|M^\alpha)$ attains the maximum value $\log|G|$ if and only if α is conjugate to $s_{G,K}^{(\lambda,\mu)}$ with $K(\alpha) = K$ and $\Lambda(\alpha) = [\lambda, \mu]$.*

2. Classification of Actions With $H(R|R^G) < +\infty$

Let G be a second countable locally compact group and K be a closed normal subgroup of G. Then, the cohomology group $\Lambda(G, K)$ is defined in a similar way to Jones' [1] or Ocneanu's [6]. For each $[\lambda, \mu] \in \Lambda(G, K)$, the space $X(\mu)$, the action $\hat{\lambda}$ of G on $X(\mu)$, and the orbit space $\Omega(\lambda, \mu)$ are canonically defined as described in the latter part of Section 1. For example, $X(\mu)$ is defined as the set of unitary equivalence classes of finite dimensional irreducible μ-representations. We note that $X(\mu)$ may be empty in general. Morover, $H^1(K)^G$ is defined as a subgroup of the character group of K in a similar way to Jones [1]. $H^1(K)^G$ acts canonically on $X(\mu)$ and we denote such action by ∂. Since the action ∂ commutes with the action $\hat{\lambda}$, the action ∂ induces an action of $H^1(K)^G$ on Ω, which is also denoted by ∂. Let $P(\Omega)$ denote the set of countable probability measures on Ω. Then, the action ∂ induces an action $\overline{\partial}$ of $H^1(K)^G$ on $P(\Omega)$. For two measures m_1 and m_2 in $P(\Omega)$, we say that m_1 is equivalent to m_2 if there exists $\eta \in H^1(K)^G$ such that $m_2 = \overline{\partial}_\eta(m_1)$. $I(\Omega)$ denotes the equivalence classes of $P(\Omega)$.

Let α be an action of G on a factor M of type II_1 with $H(M|M^\alpha) < +\infty$. Then, such an action α defines a closed normal subgroup $K(\alpha)$ $(= K)$ of G such that G/K is a finite group, the characteristic invariant $\Lambda(\alpha) = [\lambda, \mu]$ in $\Lambda(G, K)$, and the inner invariant $\iota(\alpha) = [(\tau(e_\omega)_\omega)]$ in $I(\Omega(\lambda, \mu))$ as described in Section 1. It is easily checked that these are invariant up to conjugacy of actions of G on M. When G is a finite group, V. Jones [1] succeeded in characterizing the conjugacy classes of actions of G on the hyperfinite type II_1 factor R by these invariants. The next is a slight generalization of his result.

Theorem 2.1. *Let α and β be actions of a locally compact group G on the hyperfinite type* II_1 *factor R with $H(R|R^G) < +\infty$. Then, α is conjugate to β if and only if $K(\alpha) = K(\beta)$, $\Lambda(\alpha) = \Lambda(\beta)$, and $\iota(\alpha) = \iota(\beta)$.*

The most important part to prove this theorem is the fact of the uniqueness of actions of a finite group on R up to cocycle conjugacy, which has been obtained by Jones [1]. The next task is to construct a model action corresponding to those invariants. However, Jones' technique of construction seems not to be available in our general situation.

We shall briefly explain the construction of our model action. Let K be a closed normal subgroup of G such that G/K is a finite group and let (λ, μ) be a representative of an element of $\Lambda(G, K)$. Suppose that $\Omega(\lambda, \mu) \neq \emptyset$ and the measure $\iota \in I(\Omega)$ is concentrated on one point $\{\omega\}$ in $\Omega(\lambda, \mu)$. Then, we get a finite-dimensional μ-representation V of K on $\ell^2(\omega, H)$ by $V = \sum_{x\in\omega}^{\oplus} V^x$ where V^x's are μ-representations on the same space H belonging to x in $X(\mu)$. Since $\hat{\lambda}_g(V) \cong \sum_{x\in\omega}^{\oplus} V^{x\cdot g} \cong \sum_{x\in\omega}^{\oplus} V^x = V$, there exists a Borel map W from G to $U(\ell^2(\omega, H))$ satisfying

(a) $\hat{\lambda}_g(V_k) = (\mathrm{Ad}\, W_g)(V_k) \qquad (g \in G, k \in K)$

(b) $W_k = V_k$ if $k \in K$.

(c) $W_g W_\ell = c_0(g, \ell) W_{g\ell}$ where $c_0(g, \ell) \in U(\ell^\infty(\omega))$ for $g, \ell \in G$.

We note that c_0 is $U(\ell^\infty(\omega))$-valued α^0-twisted 2-cocycle of G where $\alpha_g^0 = \mathrm{Ad}\, W_g$. Let $\mathbf{T}$ denote the one-dimensional torus of scalar operators in $U(\ell^\infty(\omega))$ and π; $G \to Q = G/K$ be the canonical quotient map.

Lemma 2.2 *There exists a Borel map c from $Q \times Q$ to $U(\ell^\infty(\omega))$ such that $c_0(g, \ell)$ is cohomologous to $c(\pi(g), \pi(\ell))$ as $(U(\ell^\infty(\omega))/\mathbf{T})$-valued α^0-twisted cocycles of G.*

Thus, we may assume that W is chosen to satisfy the above properties (a), (b), and

(c)′ $\alpha_g^0 \alpha_\ell^0 = \mathrm{Ad} c(\pi(g), \pi(\ell)) \alpha_{g\ell}^0$ on $B(\ell^2(\omega, H))$ where $\alpha_g^0 = \mathrm{Ad}\, W_g$.

(d) $c(p, q) c(pq, r) = \delta(p, q, r) \alpha_p^0(c(q, r)) c(p, qr)$ where $p, q, r \in Q$ and $\delta(p, q, r) \in \mathbf{T}$.

Let u be the left regular representation of Q and $\{e_p\}_{p\in Q}$ be the family of all atoms of $\ell^\infty(Q)$. Put $x_p = \sum_{q\in Q} c(p, q) e_{pq}$ and $c_1 1(p, q) = \sum_{r\in Q} \delta(p, q, r) e_{pqr}$ on $\ell^2(Q) \otimes \ell^2(\omega, H)$. Moreover, for each $g \in G$, define the automorphism of $B(\ell^2(Q) \otimes \ell^2(\omega, H))$ by

$$\alpha_g^1 = \mathrm{Ad}\, x_{\pi(g)} (\mathrm{Ad}\, u_{\pi(g)} \otimes \mathrm{Ad}\, W_g).$$

Then, α_g^1 satisfies

(e) $\alpha_g^1 \alpha_\ell^1 = \mathrm{Ad}\, c^1(\pi(g), \pi(\ell)) \alpha_{g\ell}^1$ on $B(\ell^2(Q)) \otimes B(\ell^2(\omega, H))$

(f) $\alpha_g^1 \alpha_\ell^1 = \alpha_{g\ell}^1$ on $A_1 = \ell^\infty(Q) \otimes B(\ell^2(\omega, H))$.

We note that $c_1(p, q)$ satisfies a condition similar to (d). Therefore, by repeating the same procedure as above, we get an action α^2 of G on $A_2 =$

$\ell^\infty(Q) \otimes B(\ell^2(Q)) \otimes B(\ell^2(\omega, H))$ and inductively an action α^n of G on

$$A_n = \ell^\infty(Q) \otimes \underbrace{B(\ell^2(Q)) \otimes \cdots \otimes B(\ell^2(Q))}_{(n-1)\text{ times}} \otimes B(\ell^2(\omega, H)).$$

The desired action α of G on R is now obtained by $\alpha = \lim_{n\to\infty} \alpha^n$ on $R = \lim_{n\to\infty} A_n$. In the general case, the desired action is also obtained by a slightly modified form of the above. We omit the details.

REFERENCES

[1] V. Jones, *Actions of finite groups on the hyperfinite type* II_1 *factor,* Memoirs of A.M.S. **237** (1980).
[2] V. Jones, *Index of subfactors,* Invention Math. **72** (1983), 1–25.
[3] S. Kawakami and H. Yoshida, *Actions of a finite group on finite von Neumann algebras and the relative entropy,* J. Math. Soc. Japan **39** (1987), 609–626.
[4] S. Kawakami and H. Yoshida, Math. Japon., *Reduction theory on the relative entropy,* **33**(1988), 975–990.
[5] G.W. Mackey, *Unitary representations of group extensions* I, Acta. Math. **99** (1958), 265–311.
[6] A. Ocneanu, *Actions of discrete amenable groups on von Neumann algebras,* Springer Lecture Notes, **1138** (1985).
[7] M. Pimsner and S. Popa, *Entropy and index for subfactors,* Ann. Sci. Éc. Norm. Sup. **19** (1986), 57–106.

Nara University of Education
Nara, 630
Japan

Jones Index Theory for C*-Algebras

YASUO WATATANI

The notion of index $[M : N]$ was introduced by Jones [13] as an invariant for subfactors N of a factor M of type II_1. Subsequently Kosaki [18] defined an index E for a conditional expectation E of an arbitrary factor M onto a subfactor N using the spatial theory of Connes [6] and the theory of operator-valued weights of Haagerup [9]. We shall define an index E for a conditional expectation E on a C^*-algebra. This index theory for C^*-algebras is a mixture of the index theory by Jones and the theory of Morita equivalence by Rieffel [24], [25]. We establish the link between transfer in K-theory and a multiplication by Index E.

Let $M \supset N$ be factors of type II_1 on a Hilbert space H. Then index $[M : N]$ is defined by

$$[M : N] = \dim_N(H)/\dim_M(H)$$

if N' is finite, where $\dim_M(H)$ is the coupling constant of M. Surprisingly, Jones shows that

$$[M : N] \in \{4\cos^2 \pi/n; n = 3, 4, 5, \ldots\} \cup [4, \infty].$$

Let $M \supset N$ be von Neumann algebras. We denote by $P(M, N)$ the set of normal semifinite faithful operator-valued weights of M into $\hat{N}_+$, the extended positive part of N. Haagerup [9] shows that there exists a bijection of $P(M, N)$ and $P(N', M')$, $E \to E^{-1}$, determined by the equation of spatial derivatives in the sense of Connes [6]:

$$d(\varphi E)/d\psi = d\varphi/d(\psi E^{-1}),$$

where φ and ψ are normal faithful semifinite weights on N and M. Given a factor M, a subfactor N of M and a normal faithful semifinite conditional expectation $E: M \to N$, the index of E, denoted by Index E, is defined by Kosaki [18] as the element $E^{-1}(1)$ of $\hat{M}_+$. Since M is a factor, $E^{-1}(1)$ is a positive number (possibly infinite). In fact Kosaki also shows that

$$\operatorname{Index} E \in \{4\cos^2 n/\pi; n = 3, \dots\} \cup [4, \infty].$$

If $M \supset N$ are factors of type II_1 and $E: M \to N$ is the canonical conditional expectation, then Index $E = [M : N]$, the Index of Jones.

On the other hand, Pimsner and Popa [22] show that if $M \supset N$ are factors of type II_1 and $[M : N]$ is finite, then there exists a certain orthonormal basis of M over N. Our definition of finiteness of Index is introduced along these lines. Classification of subfactors have been deeply studied by Jones [13], Bion–Nadal [1], M. Choda [4], [5], Kawakami, Yoshida [16], [17], [33], Ocneanu [21], Pimsner and Popa [22], Wenzl [31], Goodman, de la Harpe and Jones [8]. And the case of factors of type III has been deeply analyzed by Hamachi and Kosaki [10], [11] and Loi [19] and Hiai [12]. Later, a preprint [3] by Brillet, Green and Hauvet appeared. They consider the index of conditional expectations of von Neumann algebras and some of their results are similar to ours.

We shall discuss the case of C^*-algebras. We can give a purely algebraic definition of Index E for a conditional expectation E on a k-algebra. Our definition coincides with that of Jones and Kosaki when we consider factors.

Let k be a commutative ring with an identity element 1. Throughout this note, we consider a k-algebra B and k-subalgebra A of B with the same identity element 1.

Definition 1. A *conditional expectation* $E: B \to A$ is an onto k-linear map satisfying

$$E(ab) = aE(b), \quad E(ba) = E(b)a \quad \text{and} \quad E(a) = a$$

for $a \in A$ and $b \in B$. But when we consider operator algebras, we always assume that E is positive. We call E *non-degenerate* if $E(Bb) = 0$ or $E(bB) = 0$ implies $b = 0$, for each $b \in B$. If we consider operator algebras, then E is non-degenerate if and only if E is faithful, that is, $E(b^*b) = 0$ implies $b = 0$, for each $b \in B$ because we have the following inequality:

$$\|E(x^*y)\| \le \|E(x^*x)\|^{1/2}\|E(y^*y)\|^{1/2}.$$

See J. Tomiyama [28] and Umegaki [29] for conditional expectations in operator algebras. Throughout this note we assume that conditional expectations are non-degenerate unless we mention that they are not.

Definition 2. A finite family $\{(u_1, v_1), \dots, (u_n, v_n)\} \subset B \times B$ is called a *quasi-basis* for E if

$$\sum_i u_i E(v_i x) = x = \sum_i E(x u_i) v_i \quad \text{for } x \in B.$$

A conditional expectation $E: B \to A$ is of *index-finite type* if there exists a quasi-basis for E. In this case we define the *index* of E by

$$\operatorname{Index} E = \sum_i u_i v_i \in B.$$

Remark. (1) If E is of index-finite type, then $\operatorname{Index} E$ is, in fact, in Center B and the value, $\operatorname{Index} E$, does not depend on the choice of quasi-bases.

(2) If B and A are C^*-algebras, then we can choose a quasi-basis $\{(u_i, v_i): i = 1, \ldots, n\}$ with $v_i = u_i^*$.

(3) Sometimes we do not know that E is non-degenerate a priori. But the existence of a quasi-basis guarantees that E is non-degenerate. In fact if $E(Bb) = 0$, then $b = \sum_i u_i E(v_i b) = 0$. Similarly $E(bB) = 0$ implies that $b = 0$.

Example 3. Let G be a group and H a subgroup of G. Consider the group algebras $B = k[G]$ and $A = k[H]$ with bases $\{\lambda(g);\ g \in G\}$ and $\{\lambda(h);\ h \in H\}$ over k. Define a conditional expectation $E: k[G] \to k[H]$ by

$$E\left(\sum_{g \in G} x(g)\lambda(g)\right) = \sum_{h \in H} x(h)\lambda(h)$$

where $x(g) \in k$ for $g \in G$. Then E is of index-finite type if and only if $[G : H]$ is finite and $\operatorname{Index} E = [G : H]$.

Example 4. Let G be a countable discrete group and H a subgroup of G. Consider the reduced group C^*-algebras $B = C_r^*(G)$ and $A = C_r^*(H)$. There exists a conditional expectation $E: B \to A$ such that

$$E\left(\sum_{g \in G} x(g)\lambda(g)\right) = \sum_{h \in H} x(h)\lambda(h).$$

As with Example 3, if $[G : H]$ is finite, then E is of index-finite type and $\operatorname{Index} E = [G : H]$.

Example 5. Let $B = M_2(\mathbb{C})$ be the algebra of 2×2 matrices over the complex numbers $\mathbb{C}$ and A be $\mathbb{C}1$. Let t be a scalar with $t \neq 0$ and $t \neq 1$. Define a conditional expectation $E_t: B \to A$ by

$$E_t\begin{pmatrix} a & b \\ c & d \end{pmatrix} = ta + (1 - t)d.$$

Then E_t is of index-finite type and

$$\operatorname{Index} E_t = (1/t + 1/(1-t))I.$$

If $0 < t < 1$, then $\operatorname{Index} E_t \in [4, \infty)$. But in general we have that

$$\{\operatorname{Index} E_t \in \mathbb{C}; t \in \mathbb{C}, t \neq 0 \text{ and } t \neq 1\} = \mathbb{C}\backslash\{0\}.$$

Therefore, positivity of conditional expectations E is essential if the values of $\operatorname{Index} E$ are to lie in

$$\{4\cos^2 \pi/n; n = 3, 4, \ldots\} \cup [4, \infty].$$

Example 6. If N is a subfactor of a factor M of type II_1 and $E\colon M \to N$ is the canonical conditional expectation determined by the trace, then the Jones index $[M : N]$ coincides with the $\operatorname{Index} E$ of Pimsner and Popa [22]. More generally, suppose that M is an arbitrary factor and $E\colon M \to N$ is a faithful normal conditional expectation. Then E is of index-finite type if and only if $\operatorname{Index} E$ is finite in the sense of H. Kosaki [18], and both values coincide.

Example 7. Let X and Y be compact T_2 spaces and $\pi\colon Y \to X$ a covering map. Put $B = C(Y)$ and $A = \{f\pi \in B;\ f \in C(X)\} \cong C(X)$. Assume that the number $n_x = \#\pi^{-1}(x)$ of sheets of covering $x \in X$ is bounded. Define a conditional expectation $E\colon B \to A$ by

$$E(f)(y) = (1/n_{\pi(y)})\Sigma\{f(z); \pi(z) = \pi(y)\}.$$

Then E is of index-finite type and $\operatorname{Index} E = [y \mapsto n_{\pi(y)}] \in B$. Therefore we can regard $\operatorname{Index} E$ as the number of sheets of the covering.

Example 8. Consider the free action of $Z_2 = \{1, g\}$ on $Y = S^1$, where $g \cdot y = -y$. Let $\alpha\colon Z_2 \to \operatorname{Aut} C(S_1)$ be the corresponding action. Define a conditional expectation $E\colon C(S_1) \to C(S_1)^\alpha$ by $E(f) = (f + \alpha_g(f))/2$. Then E is of index-finite type and $\operatorname{Index} E = 2$.

Example 9. Consider the non-free action of $Z_2 = \{1, g\}$ on $Y = S^1$ given by $g \cdot y = \overline{y}$ (the complex conjugate). Define a conditional expectation as in Example 8. Then E is *not* of index-finite type.

Example 10 (Jones). Let $e_1, e_2, e_3, \ldots$ be Jones projections [13] in the hyperfinite factor M of type II_1 for a fixed constant τ. So we have that

$e_ie_{i\pm 1}e_i = \tau e_i$ and $e_ie_j = e_je_i$ for $|i-j| \geq 2$. Let $B = C^*\{1, e_1, e_2, e_3, \ldots\}$ and $A = C^*\{1, e_2, e_3, \ldots\}$. Then there exists a conditional expectation $E: B \to A$ such that

$$E(xe_1y) = \tau xy \quad \text{for} \quad x, y \in B.$$

If $\tau^{-1} = 4\cos^2 \pi/n$, then E is of index-finite type and Index $E = 4\cos^2 \pi/n$. If $\tau^{-1} \geq 4$, then E is *not* of index-finite type.

Example 11. Let $1 \in N \subset M$ be a pair of finite dimensional C^*-algebras such that

$$N = \bigoplus_{j=1}^{n} N_j = \bigoplus_{j} q_j N \quad \text{with } N_j \cong \mathrm{Mat}(\nu_j)$$

and

$$M = \bigoplus_{i=1}^{m} M_i = \bigoplus_{i} p_i M \quad \text{with } M_i \cong \mathrm{Mat}(\mu_i)$$

where $\mathrm{Mat}(n)$ is the algebra of $n \times n$ matrices over $\mathbb{C}$. Let $\Lambda = \Lambda_M^N = (\lambda_{ij})_{ij}$ be an inclusion matrix. Let tr be a faithful trace on M. We define the *row-vector* $\overleftarrow{s} \in \mathbb{C}^n$ by

$$\overleftarrow{s} = (\mathrm{tr}(\rho_1), \ldots, \mathrm{tr}(\rho_n)),$$

where ρ_i is a minimal projection in M_i. Let $\overleftarrow{t} \in \mathbb{C}^n$ be the row-vector determining the trace $\mathrm{tr} = \mathrm{tr}|_N$ restricted to N. Then $\overleftarrow{t} = \overleftarrow{s}\, \Lambda_N^M$. Let $E: M \to N$ be the conditional expectation determined by tr with $\mathrm{tr}(E(x)y) = \mathrm{tr}(xy)$ for $x \in M$ and $y \in N$. Then E is of index-finite type and Index $E = \sum_i c_i p_i$, where the scalar $c_i = s_i^{-1} \sum_j \lambda_{ij} t_j$ and p_i are the minimal central projections of M. Moreover Index E is a scalar if and only if there exists a positive number β such that $\overleftarrow{s}\,(\Lambda\Lambda^t) = \beta\, \overleftarrow{s}$ and in this case Index $E = \beta$.

Example 12. Let $B = \mathbb{C}^n$ and $A = \mathbb{C}$. Let $t_1, t_2, \ldots, t_n \in \mathbb{R}^+$ be such that $\sum_i t_i = 1$. Define $E: B \to A$ by $E(x_1, x_2, \ldots, x_n) = \sum_i t_i x_i$. Then Index $E = (1/t_1, \ldots, 1/t_n) \in B$. Define a constant $c(E)$ by

$$c(E) = \sup\{c \in \mathbb{R}^+; E(b) \geq cb \text{ for all positive } b \in B\}.$$

Then $c(E) = \min\{t_i; i = 1, \ldots, n\}$. So that $c(E)^{-1} \geq$ Index E. And Index $E = c(E)^{-1}$ if and only if $t_i = 1/n$ for $i = 1, \ldots, n$.

Example 13. Let $B = M_2$ and $A = \mathbf{C}$. Define a conditional expectation $E_t: B \to A$ for t with $0 < t < 1$ by

$$E_t \begin{pmatrix} a & b \\ c & d \end{pmatrix} = ta + (1-t)b.$$

Then $\operatorname{Index} E_t = 1/t + 1/(1-t)$ and $c(E_t) = \min\{t, 1-t\}$.

Remark. In general, let $B \supset A$ be C^*-algebras and $E: B \to A$ be a conditional expectation of index-finite type. Then

$$c(E) \geq \|\operatorname{Index} E\|^{-2}.$$

Of course, this is a crude estimate.

Let $E: B \to A$ be a conditional expectation. Let $\mathcal{E}_0 = B = B_A$. Then $\mathcal{E}_0$ is a pre-Hilbert module over A with an A-valued inner product $\langle \eta(x), \eta(y) \rangle = E(x^*y)$ for $x, y \in B$, where we use the notation $\eta(x) \in \mathcal{E}_0$ for $x \in B$. Let $\mathcal{E}$ be a completion of $\mathcal{E}_0$ with the norm $\|\eta(x)\| = \|\langle \eta(x), \eta(x) \rangle\|^{1/2} = \|E(x^*x)\|^{1/2}$. Then $\mathcal{E}$ is a Hilbert C^*-module over A. Since we assume that E is faithful, the canonical map $\eta: B \to \mathcal{E}$ is injective. Let $\mathcal{L}_A(\mathcal{E})$ be the set of all (right) A-module homomorphism $T: \mathcal{E} \to \mathcal{E}$ with an adjoint A-module homomorphism $S: \mathcal{E} \to \mathcal{E}$ such that

$$\langle T\xi, \zeta \rangle = \langle \xi, S\zeta \rangle \quad \text{for all } \xi, \zeta \in \mathcal{E}.$$

We write T^* for S. Then $\mathcal{L}_A(\mathcal{E})$ is a C^*-algebra with the usual operator norm $\|T\| = \sup\{\|T\xi\|; \|\xi\| = 1\}$. For $\eta, \zeta \in \mathcal{E}$, let $\theta_{\xi,\zeta}$ be the "rank one" operator defined by $\theta_{\xi,\zeta}(\gamma) = \xi\langle \zeta, \gamma \rangle$. Let $K(\mathcal{E})$ be the closure of the linear space of $\{\theta_{\xi,\zeta}; \xi, \zeta \in \mathcal{E}\}$, the algebra of "compact operators." For $b \in B$, define $\lambda(b) \in \mathcal{L}_A(\mathcal{E})$ by

$$\lambda(b)\eta(x) = \eta(bx) \quad \text{for } x \in B.$$

Then $\lambda: B \to \mathcal{L}_A(B)$ turns out to be an injective *-homomorphism. For $x \in B$, put $e_A\eta(x) = \eta(E(x))$. Then e_A is a projection in $\mathcal{L}_A(\mathcal{E})$.

Definition 14. Let $C_r^*\langle B, e_A \rangle$ be the closure of the linear span of

$$\{\lambda(x)e_A\lambda(y) \in \mathcal{L}_a(\mathcal{E}); x, y \in B\}.$$

Then we call the C^*-algebra $C_r^*\langle B, e_A \rangle$ the *reduced C^*-basic construction.*

Remark. For $x, y \in B$, $\lambda(x)e_A\lambda(y^*)$ is a "rank one" operator $\theta_{\eta(x),\eta(y)}$. Therefore, the C^*-basic construction $C_r^*\langle B, e_A \rangle$ is the algebra $K(\mathcal{E})$ of "compact operators."

Let $B \supset A$ be C^*-algebras and $E: B \to A$ be a conditional expectation. We define a product and an involution * on $B \otimes_A B$ by

$$(b_1 \otimes b_2) \cdot (b_3 \otimes b_4) = b_1 E(b_2 b_3) \otimes b_4 \text{ for } b_1, b_2, b_3, b_4 \in B$$

and

$$(x \otimes y)^* = y^* \otimes x^*, \qquad \text{see [15].}$$

The involution * is well-defined by considering conjugate C^*-algebras. Thus $B \otimes_A B$ turns out to be a *-algebra. For $c = \sum_i x_i \otimes y_i \in B \otimes_A B$, let

$$||c||_{\max} = \sup\{||\pi(c)||; \pi \text{ is a } ^*\text{-representation of } B \otimes_A B\}.$$

Then

$$||c||_{\max} = \left\| \sum_i x_i \otimes y_i \right\|_{\max} = ||((E(x_i^* x_j))_{ij})^{1/2}((E(y_i y_j^*))_{ij})^{1/2}||.$$

Definition 15. The completion of $B \otimes_A B$ by the norm $|| \ ||_{\max}$ (after taking the quotient by the ideal $\{c \in B \otimes_A B; ||c||_{\max} = 0\}$ if necessary) is called the *unreduced C^*-basic construction.* We denote the C^*-algebra by $C^*_{\max}\langle B, e_A\rangle$.

Lemma 16. *The canonical *-homomorphism $\varphi: C^*_{\max}\langle B, e_A\rangle \to C^*_r\langle B, e_A\rangle$ is an onto *-isomorphism.*

Definition 17. In the following, we shall identify $C^*_{\max}\langle B, e_A\rangle$ with $C^*_r\langle B, e_A\rangle$. We call it the *$C^*$-basic construction* and denote it by $C^*\langle B, e_A\rangle$.

Now we shall determine the possible values of Index E when Index E is a scalar. The proof depends essentially on the C^*-basic construction and H. Wenzl's result [32].

Theorem 18. *Let $B \supset A$ be C^*-algebras and $E: B \to A$ be a conditional expectation of index-finite type. Suppose that Index E is a scalar. Then Index E is in*

$$\{4\cos^2 \pi/n; n = 3, 4, 5, \ldots\} \cup [4, \infty].$$

Remark. If B is a simple C^*-algebra, then Index E is always scalar because Center $B = \mathbb{C}$.

Suppose that $E: B \to A$ is of index-finite type. Then $B \otimes_A B$ and $C^*\langle B, e_A\rangle$ are canonically isomorphic. Moreover, the left (and right) A-module B is a finitely generated projective generator, and A and $C^*\langle B, e_A\rangle$ are (strong) Morita equivalent [24], [25]. We need this last point here. For example, the existence of a Markov trace is deduced from the observation by Rieffel [23] that there is a correspondence between traces on C^*-algebras which are Morita equivalent. Next we consider a relation between Index E for C^*-algebras and K-theory. Let $i: A \to B$ be the canonical injection. Then i induces $i_*: K_0(A) \to K_0(B)$. We shall construct a map going in the other direction. Let $\psi: A \to C^*\langle B, e_A\rangle$ be the injection defined by $\psi(a) = ae_A$. Then $\operatorname{Im}\psi$ is a full corner in B and ψ induces the isomorphism $\psi_*: K_0(A) \to K_0(C^*\langle B, e_A\rangle)$ by Rieffel [23].

Definition 19. The *transfer map* $T_E: K_0(B) \to K_0(A)$ is defined by the composition $T_E = (\psi_*)^{-1} \circ j_*$, where $j: B \to C^*\langle B, e_A\rangle$ is the canonical embedding preserving identity.

Remark. The transfer map T_E does not depend on the choice of a conditional expectation $E: B \to A$ of index-finite type. It coincides with the restriction $\operatorname{Res}[M_B] = [M_A]$ for a finitely generated projective B-module $M = M_B$.

Example 20. Let S_n be the symmetric group of degree n. Let $E: B = C^*(S_3) \to A = C^*(S_2)$ be the usual conditional expectation. Then we have the following Bratteli diagram for $C^*(S_2) \to C^*(S_3) \to C^*\langle B, e_A\rangle$:

```
   1   1
  / \/ \
 1   2   1
  \ /\ /
   3   3
```

Then $K_0(A) = \mathbf{Z} \otimes \mathbf{Z}$ and $K_0(B) = Z \oplus Z \oplus Z$. Therefore the natural map $i_*: K_0(A) \to K_0(B)$ is given by $\begin{pmatrix} 1 & 0 \\ 1 & 1 \\ 0 & 1 \end{pmatrix}$ and the transfer $T_E: K_0(B) \to K_0(A)$ is given by $\begin{pmatrix} 1 & 1 & 0 \\ 0 & 1 & 1 \end{pmatrix}$. Hence the transfer T_E is something like a "transpose" of the natural map i. This sort of observation is essentially made in Jones [14].

We shall investigate the transfer map T_E in more detail. We restate some known facts, see Rieffel [24], Blackadaar [2].

Lemma 21. *If $E\colon B \to A$ is a conditional expectation of index-finite type and $\{(u_i, u_i^*); i = 1, \ldots, n\}$ is a quasi-basis for E. Then there exists an injective *-homomorphism $\psi\colon B \to A \otimes M_n$ such that $\psi(b) = (E(u_i^* b u_j))_{ij}$. Choose another quasi-basis $\{(s_j, s_j^*); j = 1, \ldots, m\}$ for E. By adding 0, we may assume that $n = m$. Define another *-homomorphism $\psi'\colon B \to A \otimes M_n$ by $\psi'(b) = (E(s_i^* b s_j))_{ij}$. Then there exists a partial isometry $v \in A \otimes M_n$ such that $\psi(b) = v\psi'(b)v^*$ and $\psi'(b) = v^*\psi(b)v$ for $b \in B$.*

Lemma 22. *Let ω be a trace on B with $\omega E = \omega$. Then there is a unique (non-normalized) trace ω_1 on $C^*\langle B, e_A\rangle$ such that $\omega_1(x e_A y) = \omega(xy)$ for $x, y \in B$. Moreover, we have the following:*

(1) $\omega_1(1) = \omega(\operatorname{Index} E)$.

(2) *Let ω_0 be the restriction of ω_1 to A. Then the following diagram is commutative:*

$$\begin{array}{ccc} K_0(A) & \xrightarrow{\psi_*} & K_0(C^* < B, e_A >) \\ & {\scriptstyle\hat{\omega}_0}\searrow \quad \swarrow{\scriptstyle\hat{\omega}_1} & \\ & \mathbb{R} & \end{array}$$

(3) *If $\operatorname{Index} E$ is a scalar, then $\omega_1(b) = (\operatorname{Index} E)\omega(b)$ for $b \in B$ and $(\operatorname{Index} E)^{-1}\omega_1$ is a Markov trace of ω in the sense of [8].*

Theorem 23. *Let $E\colon B \to A$ be a conditional expectation of index-finite type. Let ω be a trace on B with $\omega E = E$. Assume that $\operatorname{Index} E$ is a scalar. Define a multiplication map $V\colon \mathbb{R} \to \mathbb{R}$ by $V(t) = (\operatorname{Index} E)t$ for $t \in \mathbb{R}$. Then the following diagrams commute:*

$$(1)\quad \begin{array}{ccc} K_0(A) & \xrightarrow{T_E \circ i_A} & K_0(A) \\ {\scriptstyle\hat{\omega}_0}\downarrow & & \downarrow{\scriptstyle\hat{\omega}_0} \\ \mathbb{R} & \xrightarrow{V} & \mathbb{R} \end{array} \qquad (2)\quad \begin{array}{ccc} K_0(B) & \xrightarrow{i_* \circ T_E} & K_0(B) \\ {\scriptstyle\hat{\omega}}\downarrow & & \downarrow{\scriptstyle\hat{\omega}} \\ \mathbb{R} & \xrightarrow{V} & \mathbb{R} \end{array}$$

Corollary 24. *With the notation and assumptions of Theorem 23, we have the following:*

(1) $\operatorname{Index} E$ *is in* $\hat{\omega}_0(K_0(A)) \cap \hat{\omega}(K_0(B))$.

(2) *If $K_0(A) \cong Z^n$ (resp. $K_0(B) \cong Z^m$), then $\operatorname{Index} E$ is an eigenvalue of the integral matrix corresponding to $T_E \circ i$ (resp. $i \circ T_E$) and $\operatorname{Index} E$ is an algebraic number of degree at most n (resp. m).*

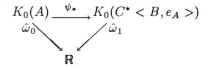

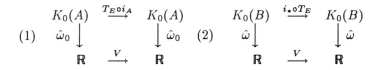

Example 25. Let $\tau^{-1} = 4\cos^2(\pi/5) = (3+\sqrt{5})/2 = \varphi^2$, where φ is the "golden ratio." Let $e_1, e_2, e_3, \ldots$ be Jones projections, as in [13], such that $e_i e_{i\pm 1} e_i = \tau e_i$ and $e_i e_j = e_j e_i$ for $|i-j| \geq 2$. Let $B = C^*\{1, e_1, e_2, e_3, \ldots\}$ and $A = C^*\{1, e_2, e_3, \ldots\}$. Then B and A are the C^*-algebras of Fibonacci in Effros–Shen [7]. Let ω be the unique trace of B. Then $K_0(B) \cong K_0(A) \cong Z^2$ and $\operatorname{Index} E = (3+\sqrt{5})/2$. Moreover $\hat{\omega}(K_0(B)) = \hat{\omega}_0(K_0(A)) = Z + \varphi Z$. Since the following diagram

$$\begin{array}{ccc} K_0(A) \cong Z^2 & \xrightarrow{T_E \circ i_*} & K_0(A) \cong Z^2 \\ \hat{\omega}_0 \big\downarrow & & \big\downarrow \hat{\omega}_0 \\ Z + \varphi Z & \xrightarrow{V} & Z + \varphi Z \end{array}$$

commutes, $T_E \circ i_*$ is represented by an integral matrix $S = \begin{pmatrix} 1 & 1 \\ 1 & 2 \end{pmatrix}$ and the spectral radius $r(S)$ of S coincides with $\varphi^2 = \operatorname{Index} E$.

Definition 26. Let B be a C^*-algebra with a unique trace tr. Define

$$\mathcal{I}(B) = \{\operatorname{Index} E \in \mathbb{R} \cup \{\infty\}; E\colon B \to A \text{ is a conditional expectation with } \operatorname{tr} E = \operatorname{tr}\}.$$

Corollary 27. *Let A_θ be the irrational rotation algebra for an irrational θ. Suppose that θ is* **not** *an algebraic number of degree 2. Then* $\mathcal{I}(A_\theta) = \{1, 2, 3, \ldots, \infty\}$.

REFERENCES

[1] J. Bion-Nadal, *Von Neumann subalgebras of type* II_1 *factors, correspondences and property* T, preprint.

[2] B. Blackadar, *K-Theory for Operator Algebras,* MSRI Publication 5, Springer-Verlag, Berlin, (1986).

[3] M. Brillet, Y. Demizeau and F.F. Hauvet, *Indice d'une espérance conditionnelle,* Compositio Math., **66** (1988), 199–236.

[4] M. Choda, *Full* II_1*-factors with non-integer index,* preprint.

[5] M. Choda, *Index for factors generated by Jones two-sided sequence of projections*, Pacific J. Math., **139** (1989), 1–16.

[6] A. Connes, *Spatial theory of von Neumann algebras,* J. Funct. Anal. **35** (1980), 153–164.

[7] E.G. Effros and C-L.Shen, *Approximately finite dimensional C^*-algebras and continued fractions,* Indiana J. Math. **29** (1980), 191–204.

[8] F.M. Goodman, P. de la Harpe and V. Jones, *Coxeter–Dynkin diagrams and towers of algebras,* MSRI Publication **14**, Springer-Verlag, Berlin (1989).

[9] U. Haagerup, *Operator valued weights in von Neumann algebras* I, J. Funct. Anal. **32** (1979), 175–206; II, J. Func. Anal. **33** (1979), 339–361.

[10] T. Hamachi and H. Kosaki, *Index and flow of weights of factors of type* III, Proc. Japan Academy, Ser. A **64** (1988), 11–13.

[11] T. Hamachi and H. Kosaki, *Inclusion of type* III *factors constructed from ergodic flows,* preprint.

[12] F. Hiai, *Minimizing indices of conditional expectations onto a subfactor,* Publ. RIMS, Kyoto Univ., **24** (1988), 673–678.

[13] V. Jones, *Index for subfactors,* Invent. Math. **72** (1983), 1–25.

[14] V. Jones, *Index for subrings of rings,* Contemp. Math. **43** (Am. Math. Soc. 1985), 181–190.

[15] V. Jones, *Braid Groups, Hecke Algebras and Type* II_1 *Factors in Geometric Methods in Operator Algebras,* Pitman Research Notes in Mathematics Series **123**, (1986), 242–273.

[16] S. Kawakami and H. Yoshida, *Finite group actions on finite von Neumann algebras and the relative entropy,* J. Math. Soc. Japan **39** (1989), 609–626.

[17] S. Kawakami and H. Yoshida, *The constituents of Jones's index analyzed from the structure of the Galois group,* Math. Japon. **33** (1988), 551–557.

[18] H. Kosaki, *Extension of Jones' theory on index to arbitrary factors,* J. Funct. Anal. **66** (1986), 123–140.

[19] P.H. Loi, *Sur la theorie de l'indice et les facteurs de type* III, C.R. Acad. Sci. Paris, **305** (1987), 423–426.

[20] M. Nagisa and G. Song, *Inheritance of the solvability of the similarity problem within a C^*-algebra and its C^*-subalgebras,* Math. Japon. **34** (1989), 73–80.

[21] A. Ocneanu, *Quantized groups, string algebras and Galois theory for algebras,* London Math. Soc., Lecture Note **136** (1988), 119–172, Cambridge Univ. Press.

[22] M. Pimsner and S. Popa, *Entropy and index for subfactors,* Ann. Sci. Ecole Norm. Sup. **19** (1986), 57–106.

[23] M.A. Rieffel, *C^*-algebras associated with irrational rotations,* Pacific J. Math. **93** (1981), 415–429.

[24] M.A. Rieffel, *Morita equivalence for operator algebras,* in Proc. Symposium Pure Math. **38**, Part 1, 285–298.

[25] M.A. Rieffel, *Applications of strong Morita equivalence to transformation group C^*-algebras,* in Proc. Symposium Pure Math. **38**, Part 1, 299–310.

[26] S. Sakai, *C^*-Algebras and W^*-Algebras,* Ergebnisse der Math. vol. 60 (1971), Berlin-Heidelberg-New York.

[27] M. Takesaki, *Conditional expectation on von Neumann algebras,* J. Func. Anal. **9** (1972), 306–321.

[28] J. Tomiyama, *On the projection of norm one in W^*-algebras,* I, Proc. Japan Acad. **33** (1957), 608–612; II, Tohoku Math. J. **10** (1958), 204–209; III, ibid., **11** (1959), 125–129.

[29] H. Umegaki, *Conditional expectation in an operator algebra* I, Tohoku Math. J. **6** (1954), 177–181; II, **8** (1956), 86–100, III, Kodai Math. Sem. Rep. **11** (1959), 51–74; IV, **14** (1962), 59–85.

[30] Y. Watatani, *L'indice d'une C^*-sous-algebre d'une C^*-algebre simple,* C.R. Acad. Sci. Paris, **305**, Serie 1, (1987), 23–26.

[31] H. Wenzl, *Representations of Hecke algebras and subfactors,* Thesis, Univ. of Pennsylvania, 1985.

[32] H. Wenzl, *On sequences of projections,* C.R. Math. Rep. Acad. Sci. Canada **9** (1987), 5–9.

[33] H. Yoshida, *On crossed products and relative entropy,* preprint

Osaka Kyoiku University
Tennoji, Osaka 543, Japan

Three Tensor Norms for Operator Spaces

VERN I. PAULSEN*

1. Introduction

The purpose of these notes is to give a survey of some of the results, questions, and applications surrounding three tensor norms which occur in the non-selfadjoint theory of operator algebras. The first norm we will focus on is the Haagerup norm on tensor products of subspaces of C^*-algebras (which we call *operator spaces*). This portion of our talk includes some joint work with R.R. Smith [18]. The other tensor norms are the min and max norms on tensor products of non-selfadjoint operator algebras, which includes some joint work with S.C. Power ([16] and [17]).

When considering a subspace or subalgebra $\mathcal{M}$ of a C^*-algebra, we have learned that it is vital to also take into account the norms on the matrix spaces over $\mathcal{M}$, $M_n(\mathcal{M})$. That is, we must regard $\mathcal{M}$ as a *matrix-normed space* in the sense of Effros [8]. For these reasons when considering tensor products of such objects we shall not just define the tensor norms on $\mathcal{M}_1 \otimes \mathcal{M}_2$ but also on the matrix spaces $M_n(\mathcal{M}_1 \otimes \mathcal{M}_2)$. We shall only be interested in tensor norms which respect all the matricial norm structure.

2. The Haagerup Tensor Norm

This tensor norm was first introduced in Effros–Kishimoto [10]. It takes its name from some quantities which appeared in several articles by Haagerup [11], [12], [13].

Let $\mathcal{A}$ and $\mathcal{B}$ be C^*-algebras and for u in $\mathcal{A} \otimes \mathcal{B}$ we set

$$\|u\|_h = \inf\{\, \|\Sigma a_i a_i^*\|^{1/2} \|\Sigma b_i^* b_i\|^{1/2} : u = \Sigma a_i \otimes b_i \}$$

where the infimum is taken over all possible ways to express u as a sum of elementary tensors. It is important to note that this norm is non-symmetric.

* Research supported in part by a grant from the NSF.

One obtains a different norm on $\mathcal{B}\otimes\mathcal{A}$. To better understand this formula, it helps to use some notation introduced by Effros. Let $M_{k,n}(\mathcal{A})$ denote the k by n matrices over $\mathcal{A}$, equipped with their usual norm. If $A = (a_{ij})$ is in $M_{k,n}(\mathcal{A})$ and $B = (b_{ij})$ is in $M_{n,m}(\mathcal{B})$, then we set $A\odot B = (a_{ij}\otimes 1)\cdot(1\otimes b_{ij})$, which is an element of $M_{k,m}(\mathcal{A}\otimes\mathcal{B})$. Now it is clear that,

$$\|u\|_h = \inf\{\|A\|\,\|B\|: u = A\odot B\}$$

where A is in $M_{l,n}(\mathcal{A})$, B in $M_{n,l}(\mathcal{B})$ and n arbitrary. For U in $M_k(\mathcal{A}\otimes\mathcal{B})$ we set

$$\|U\|_h = \inf\{\|A\|\,\|B\|: U = A\odot B\}$$

with A in $M_{k,n}(\mathcal{A})$, B in $M_{n,k}(\mathcal{B})$ and n arbitrary.

This last formula easily generalizes and allows one to define the Haagerup norm on $\mathcal{A}_1\otimes\ldots\otimes\mathcal{A}_m$, by setting, for U in $M_k(\mathcal{A}_1\otimes\ldots\otimes\mathcal{A}_m)$,

$$\|U\|_h = \inf\{\|A_1\|\ldots\|A_m\|: U = A_1\odot\ldots\odot A_m\}$$

where A_1 is in $M_{k,n_1}(\mathcal{A}_1)$, A_2 is in $M_{n_1,n_2}(\mathcal{A}_2)$, etc. We use $\mathcal{A}_1\otimes_h\ldots\otimes_h\mathcal{A}_m$ to denote the tensor product equipped with this family of norms.

If $V:\mathcal{A}\times\mathcal{B}\to\mathbb{C}$ is a bilinear functional, we set $\|V\|_h = \|L_V\|$ where L_V denotes the associated linear functional, $L_V:\mathcal{A}\otimes_h\mathcal{B}\to\mathbb{C}$. Haagerup's work on the Grothendieck inequality for C^*-algebras lead to the following representation theorem for these bilinear forms, due to Effros–Kishimoto [10].

Theorem [10]. *Let $V:\mathcal{A}\times\mathcal{B}\to\mathbb{C}$ be a bilinear functional with L_V bounded, then there exist Hilbert spaces $\mathcal{H}$ and $\mathcal{K}$, unital*-homomorphisms, $\pi:\mathcal{A}\to\mathcal{B}(\mathcal{K})$, $\rho:\mathcal{B}\to\mathcal{B}(\mathcal{K})$ vectors $\xi\in\mathcal{H}$, $\eta\in\mathcal{K}$, and an operator $T\in\mathcal{B}(\mathcal{H},\mathcal{H})$ such that $V(a,b) = \langle\pi(a)T\rho(b)\eta,\xi\rangle$ with $\|V\|_h = \|T\|\cdot\|\eta\|\cdot\|\xi\|$.*

An analogous generalization of this representation theorem for trilinear functionals was obtained later by Effros [9].

The Haagerup norm plays a role in some of the conceptual simplifications of the proof of the equivalence of injectivity and the existence of a normal virtual diagonal. See Effros–Kishimoto [10] and Haagerup [11] for a detailed account of these connections. The vital link is the following theorem of Effros–Kishimoto, which also helps to illustrate the "naturalness" of the Haagerup norm.

Theorem [10]. *Let $\mathcal{R}\subseteq\mathcal{B}(\mathcal{H})$ be a von Neumann algebra, and let $\mathcal{R}'$ be its commutant, then the space of completely bounded $\mathcal{R}'$-bimodule maps from $\mathcal{B}(\mathcal{H})$ to itself $CB_{\mathcal{R}'}(\mathcal{B}(\mathcal{H}))$ is completely isometrically isomorphic* *to a σ-weak completion of $\mathcal{R} \otimes_h \mathcal{R}$ via the map which sends $x \otimes y$ to the $\mathcal{R}'$-bimodule map, $L(b) = xby$.*

Moreover, under certain restrictions on $\mathcal{R}$ they prove that every bounded $\mathcal{R}'$-bimodule map on $\mathcal{B}(\mathcal{H})$ is automatically completely bounded.

There is another setting in which the Haagerup norm arises naturally. This is in the setting of Christensen and Sinclair's theory of *multilinear completely bounded* mappings [6].

Let $\mathcal{A}$, $\mathcal{C}$ and $\mathcal{B}$ denote C^*-algebras and let $V: \mathcal{A} \times \mathcal{B} \to \mathcal{C}$ be bilinear. Recall that the norm of a bilinear mapping is given by $\|V\| = \sup \|V(a, b)\|$ where the supremum is taken over all a, b which are of norm less than 1. If we think of V as a $\mathcal{C}$-valued "multiplication" on $\mathcal{A} \times \mathcal{B}$ then it naturally induces a bilinear map $V_n: M_n(\mathcal{A}) \times M_n(\mathcal{B}) \to M_n(\mathcal{C})$ by setting $V_n((a_{ij}), (b_{ij})) = (\Sigma_k V(a_{ik}, b_{kj}))$. A bilinear map is *completely bounded* provided that $\|V\|_{cb} = \sup \|V_n\|$ is finite.

Norming the space of bilinear maps induces norms on $\mathcal{A} \otimes \mathcal{B}$ and the corresponding norm that one obtains in this case is the Haagerup norm [8]. More precisely, if we let $L_V: \mathcal{A} \otimes_h \mathcal{B} \to \mathcal{C}$ be the associated linear map, then $\|V\|_{cb} = \|L_V\|_{cb}$.

There is an analogous definition of completely bounded multilinear maps. The definition of V_n again comes from viewing V as a generalized product.

Christensen and Sinclair obtained the following representation theorem for completely bounded multilinear maps which extends the representation of Effros and Kishimoto.

Theorem [6]. *Let $\mathcal{A}_i$ be unital C^*-algebras, $V: \mathcal{A}_i \times \ldots \times \mathcal{A}_m \to \mathcal{B}(\mathcal{H})$ a completely bounded multilinear mapping, then there exist Hilbert spaces $\mathcal{H}_i$, and unital *-homomorphisms, $\Pi_i: \mathcal{A} \to \mathcal{B}(\mathcal{H}_i)$ and operators $T_i \in \mathcal{B}(\mathcal{H}_i, \mathcal{H}_{i-1})$, $T_1 \in \mathcal{B}(\mathcal{H}_1, \mathcal{H})$, $T_{n+1} \in \mathcal{B}(\mathcal{H}_i, \mathcal{H}_n)$ such that,*

$$(*) \qquad V(a_1, \ldots, a_n) = T_1 \Pi_1(a_1) T_2 \Pi_2(a_2) \ldots T_n \Pi_n(a_n) T_{n+1}$$

with $\|T_1\| \cdots \|T_{n+1}\| = \|V\|_{cb}$.

Their proof uses Wittstock's Hahn–Banach theorem for matricial sublinear functionals.

Subsequent to this theorem Effros and Ylinen, independently, observed that by using the unitary dilation of contraction operators one could rewrite $(*)$ as

$$(**) \qquad V(a_1, \ldots, a_n) = \tilde{T}_1 \tilde{\Pi}_1(a_1) \tilde{\Pi}_2(a_2) \ldots \tilde{\Pi}_n(a_n) \tilde{T}_{n+1}$$

with $\|\tilde{T}_1\| \circ \|\tilde{T}_{n+1}\| = \|V\|_{cb}$, but the $*$-homomorphisms $\tilde{\Pi}_i$ are no longer unital (or even non-degenerate in the non-unital case). This observation leads to the following result of Christensen–Effros–Sinclair [5].

Corollary [5]. *The map $a_1 \otimes \ldots \otimes a_n \to a_1 * \ldots * a_n$ induces a completely isometric linear representation of $\mathcal{A}_1 \otimes_h \ldots \otimes_h \mathcal{A}_n$ as a subspace of the free product C^*-algebra, $\mathcal{A}_1 * \ldots * \mathcal{A}_n$ with no amalgamations.*

Thus, the Haagerup norm also yields an intrinsic description of the norm on this subspace. It would be interesting to know if variations of the Haagerup norm could be given which describes the norms on this subspace inside $\mathcal{A}_1 * \ldots * \mathcal{A}_n$ with some amalgamations. For example, if we amalgamate along the identities, then how do we describe the norm on this subspace? The corresponding theorem for the multilinear functional should have the form (**), but with the $\tilde{\Pi}_i$'s unital. This norm would most likely be equivalent to, but probably not equal to the Haagerup norm.

Perhaps the most stiking application of the Christensen–Sinclair representation theorem is to Hochschild cohomology of operator algebras. In [5], it is proved that if we define Hochschild cohomology with all mappings completely bounded, then $H^n_{cb}(\mathcal{A}, \mathcal{B}(\mathcal{H})) = 0$ for any C^*-algebra $\mathcal{A}$. The representation theorem makes the proof of this result mostly algebraic. By combining this result with the averaging techniques of Johnson, Kadison, and Ringrose, they are able to prove that the bounded Hochschild cohomology groups also vanish for many C^*-algebras.

Our work with R.R. Smith [18] generalizes the Christensen–Sinclair representation theorem to completely bounded multilinear mappings which are only defined on subspaces of C^*-algebras. Our methods also simplify their proof somewhat.

Theorem [18]. *Let $\mathcal{A}_i$ be C^*-algebras, $\mathcal{M}_i \subseteq \mathcal{A}_i$ subspaces and let $V: \mathcal{M}_i \times \ldots \times \mathcal{M}_n \to \mathcal{B}(\mathcal{H})$ be a completely bounded multilinear mapping, then*

$$V(m_1, \ldots, m_n) = T_1 \Pi_1(m_1) \ldots \Pi_n(m_n) T_{n+1}$$

where $\Pi_i: \mathcal{A}_i \to \mathcal{B}(\mathcal{H}_i)$ are *-homomorphisms and $\|V\|_{cb} = \|T_1\| \ldots \|T_{n+1}\|$.

Corollary [18]. *Let $\mathcal{M}_i \subseteq \mathcal{A}_i$, and let $V: \mathcal{M}_1 \times \ldots \times \mathcal{M}_n \to \mathcal{B}(\mathcal{H})$ be a completely bounded multilinear mapping, then there exists $V_1: \mathcal{A}_1 \times \ldots \times \mathcal{A}_n \to \mathcal{B}(\mathcal{H})$ which extends V and satisfies $\|V_1\|_{cb} = \|V\|_{cb}$.*

As before, completely bounded multilinear maps on $\mathcal{M}_1 \times \ldots \times \mathcal{M}_n$ correspond to completely bounded linear maps on $\mathcal{M}_1 \otimes_h \ldots \otimes_h \mathcal{M}_n$. It is important to note that the Haagerup norm on subspaces is defined as above with the subspace $\mathcal{M}_i$ replacing the C^*-algebra $\mathcal{A}_i$ in the definition. One would expect that for u in $\mathcal{M}_1 \otimes \mathcal{M}_2$, $\|u\|_h$ would decrease if one regarded u as an element of $\mathcal{A}_1 \otimes \mathcal{M}_2$ instead and computed the Haagerup norm there. However, it is fairly easily seen from the last corollary that this is not the case.

Corollary [18]. *The Haagerup norm is completely injective (or hereditary), that is, the natural inclusion*

$$\mathcal{M}_1 \otimes_h \cdots \otimes_h \mathcal{M}_n \to \mathcal{A}_1 \otimes_h \cdots \otimes_h \mathcal{A}_n,$$

is a complete isometry. In particular, we have that for U in $M_k(\mathcal{M}_1 \otimes \mathcal{M}_2)$, $\|U\|_h = \inf\{\|A\| \cdot \|B\| : A \in M_{k,n}(\mathcal{M}_1),\ B \in M_{n,k}(\mathcal{M}_2),\ U = A \odot B\} = \inf\{\|A\| \cdot \|B\| : A \in M_{k,n}(\mathcal{A}_1),\ B \in M_{n,k}(\mathcal{B}),\ U = A \otimes B\}$.

Question: Is it possible to give a direct proof of this equality which does not rely on the constructions in [P-S]? Is this possible in the case where the algebras are either finite dimensional or commutative?

If we look at the case of commutative C^*-algebras, $C(X)$ and $C(Y)$, then for u in $C(X) \otimes C(Y)$ the ratio of the projective tensor norm $\|u\|_p$ to the Haagerup tensor norm $\|u\|_h$ is bounded by the (commutative) Grothendieck constant, $1 \le \|u\|_p / \|u\|_h \le K_G$ and, in fact, it is the least such constant which works for all u, X, and Y. This makes the injectivity of the Haagerup norm all the more curious.

We wish to outline the main construction in the proof of Theorem [P-S]. For this we need to recall the "off-diagonalization trick" [15] from the theory of linear completely bounded maps. Let $\mathcal{M}$ be a subspace of C^*-algebra $\mathcal{A}$ and let $\varphi: \mathcal{M} \to \mathcal{B}$, where $\mathcal{B}$ is another C^*-algebra, be a linear map. Consider the operation system,

$$\mathcal{S} = \left\{ \begin{pmatrix} \lambda & a \\ b^* & \mu \end{pmatrix} : \lambda, \mu \in \mathbb{C} \cdot 1_{\mathcal{A}},\ a, b \in \mathcal{M} \right\}$$

which is contained in $M_2(\mathcal{A})$ and define $\Phi: \mathcal{S} \to M_2(\mathcal{B})$ via

$$\Phi\left(\begin{pmatrix} \lambda & a \\ b^* & \mu \end{pmatrix}\right) = \begin{pmatrix} \lambda & \varphi(a) \\ \varphi(b)^* & \mu \end{pmatrix}.$$

It is fairly easy to show that φ is a complete contraction if and only if Φ is completely positive. Using this correspondence, gives us a means of converting theorems about completely positive maps into theorems about completely bounded maps. In this way, one finds that the Hahn–Banach extension theorem, Wittstock's decomposition theorem, and the Stinespring-like representation theorem for completely bounded maps reduces to Arveson's extension theorem and the usual Stinespring theorem for completely positive maps [15].

To play the same trick for bilinear maps we wish to extend $\mathcal{M}_1 \otimes_h \mathcal{M}_2$ "off-diagonally" in an appropriate operator system. The problem is where to find one? A posteriori, we know we could use the free product $\mathcal{A}_1 * \mathcal{A}_2$, but this only comes about as a consequence of the theorem we wish to prove, and so far we have found no way to incorporate this into a proof.

The answer turned out to be to build an operator system. M.D. Choi and E.G. Effros [4] gave an axiomatic characterization of matrix ordered spaces which have a representation as an operator system (i.e., as a selfadjoint subspace of a C^*-algebra, containing 1). Using this characterization we constructed an (abstract) operator system. The operator system that we constructed had the form,

$$\mathcal{S} = \left\{ \begin{pmatrix} a & u \\ \nu & b \end{pmatrix} : a \in \mathcal{A}_1, b \in \mathcal{A}_2, u \in \mathcal{M}_1 \otimes_h \mathcal{M}_2, \nu \in \mathcal{M}_2^* \otimes_h \mathcal{M}_1^* \right\}.$$

Manipulations with this operator system were the key ingredient of our proof.

In a similar vein, it is natural to ask when a matrix-normed system has a completely isometric representation as a space of operators (an operator space)? The answer turns out to be precisely whenthe above set $\mathcal{S}$ has an order which makes it an abstract operator system, that it is compatible with the matrix norms on $\mathcal{M}$. This leads to the following theorem of Ruan [19].

Theorem [19]. *A matrix-normed space $\mathcal{M}$, with the additional property that for A in $M_n(\mathcal{M})$, and B in $M_k(\mathcal{M})$, $\|A \oplus B\| = \max\{\|A\|, \|B\|\}$, $A \oplus B$ in $M_{k+n}(\mathcal{M})$, has a completely isometric linear representation as an operator space.*

Spaces satisfying the above hypotheses are called *L^∞-matrix normed spaces* by Ruan. Thus, L^∞-matrix normed spaces are just abstract operator spaces.

We close this section with some new results on mappings into duals of operator spaces. Let $\mathcal{M} \subseteq \mathcal{A}$, $\mathcal{N} \subseteq \mathcal{B}$ and let $V: \mathcal{M} \times \mathcal{N} \to \mathbb{C}$ be a completely bounded bilinear functional. By our first corollary V can be extended to $V_1: \mathcal{A} \times \mathcal{N} \to \mathbb{C}$ and to $V_2: \mathcal{M} \times \mathcal{B} \to \mathbb{C}$. Recall that there is a natural one-to-one correspondence between bilinear functionals V, and linear maps $T_V: \mathcal{M} \to \mathcal{N}'$ where $\mathcal{N}'$ denotes the dual space of $\mathcal{N}$, via the correspondence $T_V(a)(b) = V(a,b)$. Extending V to V_1, corresponds to extending T_V to $T_1: \mathcal{A} \to \mathcal{N}'$, extending V to V_2, corresponds to lifting T_V to $T_2: \mathcal{M} \to \mathcal{B}'$, where $r: \mathcal{B}' \to \mathcal{N}'$ is the restriction map. We summarize these observations in the two diagrams below.

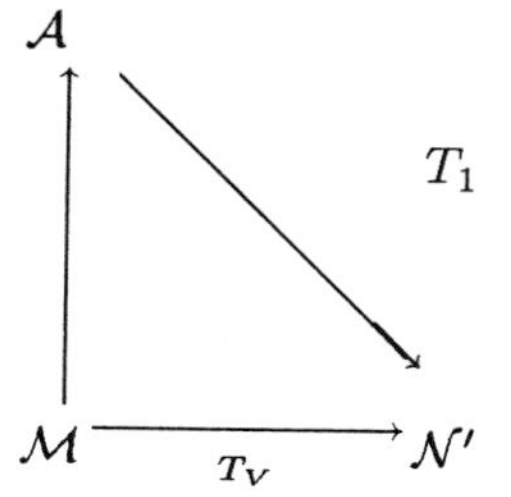

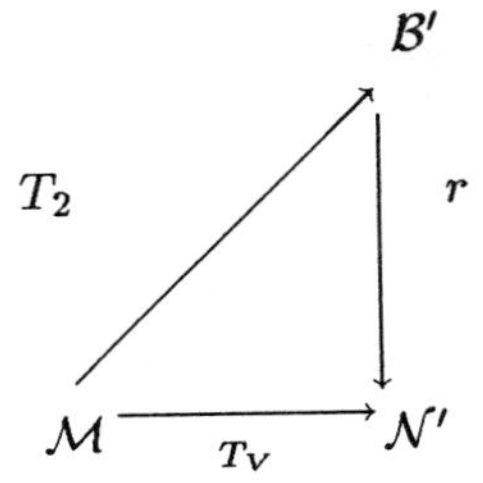

Thus, the extension theorem for completely bounded bilinear functionals, yields an extension and lifting theorem for a certain class of maps into dual spaces. The question that we wish to address is how to characterize those maps? We shall construct a family of matrix norms on $\mathcal{N}'$ such that $\|V\|_{cb} = \|T_V\|_{cb}$, so that the answer is precisely those linear maps $T: \mathcal{M} \to \mathcal{N}'$ which are completely bounded. One thing that is surprising, is that these matrix-norms on $M_n(\mathcal{N}')$ are not the usual matrix-norms which appear elsewhere, but a new family.

Given a n by n matrix of linear functionals (f_{ij}) on $\mathcal{N}'$, we endow it with a norm by regarding it as a linear map from column n tuples, i.e., $M_{n,l}(\mathcal{N})$, to $\mathbb{C}^n$. That is,

$$\|(f_{ij})\|^2 = \sup\{\Sigma_i |\Sigma_j f_{ij}(b_j)|^2 : \Sigma b_j^* b_j \leq 1\}.$$

That is, we make the identification $M_n(\mathcal{N}') = \mathcal{B}(\mathcal{N}^n, \mathbb{C}^n)$.

Lemma 1. *Let $V: \mathcal{M} \times \mathcal{N} \to \mathcal{C}$ be bilinear and let $T: \mathcal{M} \to \mathcal{N}'$ be the associated linear map, then $\|V\|_{cb} = \|T\|_{cb}$.*

Proof. Let $(a_{ij}) \in M_n(\mathcal{M})$ so that

$$\|T(a_{ij})\| = \sup\{\Sigma_i |\Sigma_j T(a_{ij})(b_j)|^2 : \Sigma b_j^* b_j \leq 1\}^{1/2}.$$

But $\Sigma_i |\Sigma_j T(a_{ij})(b_j)|^2 = \|V_n((a_{ij}), B)\|^2$, where B is the n by n matrix whose first column is $b_1, \ldots, b_n$, and whose remaining entries are 0. Since, $\|B\| \leq 1$ we have that $\|(T(a_{ij}))\| \leq \|V_n\| \cdot \|(a_{ij})\|$, and so $\|T_n\| \leq \|V_n\|$.

Conversely, given A in $M_n(\mathcal{M})$ and B in $M_n(\mathcal{N})$, we have that,

$$\|V_n(A,B)\| = \sup\{\|V_n(A,B) \cdot \lambda\| : \lambda \in \mathcal{C}^n, \|\lambda\| \leq 1\}.$$

Now note that $\|V_n(A,B)\cdot\lambda\| = \|V_n(A,B_\lambda)\|$, where

$$B_\lambda = B\cdot\begin{pmatrix}\lambda_1 & \\ \vdots & 0\\ \lambda_n & \end{pmatrix}.$$

These observations imply that $\|V_n\| \le \|T_n\|$, which completes the proof of the lemma.

Theorem 2. *If the dual of an operator space is endowed with the above matrix-norms, then every completely bounded map from an operator space into the dual space enjoys the Hahn–Banach extension property.*

The extension theorem for completely bounded multilinear functionals implies similar extension theorems into unusual matrix-normed spaces.

The way to systematically describe these norms is to introduce the concept of a *vector normed space.* A vector space E with a family of norms on n-tuples over E, E^n, will be called a vector normed space, provided that for each scalar matrix A, n by k, the linear map from E^k to E^n induced by A, has norm no larger than its norm as an ordinary scalar matrix. If E and F are two vector-normed spaces, then we may norm $M_n(\mathcal{B}(E,F))$ by identifying it with $\mathcal{B}(E^n,F^n)$. We note that this is precisely the manner in which the norms are introduced on $\mathcal{B}(\mathcal{H}) = \mathcal{B}(\mathcal{H},\mathcal{H})$ for a Hilbert space $\mathcal{H}$.

If we consider trilinear maps $V\colon \mathcal{L}\times\mathcal{M}\times\mathcal{N}\to\mathbb{C}$, then these maps can be identified with linear maps $T\colon\mathcal{M}\to\mathcal{B}(\mathcal{N},\mathcal{L}')$. Endow $\mathcal{N}^n$, $\mathcal{L}^n$ with the norms that they inherit by considering them as n by l matrices, i.e., $\|(a_1,\ldots,a_n)\|^2 = \|\Sigma a_j^* a_j\|$. We call this the *column norm.* The norm on $\mathcal{L}^n$ endows $\mathcal{L}'^n$ with a natural dual norm via the dot product

$$\|(f_1,\ldots,f_n)\| = \sup\{|\Sigma f_i(a_i)|\colon \|(a_1,\ldots,a_n)\|\le 1\}.$$

Now that $\mathcal{N}^n$ and $\mathcal{L}'^n$ are vector-normed this endows $\mathcal{B}(\mathcal{N},\mathcal{L}')$ with a matrix-norm structure.

Theorem 3. *Let $V\colon\mathcal{L}\times\mathcal{M}\times\mathcal{N}\to\mathbb{C}$, be trilinear and let $T\colon\mathcal{M}\to\mathcal{B}(\mathcal{N},\mathcal{L}')$ be the associated linear map, where $\mathcal{B}(\mathcal{N},\mathcal{L}')$ is endowed with the above matrix-norm, then $\|V\|_{cb} = \|T\|_{cb}$. Consequently, this matrix-normed space enjoys the Hahn–Banach extension property for completely bounded linear maps from operator spaces into it.*

The proof is similar to the proof of the last theorem.

As we have seen above the extension theorem for multilinear maps, implies Hahn–Banach extension theorems for many esoteric matrix-normed spaces.

In the above examples, we only considered the case where the range space was $\mathbb{C}$. If we replace $\mathbb{C}$ by an arbitrary C^*-algebra $\mathcal{C}$, then we need to modify the definitions of our matrix-norms slightly. We define a family of matrix norms on $\mathcal{B}(\mathcal{N},\mathcal{C})$ by regarding $M_n(\mathcal{B}(\mathcal{N},\mathcal{C}))$ as a subspace of $\mathcal{B}(M_n(\mathcal{N}), M_n(\mathcal{C}))$. Given (T_{ij}) in $M_n(\mathcal{B}(\mathcal{N},\mathcal{C}))$, we regard this matrix as a linear map from $M_n(\mathcal{N})$ to $M_n(\mathcal{C})$ by respecting the laws of matrix multiplication, i.e., $(T_{ij})(N_{ij}) = (\Sigma_k T_{ik}(N_{kj}))$.

Theorem 4. *Let $\mathcal{M}$, $\mathcal{N}$ be operator spaces, $\mathcal{C}$ a C^*-algebra, let $V: \mathcal{M}\times\mathcal{N} \to \mathcal{C}$ be bilinear and let $T: \mathcal{M} \to \mathcal{B}(\mathcal{N},\mathcal{C})$ be given by $T(M)(N) = V(M,N)$ then $\|V_n\| = \|T_n\|$. Consequently, completely bounded linear maps from operator spaces into $\mathcal{B}(\mathcal{N},\mathcal{C})$ enjoy the Hahn–Banach extension property, whenever $\mathcal{C}$ is injective.*

It is not hard to show that when $\mathcal{C} = \mathbb{C}$, so that $\mathcal{N}' = \mathcal{B}(\mathcal{N},\mathcal{C})$, then the norm on $M_n(\mathcal{N}')$, defined earlier coincides with the norm on $M_n(\mathcal{B}(\mathcal{N},\mathcal{C}))$ defined above.

The matrix-norms on $\mathcal{B}(\mathcal{N},\mathcal{C})$ enjoy some other unusual properties. Every completely bounded map $T: \mathcal{M} \to \mathcal{B}(\mathcal{N},\mathcal{C})$ has the property that for each m in $\mathcal{M}$, $T(m): \mathcal{N} \to \mathcal{C}$ is necessarily completely bounded and $\|T(m)\|_{cb} \leq \|T\|_{cb}\|m\|$. To see this use the fact that since $\mathcal{M}$ is an operator space, for (n_{ij}) k by k,

$$\begin{aligned} \|(T(m)(n_{ij}))\| &= \|\mathrm{Diag}(T)\cdot \mathrm{Diag}(m)\cdot (n_{ij})\| T_k(\mathrm{Diag}(m))(n_{ij})\| \\ &\leq \|T_k\|\,\|\mathrm{Diag}(m)\|\,\|n_{ij}\| \\ &\leq \|T\|_{cb}\|m\|\,\|n_{ij}\| \end{aligned}$$

where Diag(m) is used to indicate the diagonal matrix whose diagonal entry is m. Consider the completely bounded maps from $\mathcal{N}$ to $\mathcal{C}$, $CB(\mathcal{N},\mathcal{C})$, endowed with $\|\cdot\|_{cb}$ and endow $M_n(CB(\mathcal{N},\mathcal{C}))$ with a family of matrix-norms by identifying it with a subspace of $CB(M_n(\mathcal{N}), M_n(\mathcal{C}))$.

From the above calculation it follows easily that for $\mathcal{M}$, $\mathcal{N}$, $\mathcal{C}$ operator spaces, every completely bounded map $T: \mathcal{M} \to \mathcal{B}(\mathcal{N},\mathcal{C})$ is completely bounded as a map $T: \mathcal{M} \to \mathcal{CB}(\mathcal{N},\mathcal{C})$ and, moreover, the two *cb*-norms of T are equal!

The matrix-normed spaces that we have been considering above are not operator spaces, in general. For example, if $C([0,1])$ denotes the continuous functions on $[0,1]$, then $C([0,1])'$ is not an operator space. To see this let $S_t \in C([0,1])'$ be defined by $S_t(\chi) = \chi(t)$, so that $\|S_t\| = 1$. Since

$\|S_0 \oplus S_1\| = 2$, we see that $C([0,1])'$ is not L^∞-matrix-normed, and so not an operator space by Ruan's theorem.

Thus, the above considerations lead us naturally to families of matrix-normed spaces which are not operator spaces but which do enjoy the Hahn–Banach extension property for maps whose domains are operator spaces. It is clear from these considerations that a better understanding of the category of matrix-normed spaces, with morphisms the completely bounded maps would be helpful. Recall that an element $\mathcal{C}$ of a category is called injective if every morphism from a subspace into $\mathcal{C}$, extends to a morphism of the whole space into $\mathcal{C}$, i.e., if maps into $\mathcal{C}$ enjoy the Hahn–Banach extension property. We know that if we restrict the category to just operator spaces, and completely bounded maps, then, $\mathcal{B}(\mathcal{H})$ is injective. Effros and Ruan have shown that $\mathcal{B}(\mathcal{H})$ is not injective in the category of all matrix-normed spaces [22].

Problem. Is $\mathcal{B}(\mathcal{H})$ injective in some category of matrix-normed spaces which contains operator spaces and their duals? If $\mathcal{C}$ is injective in this category of matrix-normed spaces, and $\mathcal{N}$ is arbitrary is $\mathcal{B}(\mathcal{N},\mathcal{C})$, endowed with the family of matrix-norms described above, injective?

3. The Min and Max Norm on Tensor Products of Operator Algebras

In this section we will be concerned with matrix-norms on the tensor products of non-selfadjoint subalgebras of C^*-algebras, which also respect the algebraic structure. We shall always assume that our algebras contain a unit and that all maps are unit preserving. Let $\mathcal{B}_i$ be C^*-algebras and let $\mathcal{A}_i$ be subalgebras, so that $M_n(\mathcal{A}_i)$ is endowed with the norm that it inherits as a subalgebra of $M_n(\mathcal{B}_i)$, $i = 1, 2$. We shall call these matrix-normed algebras, *operator algebras.* We are interested in constructing matrix-norms on $\mathcal{A}_1 \otimes \mathcal{A}_2$ such that it has a completely isometric homomorphic representation as an operator algebra. An axiomatic characterization of operator algebras has only recently been given by Blecher–Sinclair [3].

Theorem [3]. *Let $\mathcal{A}$ be an L^∞-matrix-normed space with a completely contractive multiplication $m: \mathcal{A} \times \mathcal{A} \to \mathcal{A}$ and an identity e, with $\|e\| = 1$. Then there is a unital completely isometric homomorphism of $\mathcal{A}$ into a C^*-algebra.*

We shall be concerned with two particular tensor norms, and in particular with knowing when they are equal.

To define the max *norm,* consider pairs of commuting, completely contractive homomorphisms, $\rho_i: \mathcal{A}_i \to \mathcal{B}(\mathcal{H})$, $i = 1, 2$. Such pairs induce a homomorphism $\rho_1\rho_2: \mathcal{A}_1 \otimes \mathcal{A}_2 \to \mathcal{B}(\mathcal{H})$ via $\rho_1\rho_2(a_1 \otimes a_2) = \rho_1(a_1) \cdot \rho_2(a_2)$. For $U = (u_{ij})$ in $M_k(\mathcal{A}_1 \otimes \mathcal{A}_2)$, we set,

$$||U||_{\max} = \sup\{||(\rho_1\rho_2(u_{ij}))||\},$$

where the supremum is taken over all such commuting pairs of completely contractive representations. We let $\mathcal{A}_1 \otimes_{\max} \mathcal{A}_2$ denote the tensor product endowed with this family of matrix-norms. It is not difficult to check that it is an operator algebra by either invoking the Blecher–Sinclair theorem or by directly constructing a completely contractive homomorphism.

The second tensor norm that we are interested in, we call the min *norm,* although perhaps spatial would have been better. Given $\rho_i: \mathcal{A}_i \to \mathcal{B}(\mathcal{H}_i)$, $i = 1, 2$, completely contractive homomorphisms, define $\rho_1 \otimes \rho_2: \mathcal{A}_1 \otimes \mathcal{A}_2 \to \mathcal{B}(\mathcal{H}_1 \otimes \mathcal{H}_2)$ via $\rho_1 \otimes \rho_2(a_1 \otimes a_2) = \rho_1(a_1) \otimes \rho_2(a_2)$. For $U = (u_{ij})$ in $M_k(\mathcal{A}_1 \otimes \mathcal{A}_2)$ set,

$$||U||_{\min} = \sup\{||(\rho_1 \otimes \rho_2(u_{ij}))||\}$$

where the supremum is taken over all such pairs.

Proposition [17]. *Let $\mathcal{A}_i$ be subalgebras of $\mathcal{B}_i$, $i = 1, 2$, then the inclusion of $\mathcal{A}_1 \otimes \mathcal{A}_2$ into $\mathcal{B}_1 \otimes \mathcal{B}_2$ induces a completely isometry in the min norm.*

The above proposition was our reason for calling this norm the min norm. However, it is still not known if it is the minimal tensor norm in some appropriate sense.

Problem. What are the appropriate axioms for cross-norms on operator algebras? Is the min norm the minimal such cross-norm?

To illustrate some of the difficulties, we note that unlike the case for C^*-algebras, it is possible to construct norms on $\mathcal{A}_1 \otimes \mathcal{A}_2$ such that it is an operator algebra in this norm, and such that each inclusion map is a complete isometry, but one can still have that $||a \otimes b|| < ||a|| \cdot ||b||$ [7].

The main problem that we shall be concerned with in these notes is determining when the min and max norms agree. We give some examples of this and discuss some of the applications. Our primary interest in these two norms stems from the fact that if these two norms coincide, then any pair of commuting completely contractive representations on $\mathcal{A}_1$ and $\mathcal{A}_2$, will automatically have a dilation to $\mathcal{B}_1 \otimes_{\min} \mathcal{B}_2$, by the above proposition.

For this reason many dilation theorems are best interpreted in terms of the equality of these two norms.

Theorem [1]. *Let* $A(\mathbf{D})$ *denote the disk algebra, then the* min *and* max *tensor norms agree on* $A(\mathbf{D}) \otimes A(\mathbf{D})$ *and the completion of this algebra can be identified with the bidisk algebra* $A(\mathbf{D}^2)$.

Example (Parrott, Crabbe-Davie, Varapoulos). The min and max norms are different on $A(\mathbf{D}) \otimes A(\mathbf{D}) \otimes \mathrm{A}(\mathbf{D})$.

In addition to the dilation theorem of Ando having a tensor norm interpretation, various generalizations of the Sz.-Nagy-Foias lifting theorem can be given a tensor norm interpretation. Let $\mathcal{A}$ be a subalgebra of the C^*-algebra $\mathcal{B}$. We say that generalized *Sz.-Nagy Foias lifting* holds for this pair of algebras provided that:

Given $\rho_1: \mathcal{A} \to \mathcal{B}(\mathcal{H}_i)$, $i = 1, 2$ a completely contractive homomorphism, and a contraction $T \in \mathcal{B}(\mathcal{H}_2, \mathcal{H}_1)$, which intertwines, $\rho_1(a)T = T\rho_2(a)$, for all a in $\mathcal{A}$, there exists *-homomorphisms $\Pi_i: \mathcal{B} \to \mathcal{B}(\mathcal{K}_i)$, $\mathcal{K} \subseteq \mathcal{K}_i$, $i = 1, 2$ and an operator $\tilde{T} \in \mathcal{B}(\mathcal{K}_2, \mathcal{K}_1)$ with $\Pi_i(b)\tilde{T} = \tilde{T}\Pi_2(b)$ for all b in $\mathcal{B}$, $\|T\| = \|\tilde{T}\|$, such that, $P_{\mathcal{H}_i}\Pi_i(a)|_{\mathcal{H}_i} = \rho_i(a)$, $i = 1, 2,$, and $\rho_1(a)T = \rho_{\mathcal{H}_1}\Pi(a)\tilde{T}|_{\mathcal{H}_2} = T\rho_2(a)$, for all a in $\mathcal{A}$.

We say that *Ando's dilation* holds, provided that:

Given $\rho: \mathcal{A} \to \mathcal{B}(\mathcal{H})$, a completely contractive homomorphism, and a contraction $T \in \mathcal{B}(\mathcal{H})$ which commutes with $\rho(\mathcal{A})$, there exists a *-homomorphism $\Pi: \mathcal{B} \to \mathcal{B}(\mathcal{H})$, $\mathcal{H} \subseteq \mathcal{H}$ and a unitary $U \in \mathcal{B}(\mathcal{H})$ which commutes with $\Pi(\mathcal{B})$ such that, $T^n\rho(a) = \rho_{\mathcal{H}} U^n \Pi(a)|_{\mathcal{H}}$, for all integers $n \geq 0$ and a in $\mathcal{A}$.

These rather cumbersome conditions have fairly tidy descriptions in terms of tensor norms. Let $\mathcal{T}_n$ denote the n by n upper triangular matrices, regarded as a subalgebra of the n by n matrices, M_n.

Proposition [17]. *Ando's dilation holds for* $\mathcal{A}$, *if and only if the* min *and* max *norms are equal on* $\mathcal{A} \otimes A(\mathrm{D})$. *Sz.-Nagy-Foias lifting holds for* $\mathcal{A}$ *if and only if the* min *and* max *norms are equal on* $\mathcal{A} \otimes \mathcal{T}_2$.

In this light the results of Ball–Gohberg [2], and Sz.-Nagy-Foias [20], can be interpreted as:

Theorem [27, 20]. *The* min *and* max *norms are equal on* $\mathcal{T}_n \otimes \mathcal{T}_2$, *for all* n. *The min and max norms are equal on* $A(D) \otimes \mathcal{T}_2$.

Theorem [17]. *The* min *and* max *norms agree on* $\mathcal{A} \otimes A(\mathbf{D})$ *if and* *only if the* min *and* max *norms agree on* $\mathcal{A} \otimes \mathcal{T}_n$ *for all* n.

The proof of the above result relied on a Toeplitz like construction. Armed with this result and Ando's theorem we have:

Corollary [17]. *The* min *and* max *norms agree on* $\mathcal{A}(\mathbf{D}) \otimes \mathcal{T}_n$ *and on* $\mathcal{T}_m \otimes \mathcal{T}_n$, *for all* m *and* n.

The precise relation between Ando's dilation and Sz.-Nagy-Foias lifting is still not known. In particular, we have no example of an algebra for which the min and max norms agree on $\mathcal{A} \otimes \mathcal{T}_2$ but are different for $\mathcal{A} \otimes A(\mathbf{D})$. But we believe that such examples must exist. More interesting would be to find algebras $\mathcal{A}_n$ such that equality of the min and max norms occurs for $\mathcal{A}_n \otimes \mathcal{T}_k$, $k < n$, but are different for $\mathcal{A}_n \otimes \mathcal{T}_n$.

We close with a few examples to illustrate the difficulties of dealing with these tensor norms. Let $\mathcal{A}_2$ denote the 2-dimensional subalgebra of M_2,

$$\mathcal{A}_2 = \left\{ \begin{bmatrix} \lambda & \mu \\ 0 & \lambda \end{bmatrix} : \lambda, \mu \in \mathbb{C} \right\}.$$

Example [17]. (a) The min and max norms differ on $\mathcal{A}_2 \otimes \mathcal{A}_2$.

(b) The min and max norms differ on $\mathcal{A}_2 \otimes \mathcal{T}_2$.

(c) The min and max norms differ on $\mathcal{T}_2 \otimes \mathcal{T}_2 \otimes \mathcal{T}_2$.

(d) If $\mathcal{B}$ is a nuclear C^*-algebra, then the min and max norms agree on $\mathcal{B} \otimes \mathcal{A}$, for any operator algebra $\mathcal{A}$.

We hope that the above results demonstrate that a theory of tensor norms on non-selfadjoint spaces is beginning to emerge, and that the theory does play a role in increasing our understanding of both the selfadjoint and non-selfadjoint theory of operator algebras.

REFERENCES

[1] T. Ando, *On a pair of commutative contractions,* Acta Sci. Math. **24** (1963), 88–90.

[2] J.A. Ball and I. Gohberg, *A commutant lifting theorem for triangular matrices with diverse applications,* J. Int. Eng. and Op. Thy. **8** (1985), 205–267.

[3] D. Blecher, *The Geometry of the Tensor Product of C^*-algebras,* Ph.D. Thesis, University of Edinborough, 1988.

[4] M.D. Choi and E.G. Effros, *Injectivity and operator spaces,* J. Funct. Anal. **24** (1977), 156–209.

[5] E. Christensen, E.G. Effros, and A.M. Sinclair, *Completely bounded maps and C^*-algebraic cohomology,* Inventiones Math. **90** (1987), 279–296.

[6] E. Christensen and A.M. Sinclair, *Representations of completely bounded multilinear operators,* J. Funct. Anal. **72** (1987), 151–181.

[7] M.J. Crabb and A.M. Davie, *von Neumann's inequality for Hilbert space operators,* Bull. London Math. Soc. F (1975), 49–50.

[8] E.G. Effros, *Advances in quantized functional analysis,* Proc. I.C.M. Berkeley, 1986.

[9] E.G. Effros, *On multilinear completely bounded maps,* Contemp. Math. **62** (1987), 450–479.

[10] E.G. Effros and A. Kishimoto, *Module maps and Hochschild–Johnson cohomology,* Indiana Math. J. **36** (1987), 257–276.

[11] U. Haagerup, *All nuclear C^*-algebras are amenable,* preprint, 1981.

[12] U. Haagerup, *Injectivity and decomposition of completely maps,* Lecture Notes in Math., **1132**, Springer-Verlag, Berlin, 1983, 170–222.

[13] U. Haagerup, *The Grothenedieck inequality for bilinear forms on C^*-algebras,* Adv. in Math. **56** (1985), 93–116.

[14] S. Parrott, *Unitary dilations for commuting contractions,* Pacific J. Math. **34** (1970), 481–490.

[15] V.I. Paulsen, *Completely bounded maps and dilations,* Pitman Research Notes in Mathematics **146**, Longman, London, 1986.

[16] V.I. Paulsen and S.C. Power, *Lifting theorems for nest algebras,* J. Operator Theory, to appear.

[17] V.I. Paulsen and S.C. Power, *Tensor products of non-selfadjoint operator algebras,* Rocky Mountain Math. J., to appear.

[18] V.I. Paulsen and R.R. Smith, *Multilinear maps and tensor norms on operator systems,* J. Funct. Anal. **73** (1987), 258–276.

[19] Z.J. Ruan, *Subspace of C^*-algebras,* J. Funct. Anal. **76** (1988), 217–230.

[20] B. Sz-Nagy and C. Foias, *Harmonic Analysis of Operators of Hilbert Space,* American Elsevier, New York, 1970.

[21] N.Th. Varopoulos, *On an inequality of von Neumann and an application of the metric theory of tensor products to operator theory,* J. Funct. Anal. **16** (1974), 83–100.

[22] E.G. Effros and Z.J. Ruan, *On matricially normed spaces,* Pacific J. Math. **132** (1988), 243–264.

Department of Mathematics
University of Houston
Houston, Texas

Extension Problems for Maps on Operator Systems

R.R. SMITH

1. Introduction

Given operator systems $E \subseteq F$ and a C^*-algebra B the extension problem considers a map $\phi: E \to B$ and asks for an extension $\psi: F \to B$ which makes the following diagram commute:

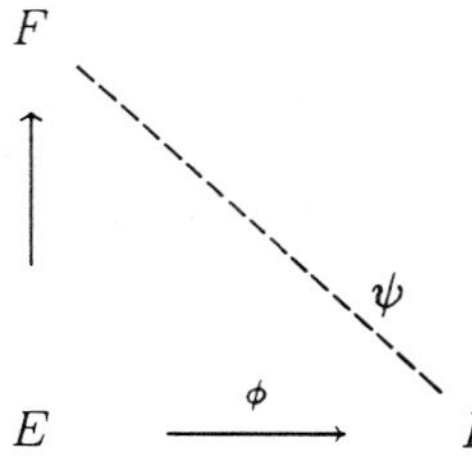

(∗)

Usually a condition is imposed on ϕ (contractive, positive, completely positive, completely bounded etc.) and a solution to (∗) is sought which possesses the same property. The earliest and most celebrated result is the Hahn-Banach theorem which, in this language, asserts that (∗) may be solved for contractive maps if B is one dimensional.

For a fixed range algebra B, if (∗) can always be solved for completely positive maps then B is said to be injective, while if E and F are required to be separable B is separably injective. Arveson's Hahn-Banach theorem asserts that $B(H)$ is injective [5], and then (∗) can be solved for completely contractive maps [17,25]. For other classes of maps the situation is more complicated. Størmer has considered positive maps [24] where conditions must be imposed on E. A counterexample to a general extension theorem appears in [19], while section 2 contains a counterexample for n-positive maps, first appearing in [20].

This example is based upon a closely related problem, that of the lifting of maps. Consider the diagram

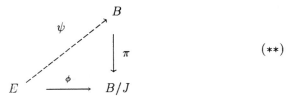

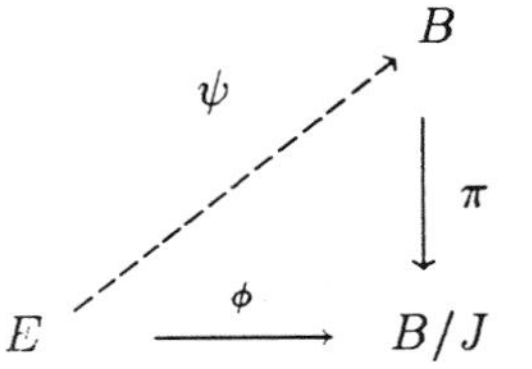

(**)

where B is a C^*-algebra, J is a closed two sided ideal and E is an operator system. If ψ exists to make the diagram commute then ϕ is liftable. Once again a condition is imposed on ϕ and liftings of the same type are sought. The known results are that (**) can be solved if

(1) E is finite dimensional, ϕ is positive [3]
(2) E is separable and has the positive approximation property, ϕ is positive [3]
(3) E is a separable nuclear C^*-algebra, ϕ is completely positive [7]
(4) E is finite dimensional, ϕ is n-positive [20]
(5) E is separable and has the n-positive approximation property, ϕ is n-positive [20]

and (**) can in general not be solved if

(6) E is finite dimensional, ϕ completely positive [4]
(7) E is a finite dimensional C^*-algebra, ϕ is completely contractive [13].

A sketch of (4) and (5) is given in the second section.

The third section of this paper is concerned with separable injectivity, which has been studied in [8,22,23]. The main result is that if A is n-homogeneous and σ-unital then the corona algebra $M(A)/A$ is separably injective. Background results may be found in [11,12,16,23].

2. Liftings

Let B be a unital C^*-algebra with a closed two sided ideal J, and let E be a finite dimensional operator system in the quotient algebra B/J. Denote by π the quotient homomorphism $B \to B/J$. By a lifting of E we mean a map $\phi: E \to B$ such that $\pi\phi = id$. Further properties are usually required of ϕ and we list the previously known results. Positive unital liftings always exist (Andersen [3]), isometric liftings always exist (Choi-Effros [8]) but completely positive liftings may not (Anderson [4]). Thus the following two results are best possible in the general context. Since full proofs will appear in [20], we given only a sketch.

Theorem 2.1. A finite dimensional operator system E in B/J has an n-positive lifting for each integer $n \geq 1$.

Proof. Fix $n \geq 1$. By [3] $E \otimes M_n$ has a positive lifting $\sigma: E \otimes M_n \to B \otimes M_n$. Now integrate with respect to Haar measure on the unitary group of M_n to obtain a modified positive lifting

$$\psi(X) = \int U\sigma(U^* X U)U^* d\mu(U)$$

for $X \in E \otimes M_n$. Then for all unitaries $V \in M_n$

$$\psi(V^* X V) = V^* \psi(X) V.$$

Such a map must have the form

$$\psi(X) = \tau \otimes id_n(X) + \lambda(\text{trace } X) \otimes I_n \ (X \in E \otimes M_n)$$

where $\tau: E \to B$ and $\lambda: E \to J$.
Select a strictly positive state θ on E and define $\phi: E \to B$ by

$$\phi(a) = \tau(a) + c\theta(a)\lambda(1)$$

where c is a suitably large positive constant. Then ϕ may be verified to be an n-positive lifting of E, and a standard argument [6] may be employed to make ϕ unital.

Corollary 2.2. A finite dimensional operator system E in B/J has an n-positive and n-isometric lifting for each integer $n \geq 1$.

Proof. Let ϕ be a $2n$-positive unital lifting of E, by Theorem 2.1. If $X \in E \otimes M_n$ then $||X|| \leq 1$ if and only if $\begin{pmatrix} I_n & X \\ X^* & I_n \end{pmatrix} \geq 0$ in $E \otimes M_{2n}$, from which it follows that $||\phi \otimes id_n(X)|| \leq 1$. The result is immediate from this.

Using Arveson's technique from [6] it is possible in certain circumstances to pass from finite dimensional operator systems to separable subalgebras of B/J.

Theorem 2.3 [20]. If $A \subseteq B/J$ is unital, separable, and has the n-positive approximation property then A has an n-positive, unital lifting.

As an application of these results we answer a question which arises from work of Størmer [24] on extension of positive maps. Let F_2 be the free group on two generators and let K be the Hilbert space $\ell_2(F_2)$. The reduced group C^*-algebra $C^*_\lambda(F_2)$ is generated on K by the group of left translation operators, and is a quotient of $C^*(F_2)$ by a certain closed two sided ideal J. The algebra $C^*_\lambda(F_2)$ has the n-positive approximation property for each integer n [10], and so there exists an n-positive unital lifting $\phi_n: C^*_\lambda(F_2) \to$

$C^*(F_2)$. If $C^*(F_2)$ is faithfully represented on some Hilbert space H, then $\phi_n: C^*_\lambda(F_2) \to B(H)$ is an example of an n-positive unital map which has no n-positive (or positive) extension to $\tilde{\phi}_n: B(K) \to B(H)$, [19,20].

Since $C^*_\lambda(F_2)$ also has the completely contractive approximation property [10] it is natural to ask whether there exists a completely contractive lifting $\phi: C^*_\lambda(F_2) \to C^*(F_2)$. The above techniques do not apply, but we suspect that the answer is negative.

3. Extension of completely positive maps

In this section we consider the problem of extending a completely positive unital map $\phi: E \to B$ to a completely positive unital map $\tilde{\phi}: F \to B$ where $E \subseteq F$ are operator systems and B is a unital C^*-algebra. If such extensions always exist then B is injective, while restricting E and F to be separable gives the definition of separable injectivity [8,23]. For von Neumann algebras these two notions coincide [9], but the Calkin algebra is not separably injective [4]. This last algebra is one example of a more general construction called a corona algebra (named after the corona set $\beta N \backslash N$ in the Stone-Cech compactification of the integers). Starting from a non-unital C^*-algebra A we first form the multiplier algebra $M(A)$ and then the corona algebra $\chi(A) = M(A)/A$. For technical reasons it is convenient to assume that A is σ-unital in the sense that it has a countable approximate unit. Corona algebras, while appearing exotic, have found some surprising applications in recent years [12,14,16].

We wish to present here the result that if A is σ-unital and n-homogeneous then $\chi(A)$ is separably injective. This would be an immediate corollary of results in [23] were it true that $\chi(A)$ is n-homogeneous. However this is not always the case [23] and the most that can be said is that $\chi(A)$ is n-subhomogeneous (has only irreducible representations of degree n or less). The proof will consist of a sequence of lemmas some of which use Choquet theory [1].

We begin with two simple observations. It will suffice to consider an extension to $\tilde{E} = E + \mathbb{C}z$ where z is a self-adjoint element. If this can be accomplished then a countable repetition of the argument will allow us to pass to any separable operator system F containing E. It will also suffice to construct an n-positive extension since n-positive maps into n-subhomogeneous algebras are automatically completely positive [21].

If L is a compact convex set in a locally convex topological space V then $A(L)$ and $A^b(L)$ denote respectively the Banach spaces of continuous and bounded real valued affine functions on L. The notation $a < b$ for functions or C^*-algebra elements will mean that there exists $\varepsilon > 0$ such that $a + \varepsilon 1 \leq b$.

Lemma 3.1. Let $L \subseteq V$ be a compact convex set with a sequence $\{K_i\}_{i=1}^{\infty}$ of compact convex subsets and let $f \in A^b(L)$. Let $\{a_{ij}\}_{j=1}^{\infty}, \{b_{ij}\}_{j=1}^{\infty} \in A(K_i)$ be sequences satisfying

$$a_{ij} < f < b_{ij} \quad \text{on} \quad K_i.$$

Then there exists a uniformly bounded sequence $\{c_r\}_{r=1}^{\infty} \in A(L)$ such that

$$a_{ij} < c_r < b_{ij} \quad \text{on} \quad K_i$$

for $1 \leq i, j \leq r$.

Proof. Let $K_0 = L$ and define

$$a_{0j} = -2\|f\| \ , \ b_{0j} = 2\|f\| \quad \text{for} \quad j \geq 0.$$

For $i, j \geq 0$ define

$$X_{ij} = \{(k, a_{ij}(k)) \colon k \in K_i\} \subseteq V \times \mathbf{R}$$

and

$$Y_{ij} = \{(k, b_{ij}(k)) \colon k \in K_i\} \subseteq V \times \mathbf{R}.$$

Fix $r \geq 1$ and let $X = \operatorname{conv}\{X_{ij}\}_{i,j=0}^{r}, Y = \operatorname{conv}\{Y_{ij}\}_{i,j=0}^{r}$. Both are compact convex subsets of $V \times \mathbf{R}$, which we claim are disjoint. If not, there exist convex combinations from X and Y respectively such that

$$\sum_{i,j=0}^{r} \lambda_{ij}(k_{ij}, a_{ij}(k_{ij})) = \sum_{i,j=0}^{r} \mu_{ij}(k'_{ij}, b_{ij}(k'_{ij})).$$

The first variable gives

$$\sum_{i,j=0}^{r} \lambda_{ij} k_{ij} = \sum_{i,j=0}^{r} \mu_{ij} k'_{ij}$$

and so, applying the hypotheses,

$$\sum_{i,j=0}^{r} \lambda_{ij} a_{ij}(k_{ij}) < \sum_{i,j=0}^{r} \lambda_{ij} f(k_{ij}) = \sum_{i,j=0}^{r} \mu_{ij} f(k'_{ij}) < \sum_{i,j=0}^{r} \mu_{ij} b(k'_{ij}).$$

This contradiction shows that X and Y are disjoint, and can thus be separated by a hyperplane. As in [1] the hyperplane is the graph of an element of $c_r \in A(L)$ which by construction satisfies the conclusion of the lemma.

If U and K are compact convex sets then $BA(U \times K)$ and $BA^b(U \times K)$ denote respectively the Banach spaces of continuous and bounded biaffine real valued functions on $U \times K$. Each is an order unit space in the uniform norm [1] and so there is a compact convex set L (the tensor product $U \otimes K$ [15]) for which $BA(U \times K) = A(L)$. If U is embedded in a finite dimensional vector space then there are simple identifications $BA^b(U \times K) = A^b(L) = A(L)^{**} = BA(U \times K)^{**}$ [2]. If $u \in U, k \in K$ then $u \otimes k \in L$ denotes the state $f \rightarrow f(u,k)$ and $u \otimes K$ represents the compact convex subset $\{u \otimes k: k \in K\}$ of L.

Let B be a nuclear unital C^*-algebra, let $E \subseteq \tilde{E} = E + \mathbb{C}z$ be separable operator systems and let $\phi: E \rightarrow B$ be a completely positive unital map. Since B^{**} is injective ϕ extends to a completely positive map $\tilde{\phi}: \tilde{E} \rightarrow B^{**}$. For ease of notation write $\theta = \tilde{\phi} \otimes id_{2n}: \tilde{E} \otimes M_{2n} \rightarrow B^{**} \otimes M_{2n} = (B \otimes M_{2n})^{**}$, and let U denote the unit ball in the self-adjoint part of M_{2n}. Fix a countable dense sequence $\{u_i\}_{i=1}^{\infty}$ in U which begins with $u_1 = 0$. Let K be the state space of $B \otimes M_{2n}$, set $L = U \otimes K$, and define $f \in BA^b(U \times K) = A^b(L)$ by

$$f(u,k) = \theta(z \otimes u)(k).$$

For each $i \geq 1$ define $K_i = u_i \otimes K \subseteq L$, and define two separable subsets of $(E \otimes M_{2n})_{\text{s.a.}}$ by

$$\Gamma_i = \{x: x \in E \otimes M_{2n}\ ,\ x < z \otimes u_i\}$$
$$\Sigma_i = \{y: y \in E \otimes M_{2n}\ ,\ y > z \otimes u_i\}.$$

Now choose dense sequences $\{x_{ij}\}_{j=1}^{\infty} \in \Gamma_i, \{y_{ij}\}_{j=1}^{\infty} \in \Sigma_i$.

Identify $(B \otimes M_{2n})_{\text{s.a.}}$ with $A(K)$ and $BA(U \times K)$ with the space of affine maps from U into $(B \otimes M_{2n})_{\text{s.a.}}$.

Lemma 3.2. There exists a uniformly bounded sequence of continuous affine maps $\psi_r: U \rightarrow (B \otimes M_{2n})_{\text{s.a.}}$ such that

$$\theta(x_{ij}) < \psi_r(u_i) < \theta(y_{ij})$$

for $1 \leq i, j \leq r$.

Proof. Setting $a_{ij} = \theta(x_{ij})$ and $b_{ij} = \theta(y_{ij})$ in the previous lemma, this result is then a mere translation of Lemma 3.1 into C^*-algebra terms.

Remark 3.3. Since $u_1 = 0$ it is clear that $\psi_r(0) = 0$ and so each ψ_r is the restriction of a real linear map, also denoted ψ_r, from $(M_{2n})_{\text{s.a.}}$ into $(B \otimes M_{2n})_{\text{s.a.}}$.

Theorem 3.4. Let A be a σ-unital n-homogeneous C^*-algebra. The $M(A)/A$ is separably injective.

Proof. We consider a completely positive unital map $\phi: E \to M(A)/A$ with associated positive map $\theta = \phi \otimes id_{2n}: E \otimes M_{2n} \to (M(A)/A) \otimes M_{2n}$ and, as a first step, extend it to a positive map $\eta: (E + \mathbb{C}z) \otimes M_{2n} \to (M(A)/A) \otimes M_{2n}$. The range algebra is nuclear and so Lemma 3.2 may be applied with $B = (M(A)/A) \otimes M_{2n}$. Let D be the separable subalgebra of B generated by $\theta(E \otimes M_{2n})$ and $\{\psi_r(U)\}_{r=1}^{\infty}$. D is nuclear and so there exists a completely positive lifting $\xi: D \to M(A) \otimes M_{2n}$ [7]. Denote by ρ the quotient map $M(A) \otimes M_{2n} \to M(A)/A \otimes M_{2n}$.

Since A is n-homogeneous and σ-unital it is possible to choose from the center $Z(A)$ a sequence of positive elements such that $\Sigma_{r=1}^{\infty} f_r = 1$. Now define $\mu: U \to M(A)/A \otimes M_{2n}$ by

$$\mu(u) = \rho(\sum_{r=1}^{\infty}(f_r \otimes I_{2n})\xi(\psi_r(u))), u \in U.$$

The infinite sum is well defined since it is calculated in $M(A) \otimes M_{2n}$.

If $x_{ij} \in \Gamma_i$ and $r \geq i, j$ then

$$\theta(x_{ij}) < \psi_s(u_i) \quad \text{for} \quad s \geq r$$

by construction. Thus

$$\xi\theta(x_{ij}) = \sum_{s=1}^{\infty}(f_x \otimes I_{2n})\xi\theta(x_{ij}) \leq \sum_{s=1}^{r-1}(f_s \otimes I_{2n})\xi\theta(x_{ij}) + \sum_{s=r}^{\infty} f_s\xi\psi_s(u_i),$$

from which it follows that

$$\theta(x_{ij}) = \rho\xi\theta(x_{ij}) \leq \mu(u_i)$$

since ρ annihilates any finite sum containing f_s's. Similarly

$$\theta(y_{ij}) \geq \mu(u_i).$$

As in the previous remark μ is the restriction of a linear map $\tilde{\mu}: M_{2n} \to M(A)/A \otimes M_{2n}$.

Now define $\eta: \tilde{E} \otimes M_{2n} \to M(A)/A \otimes M_{2n}$ as an extension of θ by setting

$$\eta(z \otimes u) = \tilde{\mu}(u) \quad \text{for} \quad u \in M_{2n}.$$

The inequalities which will verify the positivity of η are precisely

$$\theta(x_{ij}) \le \mu(u_i) \le \theta(y_{ij}), \quad u_i \in U.$$

Now regard $\tilde{E}\otimes M_{2n}$ as 2×2 matrices with entries from $\tilde{E}\otimes M_n$, and for $x \in \tilde{E}\otimes M_n$ define $\lambda(x) \in M(A)/A\otimes M_n$ to be the (1,2) entry of $\eta\begin{pmatrix} 0 & X \\ X^* & 0\end{pmatrix}$. Note that λ is real linear, contractive and extends $\phi \otimes id_n : E \otimes M_n \to M(A)/A \otimes M_n$. Now define $\tilde{\lambda}: \tilde{E} \otimes M_n \to M(A)/A \otimes M_n$ by

$$\tilde{\lambda}(x) = \frac{1}{2\pi}\int_0^{2\pi} \int\int e^{-it} v^* \lambda(e^{it} vxw) w^* \, dt\, dv\, dw$$

where the second and third integrations are with respect to Haar measure on the unitary group of M_n. It is easily verified that $\tilde{\lambda}$ is now complex linear, is still a contractive extension of $\phi \otimes id_n$, and satisfies

$$\tilde{\lambda}(vxw) = v\tilde{\lambda}(x)w$$

for all unitaries $v, w \in M_n$. Thus there exists a map $\tilde{\phi}: \tilde{E} \to M(A)/A$ such that $\tilde{\lambda} = \tilde{\phi} \otimes id_n$. This map is unital and n-contractive, from which it follows that it is n-positive. As remarked at the outset this is sufficient to prove the theorem.

Remark 3.5. The theorem is also valid if A is only assumed to be n-subhomogeneous and σ-unital. The only change in the proof would be the substitution of a quasi-central approximate unit for the central elements f_r.

REFERENCES

1. E.M. Alfsen, Compact convex sets and boundary integrals, Ergebnisse der Math., Springer Verlag, Berlin (1971).
2. T.B. Andersen, On Banach space valued extensions from split faces, Pacific J. Math., 42 (1972), 1-9.
3. T.B. Andersen, Linear extensions, projections and split faces, J. Funct. Anal., 17 (1974), 161-173.
4. J. Anderson, A. C^*-algebra for which $Ext(A)$ is not a group, Ann. Math., 107 (1978), 455-458.
5. W.B. Arveson, Subalgebras of C^*-algebras, Acta Math., 123 (1969), 141-224.
6. W.B. Arveson, Notes on extensions of C^*-algebras, Duke Math. J., 44 (1977), 329-355.
7. M.D. Choi and E. Effros, The completely positive lifting problem for C^*-algebras, Ann. Math., 104 (1976), 585-609.
8. M.D. Choi and E. Effros, Lifting problems and the cohomology of C^*-algebras, Can. J. Math., 29 (1977), 1092-1111.

9. M.D. Choi and E. Effros, Injectivity and operator spaces, J. Funct. Anal., 24 (1977), 156-209.
10. J. de Cannière and U. Haagerup, Multipliers of the Fourier algebras of some simple Lie groups and their discrete subgroups, Amer. J. Math., 107 (1985), 455-500.
11. K. Grove and G.K. Pedersen, Substonean spaces and corona sets, J. Funct. Anal., 56 (1984), 124-143.
12. K. Grove and G.K. Pedersen, Diagonalizing matrices over $C(X)$, J. Funct. Anal., 49 (1984), 65-89.
13. U. Haagerup, Injectivity and decomposition of completely bounded maps, Lecture Notes in Math., 1132, 170-222, Springer Verlag, Berlin 1985.
14. G.G. Kasparov, The operator K-functor and extensions of C^*-algebras, Math. USSR Izv., 16 (1981), 513-572.
15. I. Namioka and R.R. Phelps, Tensor products of compact convex sets, Pacific J. Math., 31 (1969), 469-480.
16. C.L. Olsen and G.K. Pedersen, Corona C^*-algebras and their applications to lifting problems, preprint (1988).
17. V.I. Paulsen, Completely bounded maps on C^*-algebras and invariant operator ranges, Proc. Amer. Math. Soc., 86 (1982), 91-96.
18. G.K. Pedersen, SAW^*-algebras and corona C^*-algebras, contributions to non-commutative topology, J. Operator Theory, 15 (1986), 15-32.
19. A.G. Robertson, A non-extendible positive map on the reduced C^*-algebra of a free group, Bull. London Math. Soc., 18 (1986), 389-391.
20. A.G. Robertson and R.R. Smith, Liftings and extensions of maps on C^*-algebras, J. Operator Theory, **21** (1989),117–131.
21. R.R. Smith, Completely bounded maps between C^*-algebras, J. London Math. Soc., 27 (1983), 157-166.
22. R.R. Smith and D.P. Williams, The decomposition property for C^*-algebras, J. Operator Theory, 16 (1986), 51-74.
23. R.R. Smith and D.P. Williams, Separable injectivity for C^*-algebras, Indiana Univ. Math. J., **37** (1988), 111–133.
24. E. Størmer, Extension of positive maps into $B(H)$, J. Funct. Anal., 66 (1986), 235-254.
25. G. Wittstock, E in operatorwertiger Hahn-Banach Satz, J. Funct. Anal., 40 (1981), 127-150.

Department of Mathematics
Texas A&M University
College Station, TX 77843

Multivariable Toeplitz Operators and Index Theory

HARALD UPMEIER*

In these notes we describe the recent progress made in the study of multivariable Toeplitz operators on domains in $\mathbb{C}^n$, the C^*-algebras generated by these operators and the index theory associated with C^*-algebra extensions of Toeplitz type. These results are important for a better understanding of multivariable complex analysis and also connect Toeplitz operators with interesting C^*-algebras not of type I, namely foliation C^*-algebras and irrational rotation algebras.

1. Solvable C^*-Algebras and Generalizations

In the study of pseudo-differential operators, Toeplitz operators or Wiener–Hopf integral operators, it is often convenient to consider the C^*-algebra A generated by these operators. In order to analyze spectral and index properties of these C^*-algebras, A. Dynin [11] introduced the notion of solvable C^*-algebra:

1.1. *Definition.* A C^*-algebra A is called *solvable* of length $r < +\infty$ if it has a composition series $I_1 \subset I_2 \subset \ldots \subset I_r$ of C^*-ideals in A such that for $0 \leq k \leq r$ (putting $I_0 := \{0\}$ and $I_{r+1} := A$) the subquotients are essentially commutative, i.e.,

$$I_{k+1}/I_k \cong \mathcal{C}_0(S_k) \otimes \mathcal{K}.$$

Here $\mathcal{K}$ denotes the C^*-algebra of all compact operators and S_k is a locally compact Hausdorff space. (For $k = r$, $\mathcal{K}$ is often replaced by $\mathbb{C}$ unless one considers the stabilization $A \otimes \mathcal{K}$ of A).

* Supported by NSF-Grant 8702371

By elementary C^*-algebra theory, it follows that the spectrum of A, i.e., the space of all irreducible representations of A, is precisely the union $\bigcup_{0\leq k\leq r} S_k$, endowed with an appropriate (non-Hausdorff) topology (cf. [11, 15]). The spaces S_k are topologically embedded in Spec (A) and are called the *spectral components* of A.

1.2 *Examples.* (i) For every compact manifold M without boundary, the C^*-algebra $\Psi(M)$ generated by all 0-order pseudodifferential operators P on M, given locally by

$$Pu(x) = (2\pi)^{-n/2} \int_{\mathbb{R}^n} p(x,\xi)\hat{u}(\xi)e^{ix\cdot\zeta}d\xi$$

for a "symbol function" $p(x,\xi)$ on the cotangent bundle $T^t(\mathbb{R}^n)$, is solvable of length 1, with spectral components $S_0 = \{\text{point}\}$ and $S_1 = S^t(M)$, the cosphere bundle of M with respect to a Riemannian metric. For manifolds with non-empty smooth boundary ∂M, one has similarly a composition series of length 2 [11].

(ii) The unilateral shift T_z on the classical Hardy space $H^1(\mathbf{T})$ on the 1-torus generates a "Toeplitz C^*-algebra" $\mathcal{T}(\mathbf{T})$ which is solvable of length 1, with spectral components $S_0 = \{0\}$ and $S_1 = \mathbf{T}$. This result is due to Gohberg and Krein (cf. [9]) and has a natural generalization to the n-dimensional case (cf. Theorem 2.1 below).

(iii) For every open convex cone Λ in $\mathbb{R}^n$ one can define *Wiener–Hopf operators*

$$(\widehat{T}_\varphi f)(x) := \int_{\overline{\Lambda}} \varphi(x-y)f(y)dy, \qquad f \in L^2(\overline{\Lambda})$$

on $L^2(\overline{\Lambda})$, with L^1-symbol function φ on $\mathbb{R}^n$. In [15], it is shown that the non-unital C^*-algebra $\widehat{\mathcal{T}}(\Lambda)$ generated by these operators is solvable of length r whenever Λ is a *polyhedral* cone of dimension r or a *self-dual homogeneous* cone of "rank" r. The spectral components of $\widehat{\mathcal{T}}(\Lambda)$ can be described in terms of the facial geometry of $\overline{\Lambda}$.

(iv) Generalizing Example (ii), the author [19, 20] has studied multi-variable Toeplitz operators

$$T_f h = P(fh), \qquad h \in H^2(S)$$

on the Hardy space $H^2(S)$ associated with a *bounded symmetric domain* D in $\mathbb{C}^n$ and its Shilov boundary S. Here $P: L^2(S) \to H^2(S)$ is the Szegö projection and $f \in \mathcal{C}(S)$ is a continuous "symbol function" on S. If the "rank" of D (cf. [14]) is r, then the Toeplitz C^*-algebra generated by these operators is solvable of length r, with spectral components S_k (described in more detail in Section 2) closely related to the boundary geometry of the underlying domain D.

Our main goal in this section is to introduce a generalization of the concept of solvable C^*-algebra which has some advantages over the notion defined in Definition 1.1. First of all, Definition 1.1 describes only C^*-algebras of type I and thus excludes many interesting classes of C^*-algebras. Also, the "geometric-analytic" operators described in the examples above can be considered in more general situations which don't give rise to type I C^*-algebras anymore. In the context of Toeplitz operators such a more general result will be described in Section 3.

In order to extend the notion of solvable C^*-algebra to a non-type I setting we use the notion of *groupoid* [17]. Let $\mathcal{G}$ be a locally compact groupoid with unit space $\mathcal{G}_0$ and (left) Haar system $\lambda = \{\lambda^u : u \in \mathcal{G}_0\}$. Let $C^*(\mathcal{G}, \lambda)$ be the C^*-algebra of $(\mathcal{G}, \lambda)$, defined as the C^*-completion of the convolution *-algebra $\mathcal{C}_c(\mathcal{G})$ with product

$$(f * g)(x) := \int_{\mathcal{G}} f(x) g(xy^{-1}) d\lambda^{xx^{-1}}(y)$$

and involution $f^*(x) := f(x^{-1})^*$, $x \in \mathcal{G}$. Here x^{-1} and xy denote the inverse and partial product in $\mathcal{G}$, respectively.

1.3 *Examples.* (i) Let X be a locally compact space. Then every (locally closed) equivalence relation $\mathcal{G} \subset X \times X$ is a groupoid with unit space $\mathcal{G}_0 = X$, inverse $(x,y)^{-1} := (y,x)$ and partial product $(x,y)(y,z) := (x,z)$. If $\mathcal{G} = \Delta \approx X$ is the diagonal, the family $\lambda^x = \delta_x$ (Dirac measure) is a Haar system with $C^*(\mathcal{G}, \lambda) \cong \mathcal{C}_0(X)$. If $\mathcal{G} = X \times X$ and μ is a measure on X with full support, we may take $\lambda^x = \delta_x \times \mu$ as a Haar system, and we have $C^*(\mathcal{G}, \lambda) \cong \mathcal{K}(L^2(X, \mu))$. Taking the cartesian product of two such groupoids, we obtain $\mathcal{C}_0(X) \otimes \mathcal{K}$ as a groupoid C^*-algebra.

(ii) A locally compact group G, acting on a locally compact space X (on the right), gives rise to a groupoid $\mathcal{G} = X \times G$ with unit space $\mathcal{G}_0 = X$ and Haar system $\lambda^x = \delta_x \times \lambda$ (product of Dirac and Haar measure) at $x \in X$. In this case, inverse and (partial) product are given by $(x,g)^{-1} := (x \rtimes g, g^{-1})$ and $(x,g)(x \rtimes g, h) := (x, gh)$ for all $x \in X$ and $g, h \in G$. The associated groupoid C^*-algebra $C^*(\mathcal{G}, \lambda)$ is the crossed product $\mathcal{C}_0(X) \rtimes G$. For $X = \{\text{point}\}$, we recover the group C^*-algebra $C^*(G)$.

(iii) Every foliation $\mathcal{F}$ on a compact manifold M gives rise to a groupoid $\mathcal{G}$ called the ***holonomy groupoid,*** having unit space M and a Haar system λ^x, $x \in M$, defined in terms of a transversal measure for $\mathcal{F}$ [7]. The corresponding groupoid C^*-algebra is independent of the choice of transversal measure and is denoted by $C^*(M, \mathcal{F})$.

Our proposed extension of Definition 1.1 is now the following

1.4 *Definition.* A C^*-algebra A is called *groupoid-solvable* of length $r < +\infty$ if it has a composition series $I_1 \subset \ldots \subset I_r$ of C^*-ideals such that for $0 \leq k \leq r$ (putting $I_0 := \{0\}$ and $I_{r+1} := A$), the subquotients are isomorphic to groupoid C^*-algebras.

$$I_{k+1}/I_k \cong C^*(\mathcal{G}_k)$$

over a locally compact groupoid $\mathcal{G}_k$ (with Haar system λ_k).

Of course, just as in Definition 1.1 (which is a special case by Example 1.3 (i)), the composition series of A should be "canonically" associated with A. Since the subquotients I_{k+1}/I_k are often stable (if $k < r$) one cannot get the lower ideals I_k, $k < r$, by taking successive commutator ideals. A promising alternative is to consider the "maximal radical series" introduced by A. Dynin and studied in detail in [12]. In the following sections we will show how the study of Toeplitz operators on domains in $\mathbb{C}^n$ fits naturally into the framework of Definition 1.4. The composition series (I_k) of A gives rise to C^*-algebra extensions

$$0 \to C^*(\mathcal{G}_{k-1}) \to I_{k+1}/I_{k-1} \xrightarrow{\sigma_k} C^*(\mathcal{G}_k) \to 0 \tag{1.1}$$

for $1 \leq k \leq r$, which (via the exact sequence of K-theory [4; Theorem 9.3.1]) carry *index information*

$$\mathrm{Ind}_k : K_1(C^*(\mathcal{G}_k)) \to K_0(C^*(\mathcal{G}_{k-1})). \tag{1.2}$$

The *index problem* is to express these analytical "partial" indices in topological terms. In the "essentially commutative" setting of Definition 1.1, the index homomorphisms

$$\mathrm{Ind}_k : K^1(S_k) \to K^0(S_{k-1}), \quad 1 \leq k \leq r \tag{1.3}$$

can be described in terms of the classical Fredholm index (for families) and play an important role for the inversion of singular integral operators (cf. [11]).

2. Toeplitz C^*-Algebras of Type I

We will now use the general concepts introduced in Section 1 to describe the Toeplitz C^*-algebras associated with important classes of bounded domains D in $\mathbb{C}^n$. The natural domains to consider are the so-called *domains of holomorphy,* which have the fundamental property that there exists a holomorphic function on D without analytic continuation across any boundary point. These domains are also called *pseudoconvex* since, in case $D = \{z \in \mathbb{C}^n : \rho(z) < 0\}$ for a smooth function $\rho(z)$, a necessary and sufficient condition is that the complex Hessian matrix (or Levi form) satisfies

$$\sum_{i,j} \overline{\xi}_i \left(\frac{\partial^2 \rho}{\partial \overline{z}_i \partial z_j} \right) \xi_j \geq 0 \tag{2.1}$$

at every boundary point, provided $\sum_j \frac{\partial \rho}{\partial z_j} \xi_j = 0$ [13]. Let $H^2(D) = \{h \in L^2(D) : h \text{ holomorphic on } D\}$ be the *Bergman space* over D (with respect to Lebesgue measure) and consider the C^*-algebra $\mathcal{T}(D)$ generated by all Toeplitz operators

$$T_f h := P(fh), \qquad h \in H^2(D)$$

on $H^2(D)$, with continuous symbol $f \in \mathcal{C}(\overline{D})$. Here $P: L^2(D) \to H^2(D)$ is the (orthogonal) *Bergman projection.*

A bounded domain with smooth boundary for which the form (2.1) is even positive definite is called *strictly pseudoconvex.* The basic example of such a domain is the "Hilbert ball"

$$\Delta_n := \{z \in \mathbb{C}^n : |z_1|^2 + \cdots + |z_n|^2 < 1\}.$$

For strictly pseudoconvex domains, the theory of Toeplitz operators closely resembles the 1-dimensional theory for the unit disc Δ (cf. [5, 6, 16, 22]):

2.1 **Theorem.** *Let D be a strictly pseudoconvex domain. Then the Toeplitz C^*-algebra $\mathcal{T}(D)$ contains the compact operators $\mathcal{K}$ and there is an exact sequence*

$$0 \to \mathcal{K} \to \mathcal{T}(D) \xrightarrow{\sigma} \mathcal{C}(\partial D) \to 0, \tag{2.2}$$

where σ is the so-called symbol homomorphism, uniquely determined by $\sigma(T_f) = f|_{\partial D}$ for all $f \in \mathcal{C}(\overline{D})$. Further, for every (matrix-valued) symbol $F \in \mathcal{C}(\overline{D}, \mathrm{GL}(N, \mathbb{C}))$ the index of T_F (a matrix of Toeplitz operators) is given by

$$\mathrm{Index}(T_F) = \chi_{\partial D}(F|_{\partial D}), \tag{2.3}$$

where $\chi_{\partial D}: K^1(\partial D) \to \mathbb{Z}$ is the (odd) index character of the contact manifold ∂D.

Note that $K^1(\partial D)$ can be identified with the group of homotopy classes of continuous mappings $F: \partial D \to \mathrm{GL}(N, \mathbb{C})$ for N arbitrary. The construction of the topological index character $\chi_{\partial D}$ is closely related to the Atiyah–Singer index character for symplectic manifolds (cf. [5]).

In the 1-dimensional case $D = \Delta$, Theorem 2.1 is essentially due to Gohberg–Krein [9] and (2.3) is the classical result

$$\mathrm{Index}(T_F) = - \text{ winding number of } \mathrm{Det}(F).$$

The case of the Hilbert ball $D = \Delta_n$, worked out in [22], gives the index formula

$$\mathrm{Index}(T_F) = (-1)^n \deg(F), \tag{2.4}$$

where $\deg(F)$ is the topological degree of $F: \mathbf{S}^{2n-1} \to \mathrm{GL}(N, \mathbb{C})$. In recent work [2], Theorem 2.1 was generalized to certain non-strictly pseudoconvex domains of "finite type," among them all pseudoconvex bounded domains with real-analytic boundary.

For domains with non-smooth boundary, Toeplitz operators are not necessarily essentially normal and the commutator ideal of $\mathcal{T}(D)$ is larger than $\mathcal{K}$. An important case are the domains with *stratified* (non-smooth) boundary, of which the so-called bounded symmetric domains are the principal examples. By definition, a circular convex bounded domain D is called *symmetric* if its holomorphic automorphism group $\mathrm{Aut}(D)$ acts transitively. These domains were classified by É. Cartan into four classical series and two exceptional types. A somewhat typical example is the *matrix unit ball*

$$D = \{z \in \mathbb{C}^{h \times k} : z^* z < \mathrm{id}\}, \tag{2.5}$$

whose automorphism group consists of all Moebius transformations

$$z \longmapsto (az + b)(cz + d)^{-1},$$

with $\begin{pmatrix} a & b \\ c & d \end{pmatrix} \in SU(h, k)$. Every symmetric domain has a characteristic invariant called its *rank* $r = \mathrm{rank}(D)$. In the previous example, we have $r = \min(h, k)$. If $r = 1$, we obtain the Hilbert ball $D = \Delta_n$.

In order to describe the stratification of the boundary of any symmetric domain, it is convenient to use a more recent algebraic characterization of these domains using Jordan algebraic methods. As shown by M. Koecher (cf. also [14]) every symmetric domain can be uniquely realized as the open unit ball

$$D = \{z \in Z : \|z\| < 1\}$$

of a *Jordan triple system* $Z \approx \mathbb{C}^n$. Let $\{uv^*w\} \in Z$ denote the triple product of $u, v, w \in Z$. The norm $\|\cdot\|$ is uniquely determined by this triple product and is called the "spectral norm" of Z. Just as a Jordan algebra is a generalized anti-commutator, a Jordan triple system can be regarded as a family of anti-commutators. For example, the matrix space $Z = \mathbb{C}^{h\times k}$ of (2.5) has the triple product

$$\{uv^*w\} = \frac{1}{2}(uv^*w + wv^*u).$$

Using the triple product structure the *faces* of the compact convex set $\overline{D}$ can be described as follows: Every face F is uniquely characterized by a *tripotent,* i.e., an element $e \in Z$ satisfying $\{ee^*e\} = e$, which is the "center" of F. The face F is given by

$$F = e + (\overline{D} \cap Z_e), \tag{2.6}$$

where $Z_e := \{z \in Z : \{ee^*z\} = 0\}$ is the "Peirce 0-space" of e. For matrices, the tripotents are just the partial isometries and the Peirce 0-space Z_e is the "opposite corner" of e. In general, the "open faces" $D \cap Z_e$ determined by (2.6) are again bounded symmetric domains (in Z_e) and thus have a "rank." A tripotent e is said to have rank k if $\text{rank}(D \cap Z_e) = r - k$. Here r is the rank of D. The set S_k of all tripotents of equal rank k, for $0 \le k \le r$, is a compact real-analytic manifold. We have $S_0 = \{0\}$ (corresponding to the single face $\overline{D}$) whereas S_r is the so-called *Shilov boundary* of D consisting of all singleton faces (i.e., extreme points) in ∂D. Of particular importance is the space S_1 of all minimal (i.e., rank 1) tripotents which geometrically correspond to the maximal faces in ∂D. The space S_k is called the *k-th* (partial) *Shilov boundary* of D.

It can be shown that the "k-th partial boundary"

$$\partial_k D = \bigcup_{e \in S_k} e + (D \cap Z_e) \tag{2.7}$$

of D is a real-analytic manifold contained in $\overline{D}$ (which is not a manifold if $r > 1$). We have $\partial_0 D = D$ and $\partial_r D = S_r$, whereas $\partial_1 D$ is a dense open subset of ∂D. Moreover, the "midpoint map"

$$\pi_k : \partial_k D \to S_k \tag{2.8}$$

sending $e + (D \cap Z_e)$ to e is a real-analytic locally trivial *fibration.* The partial boundaries satisfy

$$\overline{\partial_k D} = \bigcup_{k \le \ell \le r} \partial_\ell D$$

and are the strata of $\overline{D}$ mentioned above.

It is a remarkable fact, proved in [19, 20], that the intricate boundary geometry of symmetric domains is precisely reflected in the ideal structure of the corresponding Toeplitz C^*-algebras. More precisely, we have

2.2 **Theorem.** *Let D be a bounded symmetric domain of rank r. Then the Toeplitz C^*-algebra $\mathcal{T}(D)$ (on the Hardy space) has a canonical composition series $I_1 \subset \ldots \subset I_r$ such that for $0 \leq k \leq r$ (putting $I_0 := \{0\}$ and $I_{r+1} := \mathcal{T}(D)$) there exist isomorphisms*

$$(2.9) \qquad I_{k+1}/I_k \underset{\cong}{\overset{\sigma_k}{\longrightarrow}} \mathcal{C}(S_k) \otimes \mathcal{K}, \qquad 0 \leq k \leq r.$$

(For $k = r$, $\mathcal{K}$ is replaced by $\mathbb{C}$). The ideal I_k is the kernel of the k-th "symbol homomorphism"

$$\sigma_k: \mathcal{T}(D) \to \int_{e \in S_k}^{\oplus} \mathcal{T}(D \cap Z_e)$$

(continuous field of C^*-algebras), uniquely determined by

$$\sigma_k(T_f) = (T_{f_e})_{e \in S_k}$$

for all $f \in \mathcal{C}(\overline{D})$, where $f_e(z) := f(e+z)$ for all $z \in \overline{D \cap Z_e}$.

Theorem 2.2 implies in particular that $I_1 = \mathcal{K}$ and I_r is the commutator ideal of $\mathcal{T}(D)$. The lower ideals I_k cannot be obtained by taking successive commutator ideals but coincide with the so-called "maximal radical series" of $\mathcal{T}(D)$, as shown in [12]. Thus the composition series has an intrinsic "canonical" definition.

The index theory corresponding to the subquotients (2.9) (cf. (1.13)) was developed in [12]. For example, the lowest (integer-valued) index $\mathrm{Ind}_1: K^1(S_1) \to Z$, describing the extension

$$0 \to \mathcal{K} \to I_2 \xrightarrow{\sigma_1} \mathcal{C}(S_1) \otimes \mathcal{K} \to 0,$$

is given by

$$(2.10) \qquad \mathrm{Ind}_1(F) = \chi_{S_1}(F), \quad F \in \mathcal{C}(S_1, \mathrm{GL}(N, \mathbb{C}))$$

where, as in (2.3), χ_{S_1} is the index character of the contact manifold S_1. In [21], an explicit formula for (2.10) is given, involving the standard generators of the rational cohomology of $U(N)$ and the so-called Peirce $\frac{1}{2}$-bundles over S_k. For the special case of rank 1, i.e., $D = \Delta_n$, we recover the index formula (2.4).

3. Toeplitz C^*-Algebras not of Type I

The Toeplitz C^*-algebras described in Section 2, for the case of (i) strictly pseudoconvex domains or (ii) bounded symmetric domains, are of type I (i.e., GCR-algebras) with spectrum given, by (i) $\operatorname{Spec}\mathcal{T}(D) = \partial D \cup \{0\}$ or (ii) $\operatorname{Spec}\mathcal{T}(D) = \cup_{0\le k\le r} S_k$ (endowed with a non-Hausdorff topology). In this section we describe recent work (jointly with N. Salinas and A. Sheu [18]) which shows that, even for pseudoconvex domains with smooth boundary, non-type I Toeplitz C^*-algebras abound. Our results not only tie up multivariable Toeplitz operators with important classes of C^*-algebras such as the irrational rotation algebras but also make Toeplitz C^*-algebras an important "non-commutative" tool for studying domains with highly complicated boundary geometry. This is a typical multi-variable phenomenon since for every bounded domain in $\mathbb{C}$ (automatically pseudoconvex), the corresponding Toeplitz C^*-algebra is always of type I [1].

The non-type I Toeplitz C^*-algebras discovered in [18] (cf. also [8]) are related to the so-called Reinhardt (or poly-circular) domains. A domain $D \subset \mathbb{C}^n$ is called *Reinhardt* if it is invariant under all "poly-rotations"

$$(z_1, \ldots, z_n) \mapsto (\alpha_1 z_1, \ldots, \alpha_n z_n)$$

with $\alpha = (\alpha_1, \ldots, \alpha_n) \in \mathbb{T}^n$ (n-torus). The basic example is the polydisc

$$\Delta^n := \{z \in \mathbb{C}^n : \max_j |z_j| < 1\}.$$

(The polydisc, the Hilbert ball Δ_n and products of these domains are the only *symmetric* Reinhardt domains.) It is well-known [13] that the pseudoconvex Reinhardt domains containing the origin are precisely the domains of convergence of analytic power series. They are thus less general than the pseudoconvex domains ("local" power series versus "global" holomorphic functions) but still can be viewed as the "local building blocks" of arbitrary pseudoconvex domains.

For every pseudoconvex bounded Reinhardt domain D containing the origin, put

$$(3.1) \qquad |D^x| := \{(|z_1|, \ldots, |z_n|) : (z_1, \ldots, z_n) \in D, z_1 \ldots z_n \neq 0\}$$

and consider the "logarithmic domain"

$$(3.2) \qquad C := \{(\ln|z_1|, \ldots, \ln|z_n|) : (z_1, \ldots, z_n) \in |D^x|\}.$$

Since D is "logarithmically convex" (cf. [13]), the open subset C of $\mathbb{R}^n$ is convex.

Let F be any face of $\overline{C}$, with corresponding linear span $\mathrm{lin}(F)$, and consider the domain

$$\Omega_F := F^0 + i\,\mathrm{lin}(F) \subset \mathrm{lin}(F) \otimes \mathbb{C},$$

where F^0 is the interior of F in $\mathrm{lin}(F)$. By [18; Proposition 2.2], the sets

$$B_{F,\alpha} := \{(\alpha_1 \exp(z_1), \dots, \alpha_n \exp(z_n)) : (z_1, \dots, z_n) \in \Omega_F\} \tag{3.3}$$

for $\alpha \in \mathbb{T}^n$ are precisely the *holomorphic components* of $\overline{D}^x$, i.e., the equivalence classes under the equivalence relation

(3.4) $p \sim q$ iff there exists a finite sequence of holomorphic arcs $f_i: \Delta \to \overline{D}^x$ connecting p and q.

By (3.3), the set of holomorphic components belonging to a fixed face F of $\overline{C}$ is in 1–1 correspondence with the leaves of the foliation $\mathcal{F}_T$ of $\mathbb{T}^n$ determined by the leaves

$$L_\alpha := \{(\alpha_1 \exp(2\pi i t_1), \dots, \alpha_n \exp(2\pi i t_n)) : (t_1, \dots t_n) \in \mathrm{lin}(F)\} \tag{3.5}$$

for $\alpha \in \mathbb{T}^n$. In the special case $n = 2$ and $\dim(F) = 1$, these leaves belong to the Kronecker flow associated with the line $\mathrm{lin}(F)$ in $\mathbb{R}^2$. This flow is ergodic if and only if the slope of F is irrational, in which case the corresponding foliation C^*-algebra $C^*(\mathbb{T}^2, \mathcal{F}_F)$ is the stabilization of the irrational rotation algebra.

The main result of [18] expresses the Toeplitz C^*-algebra $\mathcal{T}(D)$ in terms of the foliation C^*-algebras associated with the holomorphic components (3.3). For simplicity, let us consider domains in $\mathbb{C}^2$ (the higher dimensional case is somewhat analogous) and suppose in the following that D is a pseudoconvex bounded Reinhardt domain in $\mathbb{C}^2$ containing the origin. For $0 \leq k \leq 2$, let $\partial_k D$ denote the *union* of all holomorphic components of $\overline{D}$ (not just of $\overline{D}^x$) having complex dimension $2 - k$. Let

$$S_k := \partial_k D / \sim \tag{3.6}$$

be the quotient space with respect to the equivalence relation (3.4), i.e., the *family* of all holomorphic components contained in $\partial_k D$. Then $\partial_k D$ is a manifold and the canonical projection

$$\pi_k : \partial_k D \to S_k \tag{3.7}$$

gives rise to a *foliation* $\mathcal{F}_k$ of $\partial_k D$, with leaf space S_k (cf. (3.3) and (3.5)). The space S_k is not necessarily a locally compact Hausdorff space.

3.1 **Theorem.** *The Toeplitz* C^**-algebra* $\mathcal{T}(D)$ *has a composition series* $I_1 = \mathcal{K}$, $I_2 = \mathcal{T}(D)'$ *(commutator ideal) such that for* $0 \leq k \leq 2$ *(putting* $I_0 := \{0\}$ *and* $I_3 := \mathcal{T}(D)$*) the subquotient is isomorphic to the foliation* C^**-algebra*

$$I_{k+1}/I_k \cong C^*(\partial_k D, \mathcal{F}_k)$$

associated with the "holomorphic component" foliation $\mathcal{F}_k$ *of* $\partial_k D$.

Comparing Theorem 3.1 with Theorems 2.1 and 2.2 we see that the main difference is the fact that, for Reinhardt domains, the holomorphic components form a *foliation* of the partial boundary $\partial_k D$ whereas for strictly pseudoconvex or symmetric domains we actually get a (locally trivial) *fibration* over a compact space S_k (cf. (2.7) and (2.8)).

The *proof* of Theorem 3.1 relies on the fact that Toeplitz operators over a Reinhardt domain D are given by *weighted shifts* (cf. [8]): Since D is poly-circular, the monomials

$$z^\nu := z_1^{\nu_1} \ldots z_n^{\nu_n},$$

for multi-indices $\nu = (\nu_1, \ldots, \nu_n) \in \mathbb{Z}_{\geq}^n$, form a pairwise orthogonal total family in $H^2(D)$. Let $e_\nu := z^\nu / \|z^\nu\|$ be the corresponding orthonormal basis. The Toeplitz C^*-algebra $\mathcal{T}(D)$ is generated by the operators

$$T^\mu := T_{z_1}^{\mu_1} \cdots T_{z_n}^{\mu_n} \tag{3.8}$$

for $\mu \in \mathbb{Z}_{\geq}^n$. Define *weights* $w_\mu(\nu)$ by putting

$$T^\mu e_\nu = w_\mu(\nu) \cdot e_{\mu+\nu}.$$

Clearly, $w_\mu(\nu) = \|z^{\mu+\nu}\| / \|z^\nu\|$ is a quotient of L^2-norms over D (which in general are not explicitly computable). The operators (3.8) have a polar decomposition

$$T^\mu = U_\mu W_\mu,$$

where $U_\mu e_\nu = e_{\mu+\nu}$ is a shift operator and W_μ is a "diagonal" operator in $H^2(D)$ with weights $W_\mu e_\nu = w_\mu(\nu) e_\nu$. As a consequence of [18; Theorem 1.3], the weight sequence $w_\mu(\nu)$ is *well-behaved* in the sense of [8] for any bounded pseudoconvex Reinhardt domain in $\mathbb{C}^n$ containing the origin. Therefore (cf. [8]) the C^*-algebra $C^*(U, W)$ generated by the operators U_μ and W_μ, for $\mu \in \mathbb{Z}_{\geq}^n$, is canonically isomorphic to the groupoid C^*-algebra

$C^*(\mathcal{G}, \lambda)$ associated with the reduction $\mathcal{G} := X \rtimes \mathbb{Z}^n || X_+$ of the transformation group groupoid $X \rtimes \mathbb{Z}^n$, endowed with its natural Haar system λ. Here X is the (non-compact) character space of the (non-unital) abelian C^*-subalgebra $\mathcal{A}$ of $\ell^\infty(\mathbb{Z}^n)$ generated by the functions

$$\tilde{w}_\mu(\nu) := \begin{cases} w_\mu(\nu), & \nu \in \mathbb{Z}^n_\geq \\ 0, & \nu \in \mathbb{Z}^n \backslash \mathbb{Z}^n_\geq \end{cases}$$

(for $\mu \in \mathbb{Z}^n_\geq$) and all its translates, and X_+ is the (compact) closure in X of the set of "elementary" characters $\nu^*(f) := f(\nu)$ of $\mathcal{A}$, for $\nu \in \mathbb{Z}^n_\geq$. In [18], the groupoid $\mathcal{G}$ is described explicitly in terms of the faces F of $\overline{C}$ and the associated foliations $\mathcal{F}_F$ with leaves given by (3.5). As a result, it follows that $C^*(U, W)$ has a composition series $(\mathcal{I}_k)$ such that, for $0 \leq k \leq 2$, there exist isomorphisms

$$\mathcal{I}_{k+1}/\mathcal{I}_k \cong \int_F^\oplus C^*(\mathbb{T}^2, \mathcal{F}_F) \tag{3.9}$$

(continuous field of C^*-algebras). Here F ranges over all faces of $\overline{C}$ of equal dimension $2 - k$ (possibly including "virtual" faces at ∞). In the presence of "virtual" faces, the Toeplitz C^*-algebra $\mathcal{T}(D)$ is a proper subalgebra of $C^*(U, W)$, but a suitable approximation argument gives Theorem 3.1 from (3.9). For details, cf. [18; Theorems 4.3 and 4.4].

Let us illustrate Theorem 3.1 by an example. For a fixed irrational number θ consider the domain

$$D := \left\{ z \in \Delta^2 : |z_1| |z_2|^\theta < \frac{1}{e} \right\}.$$

In this case, $\mathcal{T}(D)$ coincides with $C^*(\mathcal{G}, \lambda)$ (there are no virtual faces) and the logarithmic domain

$$C = \{x \in \mathbb{R}^2_< : x_1 + \theta x_2 < -1\}$$

has three 1-dimensional faces (having slope 0, $-1/\theta$ and ∞, respectively) and two extreme points. It follows from (3.9) or Theorem 3.1 that $\mathcal{T}(D)$ has a composition series $\mathcal{K} \subset I_2^{sing} \subset I_2 \subset \mathcal{T}(D)$ with subquotients

$$\mathcal{T}(D)/I_2 \cong \mathcal{C}(\mathbb{T}^2) \oplus \mathcal{C}(\mathbb{T}^2) \tag{3.10}$$

$$I_2/I_1^{sing} \cong C^*(\mathbb{T}^2, \mathcal{F}_\theta) \tag{3.11}$$

and

$$I_2^{sing}/\mathcal{K} \cong \mathcal{C}(\mathbf{T}) \otimes \mathcal{K} \oplus \mathcal{K} \otimes \mathcal{C}(\mathbf{T}).$$

Here $\mathcal{F}_\theta$ is the Kronecker foliation of $\mathbf{T}^2$ associated with θ. The index mapping

$$\mathrm{Ind}_1^{sing}: K^1(\mathbf{T}^2) \oplus K^1(\mathbf{T}^2) \to K_0(C^*(\mathbf{T}^2, \mathcal{F}_\theta))$$

associated with (3.10) and (3.11) can be described as follows [18; Theorem 4.8]

$$\mathrm{tr}_\theta[\mathrm{Ind}_1^{sing}(\varphi \oplus \psi)] := \iota_\theta \mathrm{ch}(\varphi\psi^{-1})$$

for all $\varphi, \psi \in K^1(\mathbf{T}^2)$. Here $\mathrm{tr}_\theta: K_0(C^*(\mathbf{T}^2, \mathcal{F}_\theta)) \to \mathbb{R}$ is the normalized trace, ch: $K^1(\mathbf{T}^2) \to H^1(\mathbf{T}^2, \mathbb{Z}) = \mathbb{Z}^2$ is the Chern character and $\iota_\theta: \mathbb{Z}^2 \to \mathbb{R}$ is the dense embedding given by $\iota_\theta(m,n) := m\theta - n$.

REFERENCES

[1] S. Axler, J. Conway and G. McDonald, *Toeplitz operators on Bergman spaces,* Can. J. Math. **34** (1982), 466–482.

[2] P. Baum, R.G. Douglas, M.E. Taylor, *Cycles and relative cycles in analytic K-homology,* J. Diff. Geom. (to appear).

[3] C.A. Berger, L.A. Coburn, A. Korányi, Opérateurs de Wiener–Hopf sur les sphères de Lie, C.R. Acad. Sci. Paris Sér. A-B **290** (1980), 989–991.

[4] B. Blackadar, *K-Theory for Operator Algebras,* New York, Springer, 1986.

[5] L. Boutet de Monvel, *On the index of Toeplitz operators of several complex variables,* Inventiones Math. **50** (1979), 249–272.

[6] L.A. Coburn, *Singular integral operators and Toeplitz operators on odd spheres,* Indiana Univ. Math. J. **23** (1973), 433–439.

[7] A. Connes, *A survey of foliations and operator algebras.* In: *Operator Algebras and Applications* (R.V. Kadison, ed.), Proc. Symp. Pure Math. **38**, Amer. Math. Soc., Providence, R.I., 1981.

[8] R.E. Curto, P.S. Muhly, *C^*-algebras of multiplication operators on Bergman spaces,* J. Funct. Anal. **64** (1985), 315–329.

[9] R.G. Douglas, *Banach Algebra Techniques in Operator Theory,* Academic Press, New York, 1972.

[10] R.G. Douglas, S. Hurder, J. Kaminker, *Toeplitz operators and the Eta invariant. The case of* $\mathbf{S}^1$, preprint.

[11] A. Dynin, *Inversion problem for singular integral operators: C^*-approach,* Proc. Natl. Acad. Sci. USA **75** (1978), 4668–4670.

[12] D. Handelman, H.-S. Yin, *Toeplitz algebras and rotational automorphisms associated to polydiscs,* Amer. J. Math. (to appear).

[13] S. Krantz, *Function Theory of Several Complex Variables,* New York, Wiley, 1982.

[14] O. Loos, *Bounded Symmetric Domains and Jordan Pairs,* Univ. of California, Irvine, 1977.

[15] P.S. Muhly, J.N. Renault, *C^*-algebras of multi-variable Wiener-Hopf operators,* Trans. Amer. Math. Soc. **274** (1983), 1–44.

[16] I. Raeburn, *On Toeplitz operators associated with strongly pseudoconvex domains,* Studia Math. **63** (1979), 253–258.

[17] J. Renault, *A Groupoid Approach to C^*-Algebras,* Lect. Notes in Math. **793**, New York, Springer, 1980.

[18] N. Salinas, A. Sheu, H. Upmeier, *Toeplitz operators on pseudoconvex domains and foliation C^*-algebras,* preprint.

[19] H. Upmeier, *Toeplitz operators on bounded symmetric domains,* Trans. Amer. Math. Soc. **280** (1983), 221–237.

[20] H. Upmeier, *Toeplitz C^*-algebras on bounded symmetric domains,* Ann. Math. **119** (1984), 549–576.

[21] H. Upmeier, *Fredholm indices for Toeplitz operators on bounded symmetric domains,* Amer. J. Math. (to appear).

[22] U. Venugopalkrishna, *Fredholm operators associated with strongly pseudoconvex domains,* J. Funct. Anal. **9** (1972), 349–373.

Department of Mathematics
University of Kansas
Lawrence, KS 66045

On Maximality of Analytic Subalgebras Associated with Flow in von Neumann Algebras

KICHI-SUKE SAITO

Let M be a von Neumann algebra on a Hilbert space H and let $\{\alpha_t\}_{t\in\mathbb{R}}$ be a σ-weakly continuous flow on M; i.e. suppose that $\{\alpha_t\}_{t\in\mathbb{R}}$ be a one-parameter group of *-automorphisms of M such that, for each ρ in the predual, M_*, of M and for each $x \in M$, the function of t, $\rho(\alpha_t(x))$, is continuous on $\mathbb{R}$. In this note, we consider the structure of the subspace of M, $H^\infty(\alpha)$, which is defined to be

$$\{x \in M : \rho(\alpha_t(x)) \in H^\infty(\mathbb{R}), \quad \text{for all} \quad \rho \in M_*\},$$

where $H^\infty(\mathbb{R})$ is the classical Hardy space consisting of the boundary values of functions bounded analytic in the upper half-plane. As in [2, 5, 6], $H^\infty(\alpha)$ is equal to the set of elements of M such that $\mathrm{Sp}_\alpha(x) \subset [0,\infty)$, where $\mathrm{Sp}_\alpha(x)$ is the Arveson spectrum of x with respect to $\{\alpha_t\}_{t\in\mathbb{R}}$. Thus, by [6], $H^\infty(\alpha)$ is a σ-weakly closed subalgebra of M containing the identity operator such that $H^\infty(\alpha)+H^\infty(\alpha)^*$ is σ-weakly dense in M, and such that $H^\infty(\alpha) \cap H^\infty(\alpha)^* = M^\alpha$, where M^α is the fixed point algebra of M with respect to $\{\alpha_t\}_{t\in\mathbb{R}}$. Since $H^\infty(\alpha)$ has the analyticity method in [1, 6, etc.], $H^\infty(\alpha)$ provides a very interesting generalization to the noncommutative setting of certain well-known classes of function algebras. Indeed, if M is α-finite, that is, there is a faithful family of α-invariant normal states on M, then $H^\infty(\alpha)$ is a maximal subdiagonal algebra as a noncommutative generalization of weak*-Dirichlet algebras in the sense of Arveson [1] (cf. [5], [6], [19]).

In this note, we consider the recent developments to the problem of maximality of $H^\infty(\alpha)$ as a σ-weakly closed subalgebra of M. That is, we contribute the following question.

Question. When is $H^\infty(\alpha)$ maximal among the σ-weakly closed subalgebras of M?

If $M = L^\infty(\mathbf{T})$ (resp. $L^\infty(\mathbb{R})$) and α_t is "translation" by t, then $H^\infty(\alpha)$ is $H^\infty(\mathbf{T})$ (resp. $H^\infty(\mathbb{R})$). We recall that $H^\infty(\mathbf{T})$ and $H^\infty(\mathbb{R})$ are maximal weak*-closed subalgebras of $L^\infty(\mathbf{T})$ and $L^\infty(\mathbb{R})$, respectively. Further, if M is commutative, then M may be identified with $L^\infty(\Omega, m)$ for some standard Borel space Ω with a finite measure m. In this case, $\{\alpha_t\}_{t\in\mathbb{R}}$ is implemented by a measurable action of $\mathbb{R}$ on Ω leaving m quasi-invariant:

$$(\omega, t) \to T_t\omega, \quad \omega \in \Omega, \quad t \in \mathbb{R}.$$

Then, the space $H^\infty(\alpha)$ may be considered as $\{\varphi \in L^\infty(\Omega) \mid$ for almost all ω, the function of t, $\varphi(T_t\omega)$, lies in $H^\infty(\mathbb{R})\}$. In [10], Muhly proved that if m is invariant, then $H^\infty(\alpha)$ is maximal among the σ-weakly closed subalgebras of $L^\infty(\Omega)$ if and only if m is ergodic. Since m is ergodic if and only if M^α is a factor, in this case, we conclude that $H^\infty(\alpha)$ is maximal among the σ-weakly closed subalgebras of M if and only if M^α $(= \mathbb{C})$ is a factor. In [7, 8, 9], M. McAsey, P.S. Muhly and the author introduced the notion of analytic crossed products and had the following first noncommutative results.

Suppose that N is a σ-finite von Newmann algebra and that β is a $*$-automorphism of N preserving a faithful normal state. Let M be the crossed product determined by N and β and let $\{\alpha\}_{t\in\mathbb{R}}$ be the (periodic) action of $\mathbb{R}$ on M that is dual of $\{\beta^n\}_{n\in\mathbf{Z}}$. Then $H^\infty(\alpha)$ is called the analytic crossed product determined by N and β. We sometimes write it as $N \rtimes_\beta \mathbf{Z}_+$. Then we have the following

Theorem 1 ([9, Theorem 4.4]). *Suppose N is a σ-finite von Neumann algebra and β is a *-automorphism of N preserving a faithful normal state. Then $H^\infty(\alpha)$ $(= N \rtimes_\beta \mathbf{Z}_+)$ is maximal among the σ-weakly closed subalgebras of M $(= N \rtimes_\beta \mathbf{Z})$ if and only if N is a factor.*

Subsequent results along this line were obtained by the author [14, 15] who considered almost periodic actions of $\mathbb{R}$ on finite von Neumann algebras. Let G be a compact abelian group and let $\{\alpha_t\}_{t\in G}$ be a σ-weakly continuous action on a finite von Neumann algebra M. Then we have

Theorem 2 ([15, Theorem 3.14]). *Suppose that the center $\mathcal{F}(M^\alpha)$ of M^α is contained in the center $\mathcal{F}(M)$ of M and that there is no nonzero projection $p \in \mathcal{F}(M^\alpha)$ such that $M^\alpha p = Mp$. Then $H^\infty(\alpha)$ is a maximal σ-weakly closed subalgebra of M if and only if M^α is a factor and $Sp\,\alpha$ is a subgroup (of $\hat{G}$) with an Archimedean order.*

On the other hand, B. Solel studied the maximality of $H^\infty(\alpha)$ when

$\{\alpha_t\}_{t\in\mathbb{R}}$ is periodic and found an example in which $H^\infty(\alpha)$ is maximal but M^α is not a factor. Specifically, if M is the algebra of 2×2 matrices, and if $\{\alpha_t\}_{t\in\mathbb{R}}$ is a non-trivial periodic action on M, then $H^\infty(\alpha)$ is the algebra of upper triangular matrices, which is maximal in this case, but M^α is the algebra of diagonal matrices and is not a factor. In Corollary 3.12 of [16], Solel subsumes this example in a result that gives a necessary and sufficient condition for $H^\infty(\alpha)$ to be a maximal σ-weakly closed subalgebra of M under the assumption that $\{\alpha_t\}_{t\in\mathbb{R}}$ is periodic. Let $M_n = \{x \in M: \mathrm{Sp}_\alpha(x) \subset \{n\}\}$. Put $f_n = \sup(vv^*: v$ is a partial isometry in $M_n\}$ and set $p(\alpha) = \sup\{f_n: n \geq 1\}$. Then it is clear that $p(\alpha)$ is a central projection in M^α and so we have the following.

Theorem 3 ([16, Corollary 3.12]). *Let M be a σ-finite von Neumann algebra. If $\{\alpha_t\}_{t\in\mathbb{R}}$ is periodic, then $H^\infty(\alpha)$ is a maximal σ-weakly closed subalgebra of M if and only if $p(\alpha)M^\alpha$ is a factor.*

On the other hand, Muhly and the author in [11] studied the maximality problem of $H^\infty(\alpha)$ in the case that $\{\alpha_t\}_{t\in\mathbb{R}}$ is non-periodic.

Theorem 4 ([11, Theorem 4.2]). *Let M be a von Neumann algebra with a faithful normal finite trace τ on M and let $\{\alpha_t\}_{t\in\mathbb{R}}$ be a σ-weakly continuous flow on M such that $\tau\circ\alpha_t = \tau$, for all $t\in\mathbb{R}$. If M^α is a factor, then $H^\infty(\alpha)$ is a maximal σ-weakly closed subalgebra of M.*

In the above discussion, we considered the case in which $H^\infty(\alpha)$ is a maximal subdiagonal algebra of M. Next we shall consider the case in which $H^\infty(\alpha)$ is not a maximal subdiagonal algebra, as a generalization of $H^\infty(\mathbb{R})$. As in [11], Muhly and the author studied the case of continuous analytic crossed products. Let M be a crossed product determined by a von Neumann algebra N and a σ-weakly continuous flow $\{\beta_t\}_{t\in\mathbb{R}}$ on N. Let $\{\alpha_t\}_{t\in\mathbb{R}}$ be the dual action of $\{\beta_t\}_{t\in\mathbb{R}}$. Then $H^\infty(\alpha)$ is called the analytic crossed product determined by N and $\{\beta_t\}_{t\in\mathbb{R}}$. As in [11], we denote it by $M \rtimes_\beta \mathbb{R}_+$. In this case, the analytic crossed product $M\rtimes_\beta\mathbb{R}_+$ is not a maximal subdiagonal algebra for the reason that there is not a faithful normal expectation of M onto M^α $(= N)$. We have the following theorem.

Theorem 5 ([11, Theorem 5.2]). *Let M be the crossed product determined by a von Neumann algebra N and a σ-weakly continuous action $\{\beta_t\}_{t\in\mathbb{R}}$ on N. Let $\{\alpha_t\}_{t\in\mathbb{R}}$ be the dual of $\{\beta_t\}_{t\in\mathbb{R}}$. Then $H^\infty(\alpha)$ is a maximal σ-weakly closed subalgebra of M if and only if M^α $(= N)$ is a factor.*

Let M be a von Neumann algebra and let $\{\alpha_t\}_{t\in\mathbb{R}}$ be a σ-weakly continuous flow on M. Let $\mathcal{N}$ be the set of all $x \in M$ such that there is some $y \in M$ with $y = \int_{\mathbb{R}} \alpha_t(x^*x)dt$. If the linear span of $\mathcal{N}$ is σ-weakly dense in M, then we shall say that $\{\alpha_t\}_{t\in\mathbb{R}}$ is integrable. We note that $\{\alpha_t\}_{t\in\mathbb{R}}$ is integrable if and only if $\int_{\mathbb{R}} \alpha_t(x)dt$, $x \in M_+$, is a faithful normal semifinite operator valued weight from M to M^α. Since the dual action on a crossed product is integrable, we have the following theorem as a generalization of Theorem 5.

Theorem 6 ([4, Theorem]). *If $\{\alpha_t\}_{t\in\mathbb{R}}$ is integrable on M, then the fixed point algebra M^α is a factor if and only if $H^\infty(\alpha)$ is a maximal σ-weakly closed subalgebra of M.*

Recently, B. Solel proved the general result.

Theorem 7 ([17, Theorem 3.2]). *Let M be a σ-finite von Neumann algebra and let $\{\alpha_t\}_{t\in\mathbb{R}}$ be a σ-weakly continuous flow on M. If $\mathcal{F}(M)\cap M^\alpha = CI$, then $H^\infty(\alpha)$ is a maximal σ-weakly closed subalgebra of M if and only if either $\operatorname{Sp}\alpha = \Gamma(\alpha)$ (where $\Gamma(\alpha)$ denotes the Connes spectrum of α) or there is a projection $F \in M$ such that $H^\infty(\alpha) = \{x \in M : (I-F)xF) = 0\}$.*

Finally, we give the full answer to the maximality question for $H^\infty(\alpha)$. Let M be a σ-finite von Neumann algebra and let $\{\alpha_t\}_{t\in\mathbb{R}}$ be a σ-weakly continuous flow on M. Let $H_0^\infty(\alpha)$ be the σ-weakly closed subspace generated by $\{x \in M : \operatorname{Sp}_\alpha(x) \subset (0,\infty)\}$. Then $H_0^\infty(\alpha)$ is a two-sided ideal of $H^\infty(\alpha)$. Let $f(\alpha)$ be the central support of $H_0^\infty(\alpha)$ in M, that is, $f(\alpha)$ is the least central projection in M such that $f(\alpha)x = x$ for every $x \in H_0^\infty(\alpha)$. Since $H_0^\infty(\alpha)$ is $\{\alpha_t\}_{t\in\mathbb{R}}$-invariant, we have $f(\alpha) \in \mathcal{F}(M)\cap M^\alpha$. By Theorem 7, we have

Theorem 8. *Let M be a σ-finite von Neumann algebra and let $\{\alpha_t\}_{t\in\mathbb{R}}$ be a σ-weakly continuous flow on M. Then the following assertions are equivalent:*

(1) *$H^\infty(\alpha)$ is a maximal σ-weakly closed subalgebra of M; and*

(2) (i) $\mathcal{F}(M)f(\alpha) \cap M^\alpha f(\alpha) = \mathbb{C}f(\alpha)$,

(ii) either $\operatorname{Sp}(\alpha|_{f(\alpha)} = \Gamma(\alpha|_{f(\alpha)})$ *or there exists a projection F in $Mf(\alpha)$ such that*

$$H^\infty(\alpha)f(\alpha) = \{x \in Mf(\alpha) : (f(\alpha)-F)xF = 0\}.$$

REFERENCES

[1] W.B. Arveson, *Analyticity in operator algebras,* Amer. J. Math. **89** (1967), 578–642.

[2] W.B. Arveson, *On groups of automorphisms of operator algebras,* J. Funct. Anal. **15** (1974), 217–243.

[3] A. Connes and M. Takesaki, *The flows of weight on factors of type* III, Tohoku Math. J. **29** (1977), 473–575.

[4] K. Homma and K.-S. Saito, *Analytic subalgebras associated with integrable flows on von Neumann algebras,* preprint.

[5] S. Kawamura and J. Tomiyama, *On subdiagonal algebras associated with flows in operator algebras,* J. Math. Soc. Japan **29** (1977), 73–90.

[6] R.I. Loebl and P.S. Muhly, *Analyticity and flows in von Neumann algebras,* J. Funct. Anal. **29** (1978), 214–252.

[7] M. McAsey, P.S. Muhly and K.-S. Saito, *Nonselfadjoint crossed products (Invariant subspaces and maximality),* Trans. Amer. Math. Soc. **248** (1979), 381–409.

[8] M. McAsey, P.S. Muhly and K.-S. Saito, *Nonselfadjoint crossed products* II, J. Math. Soc. Japan **33** (1981), 485–495.

[9] M. McAsey, P.S. Muhly and K.-S. Saito, *Nonselfadjoint crossed products* III *(Infinite algebras)*, J. Operator Theory **12** (1984), 3–22.

[10] P.S. Muhly, *Function algebras and flows,* Acta Sci. Math. (Szeged), **35** (1973), 111–121.

[11] P.S. Muhly and K.-S. Saito, *Analytic subalgebras in von Neumann algebras,* Canad. J. Math. **39** (1987), 74–99.

[12] P.S. Muhly, K.-S. Saito and B. Solel, *Coordinates for triangular operator algebras,* Ann. of Math. **127** (1988), 245–278.

[13] K.-S. Saito, *On noncommutative Hardy spaces associated with flows in finite von Neumann algebras,* Tôhoku Math. J. **29** (1977), 585–595.

[14] K.-S. Saito, *Invariant subspaces and cocycles in nonselfadjoint crossed products,* J. Funct. Anal. **45** (1982), 177–193.

[15] K.-S. Saito, *Nonselfadjoint subalgebras associated with compact abelian group actions on finite von Neumann algebras,* Tôhoku Math. J. **34** (1982), 485–494.

[16] B. Solel, *Algebras of analytic operators associated with a periodic flow on a von Neumann algebra,* Canad. J. Math. **37** (1985), 405–429.

[17] B. Solel, *Maximality of analytic operator algebras,* preprint.

[18] S. Stratilla, *Modular theory in operator algebras,* Abacus Press, 1981.

[19] L. Zsido, *Spectral and ergodic properties of the analytic generators,* J. Approximation Theory **20** (1977), 77–138.

Department of Mathematics
Faculty of Science
Niigata University
Japan

Reflections Relating a von Neumann Algebra and Its Commutant

RICHARD V. KADISON

1. Introduction

The initial development of the theory of von Neumann algebras, proposed by von Neumann [12] and carried out by him in collaboration with F.J. Murray [9,10,11,13] can be viewed as consisting of two parts, an "algebraic theory" and a "spatial theory." In the algebraic theory, the results refer to the von Neumann algebra $\mathcal{R}$ and make no reference to the commutant; in the spatial theory, the results involve the commutant either explicitly or implicitly. Recognizing this mathematical dichotomy, Kaplansky [7,8] studied the algebraic structure of von Neumann algebras, without reference to their action on a space, isolating and putting in sharp focus many of the natural techniques that are basic to our subject. Of course, Murray and von Neumann had taken the algebraic theory to an advanced stage in their own way [9,10,11,13].

The spatial theory was developed by Murray and von Neumann, in splendid detail, for von Neumann algebras with no central summand of type III. Just two points were left undone for such algebras: the trace-scaling (non-spatial) automorphisms of a II_∞ factor (with a II_1 commutant) [4], and the structure of the II_1 commutant in that case, when the II_∞ factor is matricial [1,3,14]. The basic element in the Murray-von Neumann arguments is the trace. Their key result in this connection is:

Theorem 0. *If $\mathcal{R}$ and $\mathcal{R}'$ are von Neumann algebras of type* II_1 *acting on a Hilbert space $\mathcal{H}$ and x_0 is a separating and generating unit trace vector for $\mathcal{R}$, then there is a *anti-isomorphism φ of $\mathcal{R}$ onto $\mathcal{R}'$ such that $Ax_0 = \varphi(A)x_0$ for each A in $\mathcal{R}$.*

In effect, Murray and von Neumann construct their "reflection" about the trace vector x_0, for each A in $\mathcal{R}$, there is a unique $\varphi(A)$ in $\mathcal{R}'$ such that

$Ax_0 = \varphi(A)x_0$, and observe that φ is a *anti-isomorphism. The burden of the argument falls on finding $\varphi(A)$ given A. With that theorem and appropriate reductions to the II_1 case, Murray and von Neumann can prove that the *anti-isomorphisms are present in all the cases where there is no central summand of type III. Their experience with specifically constructed factors of type III led them to ask whether such a *anti-isomorphism might not be present for all von Neumann algebras. This question received a spectacularly positive answer by Tomita [16,17] who associates, with a separating and generating vector u for $\mathcal{R}$, a modular structure $\{J, \Delta\}$ (cf. [6; Section 9.2]), where J is a conjugate-linear, involutory isometry of $\mathcal{H}$ onto itself and Δ is a positive, self-adjoint operator (generally, unbounded). The mapping that associates JA^*J with A in $\mathcal{R}$ is the *anti-isomorphism of $\mathcal{R}$ onto $\mathcal{R}'$ associated with u. The mapping σ_t, whose value at A in $\mathcal{R}$ is $\Delta^{it}A\Delta^{-it}$, is a *automorphism of $\mathcal{R}$ for each real t; $t \to \sigma_t$ is a one-parameter group of *automorphisms of $\mathcal{R}$.

While the subalgebra of $\mathcal{R}$ consisting of those elements A such that $Au = A'u$ for some A' in $\mathcal{R}'$ plays an important role in the deep and complicated arguments that establish the results of Tomita, just noted, this subalgebra is by no means all of $\mathcal{R}$. Tomita's work broadens "Murray-von Neumann reflection," taking it away from simple reflection about a trace vector, and deepens it significantly. It replaces it by "Tomita reflection," the mapping $A \to JA^*J$.

There is, however, another direction in which one can take Murray-von Neumann reflection, which retains the elements of simple reflection and a trace. It, too, is a reflection extending Murray-von Neumann reflection. In this context, the centralizer of a state ω on $\mathcal{R}$ is used in an essential way. Let $\mathcal{R}_\omega$ be this centralizer, that is, the set of those A in $\mathcal{R}$ such that $\omega(AT) = \omega(TA)$ for all T in $\mathcal{R}$. With ω the restriction of a vector state ω_x to $\mathcal{R}$ and ω' the restriction of ω_x to $\mathcal{R}'$, we show (Theorem 5) that there is a *anti-isomorphism φ of $\mathcal{R}_\omega E$ onto $\mathcal{R}'_{\omega'}E'$ such that $Ax = \varphi(A)x$ for each A in $\mathcal{R}_\omega E$, where E and E' are the supports of ω and ω'. Again, the burden of the argument falls on finding $\varphi(A)$ in $\mathcal{R}'_{\omega'}E'$ given A in $\mathcal{R}_\omega E$, and the main element of that process is Sakai's ingenious Proposition 1 in his proof [15] of Dixmier's Radon-Nikodým conjecture [2; p. 63].

In the last part of this paper, we show that the reflection we construct in Theorem 5 (extending Murray-von Neumann reflection from the case of a trace vector to that of an arbitrary vector) and the restriction of Tomita reflection ($A \to JA^*J$) to the centralizer are identical by the techniques of modular theory.

In the next section, we establish some results about supports and centralizers of normal states that allow us to draw conclusions about general normal states rather than just those that are faithful.

Erik Christensen and Uffe Haagerup made helpful comments at an early stage of this research. The NSF supplied partial support.

2. Centralizers and Supports

In this section, we prove three lemmas relating the support of a normal state to its centralizer. We shall use these lemmas to reduce to the case of a faithful state when proving our main results.

Lemma 1. *The support of a normal state of a von Neumann algebra lies in the center of the centralizer of that state.*

Proof. Let $\mathcal{R}$ be a von Neumann algebra, ω be a normal state of $\mathcal{R}$, E be the support of ω, and A be an element of the centralizer of ω. Since $\omega(I-E)=0$ and $0 \le I-E$, $I-E$ and E are in the centralizer of ω (for $0=\omega((I-E)B)=\omega(B(I-E))$, when $B \in \mathcal{R}$). Hence $EA(I-E)$ is in the centralizer of ω, and

$$0=\omega((I-E)A^*EA(I-E))=\omega(EA(I-E)A^*E).$$

Since E is the support of ω and $0 \le EA(I-E)A^*E$, we have, as a consequence, that $EA(I-E)A^*E=0$. Hence $EA(I-E)=0$. As A^* is also in the centralizer of ω, $EA^*(I-E)=0$ and $(I-E)AE=0$. It follows that

$$A=EAE+(I-E)A(I-E)$$

whence

$$EA=EAE=AE. \qquad \blacksquare$$

Lemma 2. *If ω is a normal state of a von Neumann algebra $\mathcal{R}$, E is the support of ω, and $\mathcal{R}_\omega$ is the centralizer of ω, then $\mathcal{R}_\omega$ is the direct sum of $(I-E)\mathcal{R}(I-E)$ and $\mathcal{R}_\omega E$.*

Proof. From Lemma 1, E is in the center of $\mathcal{R}_\omega$. Thus $\mathcal{R}_\omega$ is (isomorphic to) the direct sum of $\mathcal{R}_\omega(I-E)$ and $\mathcal{R}_\omega E$. We complete the proof by showing that $\mathcal{R}_\omega(I-E)=(I-E)\mathcal{R}(I-E)$. Since $I-E$ is in the center of $\mathcal{R}_\omega$,

$$\mathcal{R}_\omega(I-E)=(I-E)\mathcal{R}_\omega(I-E)\subseteq(I-E)\mathcal{R}(I-E).$$

Suppose S and T are in $\mathcal{R}$. Since $\omega(I-E)=0$, $I-E$ is in the left and right kernels of ω. Thus

$$0=\omega(S(I-E)T(I-E))=\omega((I-E)T(I-E)S).$$

In particular, $(I-E)T(I-E) \in \mathcal{R}_\omega$, whence $(I-E)T(I-E) \in \mathcal{R}_\omega(I-E)$. It follows that

$$(I-E)\mathcal{R}(I-E) \subseteq \mathcal{R}_\omega(I-E).$$

Combining this with the reverse inclusion, noted above, we conclude that $\mathcal{R}_\omega(I-E) = (I-E)\mathcal{R}(I-E)$. ∎

Lemma 3. *If $\mathcal{R}$ is a von Neumann algebra, ω is a normal state of $\mathcal{R}$, and E is the support of ω, then the centralizer of $\omega \mid E\mathcal{R}E$ is $\mathcal{R}_\omega E$.*

Proof. From Lemma 1, E is the center of the centralizer of ω so that

$$E\mathcal{R}_\omega E = \mathcal{R}_\omega E \subseteq \mathcal{R}_\omega, \quad \mathcal{R}_\omega E \subseteq E\mathcal{R}E.$$

Hence $\mathcal{R}_\omega E$ is contained in the centralizer of $\omega \mid E\mathcal{R}E$ $(= \omega_0)$.

Suppose T in $\mathcal{R}$ is such that ETE is in the centralizer of ω_0. With S in $\mathcal{R}$, we have that

$$\begin{aligned}\omega(SETE) &= \omega((I-E)SETE) + \omega(ESETE)\\ &= \omega(ESETE)\\ &= \omega(ETESE)\\ &= \omega(ETESE) + \omega(ETES(I-E))\\ &= \omega(ETES).\end{aligned}$$

Thus $ETE \in \mathcal{R}_\omega$ and $ETE \in \mathcal{R}_\omega E$. It follows that the centralizer of ω_0 is contained in $\mathcal{R}_\omega E$. From these inclusions, we have that the centralizer of ω_0 is $\mathcal{R}_\omega E$. ∎

3. Main Results

The theorem that follows details the construction of the reflection of an operator in the centralizer of a vector state. The argument makes crucial use of Sakai's proposition [15].

Theorem 4. *Let $\mathcal{R}$ be a von Neumann algebra acting on a Hilbert space $\mathcal{H}$ and x be a unit vector in $\mathcal{H}$. Let E be the support of $\omega_x \mid \mathcal{R}$ $(= \omega)$ and E' be the support of $\omega_x \mid \mathcal{R}'$. Then A is in the centralizer of ω if and only if $AE = EA$ and there is an A' in $\mathcal{R}'$ such that $E'A' = A'E'$ and $EAx = E'A'x$, $EA^*x = E'A'^*x$.*

Proof. Suppose, first, that for a given A in $\mathcal{R}$ commuting with E, there is an A' as described. Then,

$$\begin{aligned}\omega(AB) &= \langle ABx, x\rangle = \langle Bx, A^*x\rangle = \langle Bx, A^*Ex\rangle \\ &= \langle Bx, EA^*x\rangle = \langle Bx, E'A'^*x\rangle = \langle BA'x, x\rangle \\ &= \langle BE'A'x, x\rangle = \langle BEAx, x\rangle = \langle BAx, x\rangle \\ &= \omega(BA) \qquad\qquad (B \in \mathcal{R}).\end{aligned}$$

Thus A is in the centralizer of ω.

Suppose, now, that A is in the centralizer of ω. From Lemma 1, $AE = EA$. We begin by studying the case in which x is a separating and generating vector for $\mathcal{R}$ (and, hence, for $\mathcal{R}'$ as well). In this case, the ranges $[\mathcal{R}x]$ and $[\mathcal{R}'x]$ of E' and E are $\mathcal{H}$, so that $E = E' = I$. We define an operator R $(= R_{Ax})$ with domain $\mathcal{R}x$ by

$$RBx = BAx \qquad (B \in \mathcal{R}).$$

(See [6; p. 632].) Note that, with H self-adjoint in $\mathcal{R}$,

$$\omega_x(HAA^*) = \omega_x(AA^*H) = \overline{\omega_x(HAA^*)},$$

since AA^* is in the centralizer of ω. Thus $T \to \omega_x(TAA^*)$ is a hermitian functional on $\mathcal{R}$. By Sakai's proposition [15] (cf. [6; Lemma 7.3.4]),

$$\begin{aligned}\|RBx\|^2 &= \|BAx\|^2 = \langle A^*B^*BAx, x\rangle \\ &= \langle B^*BAA^*x, x\rangle = \omega_x(B^*BAA^*) \\ &= |\omega_x(AA^*B^*B)| \le \|AA^*\|\omega_x(B^*B) \\ &= \|A\|^2\|Bx\|^2 \qquad (B \in \mathcal{R}).\end{aligned}$$

Thus R extends uniquely to a bounded operator A' on $[\mathcal{R}x]$ (which is $\mathcal{H}$, under the present assumption). From [6; Lemma 9.2.28], $A' \in \mathcal{R}'$ (though, this is immediate in the bounded case). Moreover, $A'x = Rx = Ax$. At the same time,

$$\begin{aligned}\langle A'^*x, Bx\rangle &= \langle x, A'Bx\rangle = \langle x, RBx\rangle \\ &= \langle x, BAx\rangle = \langle A^*B^*x, x\rangle \\ &= \omega_x(A^*B^*) = \omega_x(B^*A^*) \\ &= \langle B^*A^*x, x\rangle = \langle A^*x, Bx\rangle \qquad (B \in \mathcal{R}).\end{aligned}$$

Thus $A'x = Ax$ and $A'^*x = A^*x$, under the present assumption.

We reduce the general situation to the case just studied. Note for this that $E\mathcal{R}E$ acting on $[\mathcal{R}'x]$ has $\mathcal{R}'E$ as commutant, and $E'E(\mathcal{R}'E)E'E$

$(= E'\mathcal{R}'E'E)$, acting on $[\mathcal{R}'x] \cap [\mathcal{R}x]$ $(= \mathcal{H}_0)$, has $(E\mathcal{R}E)E'E$ $(= E\mathcal{R}EE')$ as its commutant. Moreover, $E'\mathcal{R}'E'E$ and $E\mathcal{R}EE'$ acting on $\mathcal{H}_0$ have x as a joint generating and separating vector. Let ω_0 be $\omega_x \mid E\mathcal{R}EE'$. Then

$$\begin{aligned}\omega_0(EBEE'EAEE') &= \omega_0(EBEAEE') = \omega(EBEA)\\ &= \omega(AEBE) = \omega(AEB)\\ &= \omega_0(EAEE'EBEE') \qquad (B \in \mathcal{R}).\end{aligned}$$

Thus $EAEE'$ is in the centralizer of ω_0. From the case where x is a joint generating and separating vector, there is an element $E'A'E'E$ in $E'\mathcal{R}'E'E$ (with A' in $E'\mathcal{R}'E' \subseteq \mathcal{R}'$) such that

$$EAx = EAEE'x = E'A'E'Ex = E'A'x,$$

$$EA^*x = EA^*EE'x = E'A'^*E'Ex = E'A'^*x. \qquad \blacksquare$$

Theorem 5. *In the notation of Theorem 4, let ω' be $\omega_x \mid \mathcal{R}'$, $\mathcal{R}_\omega$ be the centralizer of ω, and $\mathcal{R}'_{\omega'}$, be the centralizer of ω'. With A in $\mathcal{R}_\omega E$, there is a unique A' in $\mathcal{R}'_{\omega'}E'$ such that $Ax = A'x$ and $A^*x = A'^*x$; the mapping $A \to A'$ is a *anti-isomorphism of $\mathcal{R}_\omega E$ onto $\mathcal{R}'_{\omega'}E'$.*

Proof. From Lemma 1, E and E' are in the centers of $\mathcal{R}_\omega$ and $\mathcal{R}'_{\omega'}$, respectively. Thus $\mathcal{R}_\omega E$ and $\mathcal{R}'_{\omega'}E'$ are von Neumann algebras. With A in $\mathcal{R}_\omega E$, A is in $\mathcal{R}_\omega$. From Theorem 4, there is an A' in $\mathcal{R}'$ such that

$$\begin{aligned}Ax &= EAx = E'A'x = E'A'E'x,\\ A^*x &= EA^*Ex = E'A'^*x = E'A'^*E'x.\end{aligned}$$

If we use $E'A'E'$ in place of A', we may assume that

$$Ax = A'x, \ A^*x = A'^*x, \ E'A' = A'E'.$$

Now, applying Theorem 4, again, we conclude that $A' \in \mathcal{R}'_{\omega'}$, from which $A' \in \mathcal{R}'_{\omega'}E'$.

If $B' \in \mathcal{R}'_{\omega'}E'$ and $Ax = B'x$, then $(B' - A')x = 0$. It follows that $(B' - A')\mathcal{R}x = 0$, and $B' - A' = (B' - A')E' = 0$. Thus A', as described, is unique. With A and B in $\mathcal{R}_\omega E$,

$$ABx = AB'x = B'Ax = B'A'x,$$

whence $(AB)' = B'A'$. The linearity of $A \to A'$ is evident. Moreover, $(A^*)' = (A')^*$ since $A^*x = A'^*x$. With B' in $\mathcal{R}'_{\omega'}E'$, there is a B in

$\mathcal{R}_\omega E$ such that $Bx = B'x$, $B^*x = B'^*x$, by symmetry. Thus the mapping $A \to A'$ is a *anti-isomorphism of $\mathcal{R}_\omega E$ onto $\mathcal{R}'_{\omega'}E'$. ∎

4. Relating Reflections

The relation of the reflection between centralizers, developed in Section 3, to the other reflections is established with the aid of the following proposition. Its proof requires the results and techniques of modular theory.

Proposition 6. *Let $\mathcal{R}$ be a von Neumann algebra acting on a Hilbert space $\mathcal{H}$, u be a separating and generating unit vector for $\mathcal{R}$, and ω be $\omega_u \,|\, \mathcal{R}$. With $\{J, \Delta\}$ the modular structure for $\{\mathcal{R}, u\}$, the following are equivalent:*

(i) $Au = \Delta^{1/2}Au$;

(ii) $Au = JA^*Ju$;

(iii) *A is the closure of* $\Delta^{1/2}A\Delta^{-1/2}$;

(iv) $A\Delta \subseteq \Delta A$;

(v) *A is in the centralizer of ω.*

Proof. (i) $\leftrightarrow$ (ii) Suppose $Au = \Delta^{1/2}Au$. From [6; Theorem 9.2.9], $Ju = u$ and $J^2 = I$. Thus

$$JAu = J\Delta^{1/2}Au = SAu = A^*u,$$

and

$$Au = JJAu = JA^*u = JA^*Ju.$$

Assuming that $Au = JA^*Ju$, we have that

$$Au = JA^*u = JSAu = JJ\Delta^{1/2}Au = \Delta^{1/2}Au.$$

(i) $\to$ (iii) From [6; Theorem 9.2.9], $\Delta u = u$. Thus $u = \Delta^{1/2}u = \Delta^{-1/2}u$ from [6; Remark 5.6.32], and

$$Au = \Delta^{1/2}Au = \Delta^{1/2}A\Delta^{-1/2}u.$$

Recall that $J\Delta^{-1/2} = F$ (see the discussion following [6; Remark 9.2.2]) and $J\mathcal{R}'J = \mathcal{R}$ [6; Theorem 9.2.9]. Thus with B' in $\mathcal{R}'$, we have

$$\begin{aligned}\Delta^{1/2}A\Delta^{-1/2}B'u &= \Delta^{1/2}AJFB'u = \Delta^{1/2}AJB'^*u \\ &= JSAJB'^*Ju = J(JB'^*J)^*A^*u \\ &= J(JB'J)A^*u = B'JA^*u \\ &= B'JA^*Ju = B'Au = AB'u.\end{aligned}$$

Thus $\mathcal{R}'u \subseteq \mathcal{D}(\Delta^{1/2}A\Delta^{-1/2})$ and

$$\Delta^{1/2}A\Delta^{-1/2} \,|\, \mathcal{R}'u = A \,|\, \mathcal{R}'u.$$

Since $\mathcal{R}'u$ is a core for $\Delta^{-1/2}$, if $x \in \mathcal{D}(\Delta^{1/2}A\Delta^{-1/2})$, there is a sequence $\{A'_n\}$ in $\mathcal{R}'$ such that $A'_n u \to x$ and $\Delta^{-1/2}A'_n u \to \Delta^{-1/2}x$. But then $A\Delta^{-1/2}A'_n u \to A\Delta^{-1/2}x$ and $\Delta^{1/2}A\Delta^{-1/2}A'_n u = AA'_n u \to Ax$. Since $\Delta^{1/2}$ is closed,

$$\Delta^{1/2}A\Delta^{-1/2}A'_n u \to \Delta^{1/2}A\Delta^{-1/2}x.$$

Thus $\Delta^{1/2}A\Delta^{-1/2}x = Ax$ and $\Delta^{1/2}A\Delta^{-1/2} = A \,|\, \mathcal{D}(\Delta^{1/2}A\Delta^{-1/2})$. It follows that A is the closure of $\Delta^{1/2}A\Delta^{-1/2}$.

(iii) $\to$ (iv) By assumption, $\Delta^{1/2}A\Delta^{-1/2} \subseteq A$. From [6; 5.6.(13)], $\Delta^{-1/2}\Delta^{1/2}A\Delta^{-1/2} \subseteq \Delta^{-1/2}A$. Thus

$$A\Delta^{-1/2} \,|\, \mathcal{D}(\Delta^{1/2}A\Delta^{-1/2}) \subseteq \Delta^{-1/2}A.$$

We show, next, that $\mathcal{R}'u \subseteq \mathcal{D}(\Delta^{1/2}A\Delta^{-1/2})$. If $B' \in \mathcal{R}'$,

$$\Delta^{-1/2}B'u = JFB'u = JB'^{*}u = JB'^{*}Ju.$$

Now $JB'^{*}J \in \mathcal{R}$ from [6; Theorem 9.2.9], whence

$$A\Delta^{-1/2}B'u = AJB'^{*}Ju \in \mathcal{D}(\Delta^{1/2}).$$

Thus $B'u \in \mathcal{D}(\Delta^{1/2}A\Delta^{-1/2})$ and $\mathcal{R}'u \subseteq \mathcal{D}(\Delta^{1/2}A\Delta^{-1/2})$. It follows that $A\Delta^{-1/2} \,|\, \mathcal{R}'u \subseteq \Delta^{-1/2}A$. With x in $\mathcal{D}(\Delta^{-1/2})$, we can choose B'_n in $\mathcal{R}'$ such that $B'_n u \to x$ and $\Delta^{-1/2}B'_n u \to \Delta^{-1/2}x$. Since A is bounded, $A\Delta^{-1/2}B'_n u \to A\Delta^{-1/2}x$. But $A\Delta^{-1/2}B'_n u = \Delta^{-1/2}AB'_n u$ and $\Delta^{-1/2}A$ is closed. (With B bounded and T closed, BT need not be closed, but TB is closed [6; Example 5.6.33].) Thus $x \in \mathcal{D}(\Delta^{-1/2}A)$, $\Delta^{-1/2}Ax = A\Delta^{-1/2}x$, and $A\Delta^{-1/2} \subseteq \Delta^{-1/2}A$. From [6; Lemma 5.6.17], A commutes with the spectral resolution of $\Delta^{-1/2}$. Hence $A \in \mathcal{A}'$, where $\mathcal{A}$ is the abelian von Neumann algebra generated by $\Delta^{-1/2}$ (cf. [6; Theorem 5.6.18]). We have, from [6; Theorem 5.6.26], that $\Delta \,\eta\, \mathcal{A}$. Thus $A\Delta \subseteq \Delta A$.

(iv) $\to$ (v) If $A\Delta \subseteq \Delta A$, then A commutes with the abelian von Neumann algebra generated by Δ (as at the end of the preceding argument). That algebra contains Δ^{it} for each real t. Thus, $\Delta^{it}A\Delta^{-it} = A$. From [6; Theorem 9.2.13, Proposition 9.2.14], A is in the centralizer of ω.

(v) $\to$ (i) If A is in the centralizer of ω, then A commutes with Δ^{it} for each real t from [6; Proposition 9.2.14]. Using the formula from [6; Theorem 5.6.36],

$$\langle \hat{f}(H)x, y\rangle = \int_R f(t)\langle e^{itH}x, y\rangle dt,$$

with $\log\Delta$ in place of H and Ax in place of x, we have that

$$\begin{aligned}\langle \hat{f}(\log\Delta)Ax, y\rangle &= \int_R f(t)\langle \Delta^{it}Ax, y\rangle dt\\ &= \int_R f(t)\langle \Delta^{it}x, A^*y\rangle dt\\ &= \langle \hat{f}(\log\Delta)x, A^*y\rangle;\end{aligned}$$

whence $\hat{f}(\log\Delta)A = A\hat{f}(\log\Delta)$ for each f in $L_1(\mathbb{R})$. Pursuing this reasoning with appropriate choices for f, one can conclude that A commutes with the (abelian) von Neumann algebra generated by $\log\Delta$, and since Δ ($= \exp(\log\Delta)$) and $\Delta^{1/2}$ are affiliated with this algebra, that $A\Delta^{1/2} \subseteq \Delta^{1/2}A$. But careful use of Fourier transform arguments and formulae are needed for a complete proof that $A\Delta^{1/2} \subseteq \Delta^{1/2}A$. It is, perhaps, more convincing to employ a lemma from [5], which assures us that $\{\Delta^{it} : t \in \mathbb{R}\}$ and $(\Delta + I)^{-1}$ generate the same von Neumann algebra. Of course, Δ and $\Delta^{1/2}$ are affiliated with this algebra, so that $A\Delta^{1/2} \subseteq \Delta^{1/2}A$. Thus $Au = A\Delta^{1/2}u = \Delta^{1/2}Au$. ∎

The theorem that follows describes the relations among the various reflections.

Theorem 7. *Let $\mathcal{R}$ be a von Neumann algebra acting on a Hilbert space $\mathcal{H}$, u be a generating and separating unit vector for $\mathcal{R}$, ω be $\omega_u \mid \mathcal{R}$, ω' be $\omega_u \mid \mathcal{R}'$, $\mathcal{R}_\omega$ be the centralizer of ω, and $\mathcal{R}'_{\omega'}$ be the centralizer of ω'. Let φ be the reflection of $\mathcal{R}_\omega$ onto $\mathcal{R}'_{\omega'}$ (about u) described in Theorem* 5.

(i) *If u is a trace vector for $\mathcal{R}$, then u is a trace vector for $\mathcal{R}'$, and $Au = \varphi(A)u$ for each A in $\mathcal{R}$. The mapping φ is Murray-von Neumann reflection in this case.*

(ii) *$A \in \mathcal{R}_\omega$ if and only if $Au = JA^*Ju$; and when $A \in \mathcal{R}_\omega$, $\varphi(A) = JA^*J$.*

Proof. (i) That u is a trace vector for $\mathcal{R}'$ is a consequence of [6; Lemma 7.2.14]. It follows that $\mathcal{R}_\omega = \mathcal{R}$ and $\mathcal{R}'_{\omega'} = \mathcal{R}'$. Since u is generating and separating, the supports of ω and ω' are both I. From Theorem 5, $Au = \varphi(A)u$ for each A in $\mathcal{R}$. Thus φ is Murray-von Neumann reflection about the trace vector u.

(ii) The first assertion of this part follows at once from the equivalence of (ii) and (v) of Proposition 6. With A in $\mathcal{R}_\omega$, $Au = \varphi(A)u$ from Theorem 5. But $Au = JA^*Ju$ so that $(\varphi(A) - JA^*J)u = 0$. Since $\varphi(A)$ and JA^*J are in $\mathcal{R}'$ and u is separating for $\mathcal{R}'$, $\varphi(A) = JA^*J$. ∎

REFERENCES

[1] A. Connes, *Classification of injective factors,* Cases II_1, II_∞, III_λ, $\lambda \neq 1$, Ann. of Math. **104** (1976), 73–115.

[2] J. Dixmier, *Les Algèbres d'Opérateurs dans l'Espace Hilbertien,* Gauthier-Villars, Paris, 1957.

[3] U. Haagerup, *A new proof of the equivalence of injectivity and hyperfiniteness for factors on a separable Hilbert space,* J. Fnal. Anal. **62** (1985), 160–201.

[4] R. Kadison, *Isomorphisms of factors of infinite type,* Canad. J. Math. **7** (1955), 322–327.

[5] R. Kadison, *Centralizers and diagonalizing states,* in preparation.

[6] R. Kadison and J. Ringrose, *Fundamentals of the Theory of Operator Algebras,* Academic Press, Orlando, Vol. I, 1983, Vol. II, 1986.

[7] I. Kaplansky, *Projections in Banach Algebras,* Ann. of Math. **53** (1951), 235–249.

[8] I. Kaplansky, *Algebras of type I,* Ann. of Math. **56** (1952), 460–472.

[9] F. Murray and J. von Neumann, *On rings of operators,* Ann. of Math. **37** (1936), 116–229.

[10] F. Murray and J. von Neumann, *On rings of operators,* II, Trans. Amer. Math. Soc. **41** (1937), 208–248.

[11] F. Murray and J. von Neumann, *On rings of operators,* IV, Ann. of Math. **44** (1943), 716–808.

[12] J. von Neumann, *Zur Algebra der Funktionaloperationen und Theorie der normalen Operatoren,* Math. Ann. **102** (1930), 49–131.

[13] J. von Neumann, *On rings of operators,* III, Ann. of Math. **41** (1940), 94–161.

[14] S. Popa, *A short proof of "injectivity implies hyperfiniteness" for finite von Neumann algebras,* J. Operator Theory **16** (1986), 261–272.

[15] S. Sakai, *A Radon-Nikodym theorem in W^*-algebras,* Bull. Amer. Math. Soc. **71** (1965), 149–151.

[16] M. Takesaki, *Tomita's Theory of Modular Hilbert Algebras and Its Applications,* LNM Vol. 128, Springer-Verlag, Heidelberg, 1970.

[17] M. Tomita, *Standard forms of von Neumann algebras,* Fifth Functional Analysis Symposium of the Math. Soc. of Japan, Sendai, 1967.

Department of Mathematics
University of Pennsylvania
Philadelphia, PA 19104

Normal AW*-Algebras

KAZUYUKI SAITÔ and J.D.M. WRIGHT

Let us recall that a C^*-algebra A is an AW^*-algebra if (1) each maximal abelian *-subalgebra of A is generated by its projections and (2) each family of orthogonal projections $\{e_\alpha\}$ in A has a supremum $\Sigma_A e_\alpha$ in $\mathrm{Proj}(A)$ (the set of all projections in A).

Kaplansky [3] and [4] introduced AW^*-algebras and obtained their theory. In particular, by an elegant algebraic method, he extended the Murray-von Neumann type theory classification of von Neumann algebras to these more general C^*-algebras. He also showed that these conditions (1) and (2) imply that $\mathrm{Proj}(A)$ is a complete lattice.

One of the difficulties in treating AW^*-algebras is that because of the lack of the strong topology, as in von Neumann algebras, there are no guarantees for the fact that whenever $\{f_\beta\}$ is an increasing net of projections in A with the supremum $f(= \mathrm{LUB}_{\mathrm{Proj}(A)} f_\beta)$ in the complete lattice $\mathrm{Proj}(A)$, then f is the supremum of $\{f_\beta\}$ in the partially ordered set A_h of all hermitian elements in A. (We call such a family *a well-behaved family*).

An AW^*-algebra is said to be normal ([8]) if every increasing net of projections is well-behaved.

Are all AW^*-algebras normal?

It is known that monotone complete C^*-algebras (for example, von Neumann algebras and AW^*-algebras of type I) are normal.

In 1980, J.D.M. Wright [8] and M. Hamana [2] independently showed that finite AW^*-algebras are normal (K. Saitô gave a simple proof [6]).

When an AW^*-algebra is properly infinite and *σ-finite*, then, by using their ingenious arguments, Christensen and Pedersen proved that it is normal ([1]).

In this note, we would like to present the following:

Theorem. *Let A be a properly infinite AW^*-algebra whose centre Z_A is σ-finite. Then A is normal.*

We lean heavily on the results and methods of Christensen and Pedersen [1]. In fact, we can squeeze out the following proposition from [1].

Proposition. *Let A be any properly infinite AW^*-algebra and suppose that the center of A is σ-finite. Then, for every orthogonal family $\{e_\alpha\}$ of projections in A, the net, which comes from all finite partial sums of $\{e_\alpha\}$, is well-behaved.*

Let $\hat{A}$ be the regular completion of A ([2] and [7]). Note that $\hat{A}$ is a monotone complete C^*-algebra and by making use of the above result, we can show that A is an *AW^*-subalgebra* of $\hat{A}$, that is, for every orthogonal family of projections $\{f_\beta\}$ of A,

$$\Sigma_A f_\beta = \Sigma_{\hat{A}} f_\beta.$$

By a result of Saitô and Pedersen, for every increasing net $\{g_\alpha\}$ in Proj(A),

$$\mathrm{LUB}_{\mathrm{Proj}(A)} g_\alpha = \mathrm{LUB}_{\mathrm{Proj}(\hat{A})} g_\alpha.$$

Since $\hat{A}$ is monotone complete and $\{g_\alpha\}$ is well-behaved in $\hat{A}$, it follows that $\{g_\alpha\}$ is well-behaved in A.

The first author wishes to thank the SERC who financed his extended visit to the University of Reading.

REFERENCES

[1] E. Christensen and G.K. Pedersen, *Properly infinite AW^*-algebras are monotone sequentially complete,* Bull. London Math. Soc., **16** (1984), 407–410.

[2] M. Hamana, *Regular embeddings of C^*-algebras in monotone complete C^*-algebras,* J. Math. Soc. Japan **33** (1981), 159–183.

[3] I. Kaplansky, *Projections in Banach algebras,* Ann. of Math., **53** (1951), 235-249.

[4] I. Kaplansky, *Algebras of type I,* Ann. of Math. **56** (1952), 460–472.

[5] K. Saitô, *On the embedding as a double commutator in a type I AW^*-algebra II,* Tôhoku Math. J., **26** (1974), 333–339.

[6] K. Saitô, *On normal AW^*-algebras,* Tôhoku Math. J., **33** (1981), 567–572.

[7] J.D.M. Wright, *Regular σ-completions of C^*-algebras,* J. London Math. Soc., **12** (1976), 299-309.
[8] J.D.M. Wright, *Normal AW^*-algebras,* Proc. Roy. Soc. Edinburgh **85A** (1980), 137–141.

Kazuyuki Saitô
Mathematical Institute
Tôhoku University
Sendai, Japan

J.D.M. Wright
Dept. of Mathematics
University of Reading
Whiteknights, Reading
United Kingdom

Progress in Mathematics

Edited by:

J. Oesterlé
Département de Mathématiques
Université de Paris VI
4, Place Jussieu
75230 Paris Cedex 05
France

A. Weinstein
Department of Mathematics
University of California
Berkeley, CA 94720
U.S.A.

Progress in Mathematics is a series of books intended for professional mathematicians and scientists, encompassing all areas of pure mathematics. This distinguished series, which began in 1979, includes authored monographs and edited collections of papers on important research developments as well as expositions of particular subject areas.

All books in the series are "camera-ready," that is they are photographically reproduced and printed directly from a final-edited manuscript that has been prepared by the author. Manuscripts should be no less than 100 and preferably no more than 500 pages.

Proposals should be sent directly to the editors or to: Birkhäuser Boston, 675 Massachusetts Avenue, Suite 601, Cambridge, MA 02139, U.S.A.

A complete list of titles in this series is available from the publisher.

37 Boutet de Monvel/Douady/Verdier. Mathématique et Physique
38 Bertin. Séminaire de Théorie des Nombres, Paris 1981–82
39 Ueno. Classification of Algebraic and Analytic Manifolds
40 Trombi. Representation Theory of Reductive Groups
41 Stanely. Combinatorics and Commutative Algebra
42 Jouanolou. Théorèmes de Bertini et Applications
43 Mumford. Tata Lectures on Theta II
44 Kac. Infinite Dimensional Lie Algebras
45 Bismut. Large Deviations and the Malliavin Calculus
46 Satake/Morita. Automorphic Forms of Several Variables, Taniguchi Symposium, Katata, 1983
47 Tate. Les Conjectures de Stark sur les Fonctions L d'Artin en s = 0
48 Fröhlich. Classgroups and Hermitian Modules
49 Schlichtkrull. Hyperfunctions and Harmonic Analysis on Symmetric Spaces
50 Borel, et al. Intersection Cohomology
51 Bertin/Goldstein. Séminaire de Théorie des Nombres, Paris 1982–83
52 Gasqui/Goldschmidt. Déformations Infinitésimales des Structures Conformes Plates
53 Laurent. Théorie de la Deuxième Microlocalisation dans le Domaine Complexe
54 Verdier/Le Potier. Module des Fibres Stables sur les Courbes Algébriques: Notes de l'Ecole Normale Supérieure, Printemps, 1983
55 Eichler/Zagier. The Theory of Jacobi Forms

56 SHIFFMAN/SOMMESE. Vanishing Theorems on Complex Manifolds
57 RIESEL. Prime Numbers and Computer Methods for Factorization
58 HELFFER/NOURRIGAT. Hypoellipticité Maximale pour des Opérateurs Polynomes de Champs de Vecteurs
59 GOLDSTEIN. Séminaire de Théorie des Nombres, Paris 1983–84
60 PROCESI. Geometry Today: Giornate Di Geometria, Roma. 1984
61 BALLMANN/GROMOV/SCHROEDER. Manifolds of Nonpositive Curvature
62 GUILLOU/MARIN. A la Recherche de la Topologie Perdue
63 GOLDSTEIN. Séminaire de Théorie des Nombres, Paris 1984–85
64 MYUNG. Malcev-Admissible Algebras
65 GRUBB. Functional Calculus of Pseudo-Differential Boundary Problems
66 CASSOU-NOGUÈS/TAYLOR. Elliptic Functions and Rings and Integers
67 HOWE. Discrete Groups in Geometry and Analysis: Papers in Honor of G.D. Mostow on His Sixtieth Birthday
68 ROBERT. Autour de L'Approximation Semi-Classique
69 FARAUT/HARZALLAH. Deux Cours d'Analyse Harmonique
70 ADOLPHSON/CONREY/GHOSH/YAGER. Number Theory and Diophantine Problems: Proceedings of a Conference at Oklahoma State University
71 GOLDSTEIN. Séminaire de Théorie des Nombres, Paris 1985–1986
72 VAISMAN. Symplectic Geometry and Secondary Characteristic Classes
73 MOLINO. Riemannian Foliations
74 HENKIN/LEITERER. Andreotti–Grauert Theory by Integral Formulas
75 GOLDSTEIN. Séminaire de Théorie des Nombres, Paris 1986–87
76 COSSEC/DOLGACHEV. Enriques Surfaces I
77 REYSSAT. Quelques Aspects des Surfaces de Riemann
78 BORHO/BRYLINSKI/MACPHERSON. Nilpotent Orbits, Primitive Ideals, and Characteristic Classes
79 MCKENZIE/VALERIOTE. The Structure of Decidable Locally Finite Varieties
80 KRAFT/PETRIE/SCHWARZ. Topological Methods in Algebraic Transformation Groups
81 GOLDSTEIN. Séminaire de Théorie des Nombres, Paris 1987–1988
82 DUFLO/PEDERSEN/VERGNE. The Orbit Method in Representation Theory: Proceedings of a Conference Held in Copenhagen, August to September 1988
83 GHYS/DE LA HARPE. Sur les Groupes Hyperboliques d'après Mikhael Gromov
84 ARAKI/KADISON. Mappings of Operator Algebras: Proceedings of the Japan-U.S. Joint Seminar, University of Pennsylvania, Philadelphia, Pennsylvania, 1988
85 BERNDT/DIAMOND/HALBERSTAM/HILDEBRAND. Analytic Number Theory: Proceedings of a Conference in Honor of Paul T. Bateman
86 CARTIER ET AL. The Grothendieck Festschrift, Volume One
87 CARTIER ET AL. The Grothendieck Festschrift, Volume Two
88 CARTIER ET AL. The Grothendieck Festschrift, Volume Three
89 VAN DER GEER/OORT/STEENBRINK. Arithmetic Algebraic Geometry
90 SRINIVAS. Algebraic K-Theory
91 GOLDSTEIN. Seminaire de Theorie des Nombres, Paris 1988-89
92 CONNES/DUFLO/JOSEPH/RENTSCHLER. Operator Algebras, Unitary Representations, Enveloping Algebras, and Invariant Theory: Actues du Colloque en l'honneur de Jacques Dixmier